Green Sense

Green Sense

The aesthetics of plants, place and language

John Charles Ryan

TrueHeart *academic*
– Oxford –

TrueHeart Press (Oxford) Ltd.
Hexagon House
Avenue 4, Station Lane
Witney
Oxford OX28 4BN

www.trueheartpress.com
please contact: info@trueheartpress.com

First published in 2012 by TrueHeart Press (Oxford) Ltd.

ISBN: 978-0-9573017-0-2
E-ISBN: 978-0-9573017-5-7

British Library Cataloguing-in-Publication Data: A catalogue record for this book is available from the British Library.

Typeset in the UK by Concerto.

Printed in the UK, USA, and Australia.

Published by TrueHeart Press (Oxford) Ltd., 2012
www.trueheartpress.com

Contents

List of Figures and Tables

Figures

Tables

Images with an asterisk have been reprinted with permission. All other images are the author's.

Acknowledgments

I thank Associate Professor Rod Giblett and Dr Lekkie Hopkins of Edith Cowan University, Western Australia, for their supportive criticism and generous readings of drafts along the way. They have been receptive to alternative—arts-based, syncretic and even transdisciplinary—approaches to environmental scholarship since the beginning of the project. I acknowledge, in particular, Dr Giblett's significant and generous contribution of ideas, stemming from his years of research in cultural studies of nature. I also recognise the support of the International Centre for Landscape and Language (ICLL), including the literary suggestions of Professors Glen Phillips and Andrew Taylor, as well as field trips to the bush offered by the Centre. Notably, Professor Phillips guided me to the works of the poets featured in chapters 6 and 10 through his considerable knowledge of Australian literature. My thanks to Dr John Hall for editing and encouragement and to Australian garden historian Richard Aitken for critical feedback on chapter 4. The staff of the Edith Cowan University library, particularly the Virtual Document eXchange team, have secured resources from overseas libraries and given patient assistance to me in locating obscure regional documents. I also thank the passionate interviewees—the "plant people"—who contributed generously to my understanding of their perceptions of wildflowers and flora in general and who have given me permission to quote from their interview transcripts and interpret the larger significance of their statements. They include Merle Bennett, Allan Tinker, Ayleen Sands, Stephen Hopper, Lyn Alcock, Noel Nannup, David James, Kevin Collins and Don Williams. Any errors of context or content are my own. And to the astonishing plants of the South-West of Western Australia that inspire me continually and that offer their many gifts to us—home, landscape, beauty, region, nature, air, intrigue.

Prologue

Seeing-Feeling-Thinking Flora

*The Christmas trees (*Nuytsia floribunda*) were now bursting into colour again; yellow buds were already opening into acrid-scented, brilliant orange blossom, contrasting with the olive green of the coarse foliage.*

Barbara York Main (1967, p. 119)

Warm thoughts of the wilderness. Am I, perhaps,
A woman loved. All night in this lonely world,
In my scorched kitchen, the gems of my becoming sing.

Alan Alexander (1979, p. 54) "Nuytsia Floribunda"

In the small town of Kojonup, Western Australia, a regional hub for the wool industry, on a winter afternoon in 2009, I decided to enquire about the local plants. Surprisingly, at the tourism office, my curiosity was greeted with a brusque reply: "The show hasn't started. The wildflowers aren't out yet. You won't see anything." Venturing into the small bush reserves around the town, my guide and I nevertheless uncovered a world of sensory immersion. Selecting nuts from the base of a quandong (*Santalum acuminatum*), we cracked open its convoluted outer shell to expose the crisp, white flesh reminiscent of macadamia. Despite it being the "off-season," the plants offered memorable moments of interaction. We scrambled across the highway to a marri tree (*Corymbia calophylla*) to sample the medicinal *kino*, or gum resin, oozing from the bark. Closer to town, we spotted a plant with small cylindrical fruits. My guide told me coyly that varieties with white flowers are considered more effective than pharmaceutical drugs to combat male impotence. As the sun descended below the horizon, we crushed fragrant gum leaves in our hands and rubbed the resin along our forearms to assuage the sting of ant bites. Yes, there were only a few flowers in Kojonup that day. Instead, there were a myriad of textures, tastes, smells and sounds. Contrary to the opinion at the tourism centre, the botanical show had never stopped. Indeed, our foray into the bush was guided by our sensory curiosity.

Since my experience at Kojonup, I have begun to regard sensory openness to flora as integral to the formation of a "sense of place." Yet, I can trace the beginnings of this interest to an unlikely place on the other side of the globe. In 2007, I spent two months walking a south-to-north transect of the Canadian island of Newfoundland; along the western coast from Port aux Basques, where the ferry from Nova Scotia arrives, to the northern tip at L'Anse aux Meadows. Norse sailors first established a settlement in the Americas at this windswept northern outpost. Intermittently throughout my adult life, I have taken ambitious foot journeys into places inaccessible to vehicles, places where ancient populations of plants still enact ancient symbioses, where the passionate dramas of seeding and pollination delicately unfold. This walk took me both into the heart of the coastal wilderness and into the belly of Newfoundland culture. Stumbling fatigued and mosquito-bitten from a fog-ensnared foot track, I crossed a bitumen highway and found myself in the midst of the tourist season. At that moment of emergence from sodden hinterlands, I realised something from eco-philosophy, landscape aesthetics, nature writing and field botany: the cultural significance of plant aesthetics.

The penetrating gaze of a passing motorist mixed indecipherably with the glint of the apricot-coloured fruits of the cloudberry (*Rubus chamaemorus*), twinkling mesmerisingly at the side of the highway. A composite sensory world incorporated me. I felt a sense of being simultaneously nature and culture, realising that such a split is a construct of the imagination. I had conversations—at the road verge, in the front of rural markets and in trucks—with fourth-generation Newfoundlanders, as well as fresh-eyed vacationers from far-flung parts of North America and Europe. Through these dialogues, a theme emerged: the attractiveness of the coastline comprised the colour and form of its geological features, but also the visual beauty, edibility and aromas of its plants. For many of us, the Newfoundland landscape captivated our senses enough to travel great distances to behold. The *plantscape* had as much to do with the pleasurable experience of the island as the dramatic legacies of its topography.

Through the commonalities of language used to describe the botanical landscape—as beautiful, pretty, stunning, handsome or striking—the importance of an aesthetics of flora stuck with me after my island walk. However, the diction of the people I spoke with tended to construct an array of artistic qualities: forms, colours, symmetries, harmonies and balance, all visually announced. As such, aesthetics seemed to promulgate a distance between visitors and place; a distance as abstract and untenable as the divide between the Newfoundland wilderness and its small towns and bitumen thoroughfares.

I knew plants cognitively as an amateur botanist, and as an appreciator of beauty perceived them visually, but also recognised their multi-sensoriality: as effusers of intoxicating scents distinct as fingerprints; conveyors of nutriment through spicy or sweet flavours; instruments of sound resonating as vibrantly as a harpsichord; and delicacies of touch, palpable as the smooth surface of a vase.

In contrast to a visual aesthetics, bodily engagement with flora generates sense of place through the appreciation of plant appearances in dynamic connection with body sensations.

Bodily experience positioned me within the plant world, and problematised the fissure between the observer and the plant as an object of appreciation. These experiences in Newfoundland, and later in Kojonup, presented a philosophical quandary and the questions guiding this book.

Despite the possibility of a complex sensory experience of nature, there is an undeveloped corporeal aesthetics of wild plants in environmental philosophy and an equally nascent language for expressing bodily interactions with flora through the senses. Multi-sensorial experience of uncultivated plants is an undeniable part of the human encounter with place. Recounting her landing at Albany in Western Australia during the early twentieth century, former opera singer turned travel writer May Vivienne (1901, p. 3) wrote of the immediacy of "the scent of the sweet boronia wafting on the breeze from the gullies, where it grows in such luxuriance that one wonders no scent farm has been started to distil the exquisite perfume."

Could it be that, in a hyper-industrial era, we no longer interact significantly enough with wild plants through non-visual senses? Or is it part of a culture of perception encompassing a philosophy of the non-human world that maintains distance, or an attitude towards flora as potentially dangerous or mere ornamentation?

Interlude I: Agoraphobia in the Garden State

I was born in the densely populated and highly suburbanised American state of New Jersey on the eastern seaboard of the continent. Although transected by parkways and turnpikes, New Jersey is known affectionately as the "Garden State" and less affectionately as the "Garbage State." Since childhood, my sense of place for my native territory has been moulded by its flora, but not so much by the state's economic importance within the megapolis between Boston and Washington D.C. Childhood memory is laden with sheltered oak woods, pitch pines bleeding sap, lady slipper orchids appearing impetuously in the spring, inedible wild blueberries, and the cultivated bindweed or convolvulus trumpeting its flower each morning outside my bedroom window. "Agoraphobia in the Garden State" anticipates my anxieties in coming to Australia, a country whose landscapes are characteristically open expanses with congregations of plants that were almost totally unfamiliar to me. The ancient Greek *agora* was an open place of assembly. The root *agora* contrasts to *claustro*, referring to closed spaces, as in *cloisters*. For me, New Jersey represents claustrophobic sensitivity, whereas the Western Australian landscape offers agoric expansiveness. In addition to its spatial dimensions, the poem employs bodily diction, such as the "ruddy cheeks" (l. 26) of the dryads inhabiting my Northern *genius loci*.

Agoraphobia in the Garden State

For New Jersey

these woods are not open
but hemmed in with all manner
of tree and brush and herb;
oaken trunks are wishbones
slowly straining at the crotch,
a gash in a pitch pine exposes
tawny-brown flesh sealed in
a rosary of golden sap—
its untalking lips closed and
pursed prayerfully upward;
I gouged it there years back
wounding the pitted bark
with a meat cleaver;
lady slipper orchids will come
if I wait for the veiny pink
testes to emerge and droop
and—in truth—ivory lantern
bunches of wild blueberries
will call rhomboidal light to the
duff, showering hard inedible fruits
upon strata of decaying things:
samara, rhizome, flung sepals;
one morning, crouching among
convolvulus, I will decide
to kiss the dryads
on their ruddy cheeks
board the plane to Sydney,
upchuck my guts into the sea.

I reflected on fragments of contemporary eco-aesthetic philosophy, but found them largely unsatisfying. The picturesque implies a treatment of landscape as if it were a painting (Gilpin, 1786). The more recent "natural environmental" model argues against the visual framing of the landscape I encountered in Newfoundland, seeming to allow for multi-sensorial appreciation of the botanical world (Carlson, 1993, 2000). However, in broadening the cult of the scenic and opening space for the emergence of sensation, the "natural environmental" model designates natural science as the master narrative of appreciation. Carlson's model in particular contends that one should be scientifically informed about plants in order to make appropriate judgments of aesthetics. In such a model, bodily experience is characterised as unstructured—perhaps idiosyncratic—personal experience, and hence no basis for a system of perception. How could botanical science—a visually focused schema for ordering life—be integral to an aesthetics of multi-sensorial immanence?

Consequently, I turned away from current movements in continental philosophy. Instead, I embraced two philosophers who, though not necessarily classified as writers of aesthetics, have much to say about the sensuous relationship between people and place, especially as mediated by language: Martin Heidegger and Henry David Thoreau.

Martin Heidegger's interpretations of the Greek concepts *poiēsis*, *aesthesis* and "the Open," in conjunction with his valuation of poetic thought, were fertile places to begin (see chapters 2 and 9). However, the botanical writings of the nineteenth century American philosopher and "botanophilist" Henry David Thoreau, presented a philosophy of plants fundamentally linked to sensuous encounter and articulated within a distinctive habitat poetics (see chapter 3). His posthumous *Faith in a Seed* (1993) and *Wild Fruits* (2000) prove embedded with possibility. His writings trace a corporeal aesthetics of plants in which language becomes a mode of participation in nature. Thoreau also enters into dialogue with science while retaining a sensuous underpinning. He exhibits a syncretic approach to plants and place not circumscribed entirely by empiricism.

I will consider the dialogue—commonalities, divergences and points of contention—between these two philosophers in chapter 9. I return to their writings as conceptual frameworks for understanding the written and spoken resources of the South-West of Western Australia, where this book is situated, in chapter 2. The conceptual approach I adopt might be termed a hybrid Heideggerean-Thoreauvian framework. Moreover, I draw from Heidegger and Thoreau in proposing cultural botany as a context for enlivening multi-sensorial engagements with plants in chapter 1. In Thoreau, I found an exemplary figure of cultural botany: an amateur botanist, a proto-ecologist and proto-environmental ethnographer, a philologist, a writer of poetic prose and one interested in the aesthetics of plants, place and language.

Green Sense is divided into five parts: Part I: Restoring Sense to Plant Research; Part II: Botanical Histories; Part III: Botanical Cultures; Part IV: Botanical Languages; and Part V: Botanical Futures.

Part I suggests that the foregoing questions are best responded to in a context comprising methods of embodiment drawn from the humanities. Hence, chapter 1 sketches a transdisciplinary context I call *cultural botany* that extends recent research into cultural ecology and the ecological humanities, and counterbalances empiricist modes of quantifying flora. Chapter 2 provides a theoretical synopsis of the field-based methodology I call *botanic field aesthetics* for researching engagements with plants through the methods of poetic practice, ethnographic interviewing and *gestural walking*. Part II focuses on the interstices between aesthetics and the many histories of South-West flora. Chapter 3, "Narrative," argues that the writing of botanical histories are made polyvocal through the interweaving of Nyoongar, scientific, colonial and post-colonial perspectives. Chapter 4, "Values," offers a close analysis of the nineteenth century botanist John Lindley's *A Sketch of the Vegetation of the Swan River Colony* in exemplifying some colonial-era aesthetic views of South-West plants. Through a reading of the works of contemporary environmental philosophers Allen Carlson and Arnold Berleant, chapter 5 outlines the limitations of contemporary ecophilosophy in providing the multi-sensorial groundwork for an embodied aesthetics of flora.

Parts II and IV transition from previous concepts of floral aesthetics to current ones through discussion of cultural practices and language. Chapter 6 first asserts that *greenness* implies visual categorisation that disfavours arid places and plants. The chapter highlights the genesis of botanical vocabularies specific to the South-West. Chapters 7 and 8 provide readings of ethnographic interviews. In the context of a "culture of flora," chapter 7 explores the aesthetic modes at work during the South-West wildflower season. Chapter 8 explores the relation between sensory memory and local plants, as well as the contribution of memory studies to cultural botany. Part IV is focused on language, and argues that an embodied aesthetics depends on the capacity of language to open perception towards multi-sensorial entanglement. Chapter 9 proposes a language theory of sensation through an integrative reading of Thoreau and Heidegger. Chapter 10 argues that landscape poetry, which focuses on the appearances of places and plants, differs from habitat poetry, which attends to the relationship between habitat processes, felt experiences and ecological places.

Part V is future-oriented. Chapter 11 contextualises South-West plants in the global extinction crisis and conceptualises mourning in terms of sense connectivity between people, plants and places. Theorising the idea of gestural walking through Marcel Mauss's concept of body techniques, chapter 12 argues that some forms of walking offer a greater experience of corporeality. Chapter 13 comes full circle, theorising an aesthesis of plants, reflecting a growing interest in sensation in critical theory and restoring multi-sensoriality to a floral aesthetics. Identifying aesthetic themes and interconnections, the Coda suggests directions for cultural botany as the ongoing dialogue between the humanities and the sciences, nature and culture, and people and plants in the South-West region of Australia and elsewhere.

Interlude II: Understanding Parrot Bush

A jaunt through the knee-high vegetation atop Mount Lesueur north of the city of Perth, Western Australia, was my first excursion into the rich *kwongan* heathlands (for example, Beard & Pate, 1984). "Understanding Parrot Bush" narrates an induction. My body-based investigation of the South-West flora began with *Dryandra sessilis*, also known controversially as of 2007, as *Banksia sessilis* (K. Collins, pers. comm., September 9, 2009). Through the shifts between discord and ease in the structure of the writing, the poem expresses acculturation to a place through a curiosity about its flora. An initiate to the South-West, I have needed to learn the languages of its plants as a second tongue. I grapple with the multiple narratives used to describe Parrot Bush, including the master narrative of taxonomy (see chapter 3). As suggested by the enumeration of plant names in the poem, the process has been characterised by an awareness of the disjunctions and convergences of naming. *Budjan*, as Parrot Bush is known in the Aboriginal language Nyoongar, is also designated by an abundance of ever-evolving technical and colloquial names that reflect histories (see chapters 3 and 4). Naming parallels the history of human interactions with Parrot Bush. Aboriginal, scientific and colonial appellations retain important cultural and aesthetic meanings. Additionally, as a newcomer to the South-West, the quality of harshness for which Australia's plants are known, strikes me as the obverse of the silken texture of the flower itself (see chapter 6):

> turret of petals, stamens silky
> helter-skelter inside an armamentarium
> :*josephia in early taxonomy*
> you adapt your downy insides
> softer in hardness, more loving in
> the hardnesses, this land— (ll. 10-15)

The land is characterised more accurately as a confluence of extremes: softness and hardness, distance and proximity, and scarcity and density. As a crucible for these heterogeneous meanings, the poem provides a deliberately disjointed—somewhat unnerving—rumination (to read "Understanding Parrot Bush" in full, see Ryan, *in press*).

Figure 1 Survey Corps Tower Atop Mount Lesueur, Western Australia. *Image courtesy of the author.*

References

Alexander, A. (1979). Nuytsia Floribunda. In B. Bennett & W. Grono (Eds.), *Wide domain: Western Australian themes & images*. London, England: Angus & Robertson.

Beard, J., & Pate, J. (1984). Foreword. In J. S. Beard & J. S. Pate (Eds.), *Kwongan, plant life of the sandplain: Biology of a south-west Australian shrubland ecosystem* (pp. xvii-xxi). Nedlands, Australia: University of Western Australia Press.

Carlson, A. (1993). Appreciating art and appreciating nature. In S. Kemal & I. Gaskell (Eds.), *Landscape, natural beauty and the arts* (pp. 199-227). Cambridge, England: Cambridge University Press.

Carlson, A. (2000). *Aesthetics and the environment: The appreciation of nature, art, and architecture*. New York, NY: Routledge.

Gilpin, W. (1786). *Observations relative chiefly to picturesque beauty, made in the year 1772* [Google Books]. Retrieved from http://books.google.com/

Main, B. Y. (1967). *Between wodjil and tor*. Melbourne, Australia: Jacaranda.

Ryan, J. (*in press*). John Charles Ryan. In S.-P. Mitchell (Ed.), *Fremantle poets 3: Performance Poets*. Fremantle, Australia: Fremantle Press.

Thoreau, H. D. (1993). *Faith in a seed: The dispersion of seeds and other late natural history writings*. Washington, DC: Island Press.

Thoreau, H. D. (2000). *Wild fruits: Thoreau's rediscovered last manuscript.* New York, NY: W.W. Norton.

Vivienne, M. (1901). *Travels in Western Australia: Being a description of the various cities and towns, goldfields, and agricultural districts of that state.* London, England: William Heinemann.

Part I

Restoring Sense to Plant Research

Part I introduces context and approach. Through a review of literature addressing potential rapprochements between the two cultures of the sciences and humanities, chapter 1 outlines a transdisciplinary research approach to the study of flora. The first chapter will introduce the frameworks of Thoreau and Heidegger that will serve as traction throughout *Green Sense*. Cultural botany will be theorised as an extension of recent work in cultural ecology and the ecological humanities. Chapter 2 goes on to propose a qualitative suite of methods organised as botanic field aesthetics and including poetic enquiry, ethnographic practice and gestural walking for researching sense-rich interactions between plants and people in the South-West of Western Australia.

In "Swamp Plant," the Australian poet Judith Wright (1994, p. 367) sums up poignantly the convergences between flora, poetry, science, embodiment and appreciation at the heart of *Green Sense*:

> Only science, then, has noticed you,
> not poetry.
> It's that way round in this country,
> upside-down as ever.
> Living on swamp-edges
> turning your face to the ground, shyer than
> Wordsworth's violet,
> no words but dog-Latin
> have tagged you.

1

Cultures

Abstract: Since the eighteenth century, the study of plants has reflected an increasingly mechanised and technological view of the natural world that splits humanities and sciences into separate knowledge fields. In broad terms, this chapter will propose a plant research context by interrogating existing literature, in order to allow a rapprochement between knowledge forms. The nature-culture dichotomy, and more specifically the plant-human disjunction, follows a course of resolution that runs parallel to the schism between objective (technical, scientific, reductionist, visual) and subjective (emotive, artistic, relational, multisensory) forms of knowledge. The foundations of taxonomic botany, as well as the allied fields of environmental studies, ethnobotany and economic botany, reflect a universalising, sensory-limited and visually-privileged structuring of the natural world. Consisting of the study of everyday embodied interactions between humans and flora, expanding upon the lens of cultural ecology and environmental humanities, "cultural botany" provides a transdisciplinary research framework.

> *It may appear singular, but yet it is not the less correct, to attempt to connect poetry, which rejoices every where in variety of form, color, and character, with the simplest and most abstract ideas. Poetry, science, philosophy, and history are not necessarily and essentially divided; they are united wherever man is still in unison with the particular stage of his development, or whenever, from a truly poetic mood of mind, he can in imagination bring himself back to it.*
>
> Wilhelm von Humboldt cited by Alexander von Humboldt (1850, p. 31)

Since the eighteenth century, the study of plants has reflected an increasingly mechanised view of the botanical world that divides the humanities from the sciences. In broad terms, I propose a context for researching flora through an interrogation of existing literature that examines rapprochements between scientific epistemologies and humanities-based ways of knowing, as well as between modernist and postmodernist notions of objectivity and subjectivity. The nature-culture dichotomy parallels the schism between objective and subjective knowledge forms. Yet, in order to revise Cartesian subjectivity without negating the centrality and richness of experience itself, cultural botany will need to distinguish between the subject as the living centre of sensation and interaction with the plant world, and subjectivity as a state of ego-focused closure.

Heidegger critiqued the dialectic between subjectivity and objectivity by contending that high modernist subjectivity has its origins in the Cartesian reduction of human beingness to an individuated subject: a person, a body, an ego. According to Heidegger's argument, "everything is either the object of the representation of a subject or a subject that represents the world as a picture" (Schalow & Denker, 2010, pp. 262-263). Similarly, Lash and Friedman (1992, p. 5) propose that modernist subjectivity privileges judgment and colonises vision through cognition, further creating hierarchies such as "the ego over the id, the visual over touch…discursive over figural communication…culture over nature… individual over community…[allowing] the individual to be somehow 'closed', instead of open."

Taxonomic botany, as well as environmental studies, ethnobotany and economic botany, operates according to a sense-constrained visual structuring of living things fixed in a Cartesian dialectic between subjectivity and objectivity. In these disciplines, plant life is seen as a series of objects represented and studied by a subject or subjects, such as a botanist, taxonomist, ethnobotanist or other specialist. Towards transdisciplinarity, cultural botany, as the investigation of embodied interactions between plants and people, requires the shared perspectives of ecological humanities, cultural ecology and ecocriticism. The transdisciplinary context provides an approach for researching engagements with flora through diverse embodied methods and beyond the 'closed' systems of objectivist science (see chapter 2).

Cultural botany presents an alternative to an empiricist paradigm. It mediates the "two cultures" split, the divorce of the humanities from the sciences identified by Snow (1964/1993), philosophers Heidegger (1977), Berlin (1979), Prigogine and Stengers (1984), Serres (1982, 1995), Serres and Latour (1995), Serres and Zournazi (2002), and ecological thinkers Giblett (2011), Leopold (1949/1987), Thoreau (1993, 2000) and Seddon (1972, 2005). Due to their technical orientations, environmental studies, ethnobotany and economic botany offer limited theoretical promise for embodied and poetic research into plants.

I expand upon contemporary literature in ecocriticism, cultural ecology, the ecological humanities and transdisciplinarity to present cultural botany as a context for theorising bodily engagements with flora. Through the prism of cultural botany, specific philosophical and poetic knowledges represent commonplace interactions with flora outside of the limitations of the objective/subjective divide.

The Technicised Plant in Nature's Laboratory

This, scholars of all countries prize,—
Yet 'mong themselves no weavers rise.—
He who would know and treat of aught alive,
Seeks first the living spirit thence to drive:
They are the lifeless fragments in his hand,
There only fails, alas! the spirit-band.
This process, chemists name, in learned thesis,
Mocking themselves, *Naturae encheiresis.*

Johann Wolfgang von Goethe (1828/2010, p. 74)

Translated as "nature's laboratory," Goethe's *naturae encheiresis* from *Faust*, first published in 1808, expresses nineteenth century European disenchantment with the increasingly reductionist view of nature in which the living body is dissected into constituent parts, each analysed and compartmentalised into new disciplines of knowledge. In the last line of the excerpt, "mocking themselves" refers to the ironies in the separation of intellectual investigation and bodily presence. The segregation of epistemologies corresponds to the evisceration of bodies in the laboratory.

In 1790, Goethe, a polymath accomplished in both plant poetics and botanical science, published the long poem, *The Metamorphosis of Plants*, prior to his more acclaimed *Faust* (Goethe, 1790/2009). In *Metamorphosis*, Goethe proposed what Miller (2009, p. xi) describes as "a fuller integration of poetic and scientific sensibilities that would provide a way of experiencing nature both symbolically and scientifically, simultaneously." In these lines from *Faust*, Goethe critiques the structures of relation between human knowledge-making and the living objects of study that have been systematised by science, especially since eighteenth century Swedish botanist Carl Linnaeus formulated his hierarchy of plants.

Scientific taxonomy establishes standardised methods of nomenclature to reference plants worldwide and to show evolutionary relationships between species (Clarke, 2008, p. 57). Goethe's verses provoke a critical question. How does Linnaean taxonomy affect the sensuous relationship between people and plants, when, previously, the visible parts of plants, along with their gustatory, auditory, tactile and olfactory qualities, characterised human perception of flora? Bearing Linnaean descent, a modern botanist engages with plants through taxonomic keys and tools of magnification. In contemporary botanical science, DNA technology further ensures that the code of plant knowledge is transmissible to a worldwide audience of specialists (for example, Berg, 2008, pp. 292-309). As technical research, the rigorous investigation of flora tends to engage the structuring of taxonomy. Science, and, more specifically, knowledge in service to technology, provide the empirical foundation for contemporary research into plant life.

Before the eighteenth century, knowledge of plants was intimately related to the human body through herbal medicine (Arber, 1912/1938). As multi-sensorial phenomena, plants were studied for—and classified by—their curative virtues, which bore directly on human wellbeing (see chapter 13). Therapeutic properties of roots, leaves or flowers constituted a sensuous system of corporeal entanglement with flora. Before the systematisation forged by Linnaean taxonomy, herbal texts categorised plants, but according to their uses, specific locations, physical properties, the season at which their optimal therapeutic value could be attained, and their method of preparation and administration. Schiebinger (2004, p. 14) stresses that, "knowledge of plants at this time was local and particular, derived from direct experience with plants." Pre-Linnaean knowledge of flora was related syncretically to cultures and the senses. In *The Order of Things*, Michel Foucault (1966/2002) postulates that, after the eighteenth century in particular, natural observation corresponded to visual perception, excluding taste, smell, touch and "hearsay" due to their personal variability. Earlier it had been that:

> To write the history of a plant or an animal was as much a matter of describing its elements or organs as of describing the resemblances that could be found in it, the virtues that it was thought to possess, the legends and stories with which it had been involved, its place in heraldry, the medicaments that were concocted from its substance, the foods it provided, what the ancients recorded of it, and what travellers might have said of it. The history of a living being was that being itself, within the whole semantic network that connected it to the world. (140)

In 1653, the physician Nicolas Culpeper published *The Complete Herbal,* a heterogeneous text about flora, preceding Linnaean classification but proving commensurate vigour in its attention to discerning between plants through a practical taxonomy (1653/1981). The text is a compendium of the medicinal virtues of European flora and relates the methods of preparation for various species. Its descriptions are laden with multi-sensorial information, aligning embodied cognition to the attainment of practical knowledge. For instance, Culpeper (1653/1981, p. 313) cautions the user of herbs to exercise sensory powers in discerning between beneficial and deleterious roots: "Of roots choose neither such as are rotten or worm-eaten, but proper in their taste, colour and smell, such as exceed neither in softness nor hardness." Moreover, visceral cues signify unity between the powers of the senses and the therapeutic value of the plants: "Yet you may know when they are corrupted by their loss or colour, or smell, or both: and, if they be corrupted, reason will tell you that they must needs corrupt the bodies of those people that take them" (Culpeper, 1653/1981, p. 312). Enfolded within physical needs, plant knowledge systems emerged from the senses, along with stories and "hearsay" of regional locales and the seasonal particularities of people and place. Awareness of flora was diachronic. Botanical epistemologies were situated, variable, self-determined and corporeally affective.

In 1637, René Descartes in *Discourse on Method* advanced an approach to scientific enquiry based on deduction and reduction, the former involving the progression towards logical conclusions and the elimination of all illogical assumptions, and the latter involving the breaking up of the world into its constituents (Moran, 2010, p. 137). In the sixteenth and seventeenth centuries, mechanical metaphors of the body proliferated along with discrete scientific disciplines.

The post-Renaissance botany of the eighteenth century ushered in universalised methods of plant classification based on gendered power-relations. Linnaeus first outlined his sexually based system in *Systema Naturae* (1735), *Fundamenta Botanica* (1736) and *Classes Plantarum* (1738) by identifying differences between the male and female parts of the flower (Blunt, 2004, p. 258). This organisational system, known as binomial nomenclature (or genus-species designation), compartmentalises plants according to differences in sexual organs (Schiebinger, 2004). Linnaeus's emphasis on sexual morphologies, where the male parts of the flower determine higher classification categories along the taxonomic chain, encipher and reinscribe the gender hierarchies of eighteenth century Europe (Schiebinger, 2004). Linnaeus's system also served his "physico-theological" ambition to promoting the development of Swedish nationalism through natural history (D. Miller & Reill, 1996, p. 8).

Botanical science universalises plants by dismantling organic cohesion into coded blocks of information transcending regional cultures and languages. Elements of taxonomic science, such as Latinate names for genus and species and the modern usage of biochemical assays, technicise flora as an object of study. Georges-Louis Leclerc, a contemporary of Linnaeus, criticised binomial nomenclature for its abstraction and its basis in miniscule details that require a microscope (Schiebinger, 2004, p. 28).

Linnaean botany operates successfully on a global scale because it formalises plants, removing living beings from the spatio-temporal and sensory conditions of complex habitats. Another Linnaean contemporary, Swiss naturalist Albrecht von Haller, argued that temporal changes are as crucial as morphologies fixed in the isolation of a moment (Schiebinger, 2004, p. 16). The technical abstraction of plants is a-temporal in character.

Through removal from the temporal flux of biotic processes, the universalisation of flora is allied to the framing of plants in time and space. Taxonomy, the major legacy of Linnaeus, structures life into visual arrays. Multi-sensorial features are excised to create exportable images for worldwide circulation. These core practices exemplify the ordering power of what Latour (1999, p. 38) refers to as the "synoptic tableau." Latour (1999, p. 38) asserts that "once classified, specimens from different locations and times become contemporaries of one another on the flat table, all visible under the same unifying gaze." Scientific images and nomenclature are signifiers of living nature. These forms migrate as "circulating references," enabling the global construction of knowledge (Latour, 1999, p. 38). The locality, particularity and materiality of a plant in its environment are reduced by a desire for compatibility, standardisation and circulation. Visual representations, linked to classificatory hierarchies, can obscure actual mutable plants in the field, as well as complex sensory experience of plants. Rather than flora's multi-sensorial manifold, form and colour come to dictate the structure of authentic knowledge.

In Heidegger's terms, science and philosophy both constitute knowledge of the world. He interrogates the epistemological exceptionalism of scientific knowledge production as separate from poetic forms of knowing. Heidegger problematises the technical preoccupation of modern scientific enterprise. The concept of enframing (*Ge-stell*) maintains the imagistic rationality of science by correlating the systematic domination of the natural world to scientific objectivity and visual knowledge production (Prigogine & Stengers, 1984, p. 32). Glazebrook (2000, p. 246) characterises *Ge-stell* as the "challenging of nature to reveal itself in a determined way" through *a priori* assertions about reality. Scientific objectivity determines the "age of the world picture," the privileging of visuality (Glazebrook, 2000, p. 246). Classical science enframes the natural world as a two-dimensional portrait, inducing a snapshot of a plant and instantiating a living organism in space and in time.

In "The Question Concerning Technology," Heidegger (1977) describes enframing as the ordering of the visible, standing in contrast to *poiēsis*, which broadens the possibility of sense unfolding. On the one hand, enframing sets forth rigorous ordering through visual denomination of structures. On the other, the *poiētic* revealing of the world entails the culmination of the senses in temporal movement: seasonal, specific, relational and somatic (see chapter 2). Heidegger (1977, p. 311) contends that "enframing, in a way characteristic of a destining, blocks *poiēsis*." As a predominant empirical mode, taxonomic Linnaean science constrains the possibilities of plants through representation and appearance.

Figure 2 This genus of plant belongs to the ancient Proteaceae family. About 75% of Australia's *Isopogon* spp. occur in Western Australia. *Image courtesy of the author.*

Interlude III: Balance Lost and Found Whilst Stretching for Wattle Pods

Through simple rhyming verse, "I Lose My Balance" conveys ideas central to cultural botany. One of my premises is that the culture of flora is quintessentially a culture of flowers, but a visual focus on blossom may overshadow multi-sensorial possibilities for physically engaging with the whole plant through its life cycles (see chapter 7). The poem concerns the pods of wattle, the Australian *Acacia* genus, and the nourishing jewel-like seeds sequestered within each pod. The potentially undisclosed narrative of wattle seed as a staple food in Aboriginal cultures forms a backdrop for my appreciation of the plant (Daw, Walley, & Keighery, 1997; Meagher, 1974). Throughout Australia, various methods have been used to process wattle for consumption, including grinding the seeds into a meal, baking seed flour into a damper or eating the roasted seeds directly (for example, Clarke, 2007, p. 78). Writing in the 1880s on Aboriginal Australian usage of wattle, Ethel Hassell (1975, p. 19) reported that "when the seeds of the wattle tree were ripe, they were collected by the women and ground into meal." Not only by taste, wattle may be appreciated through sight, sound, smell and touch:

> the damper in the fire,
> its scent of earthen bread

would stet my hungry ire
and leave my digits bled (ll. 17-20)

Eating a wattle seed involves contemporary participation in the long-standing tradition of deriving bodily sustenance from wild foods (to read "Balance Lost and Found" in full, see Hickman & Ryan, *in press*).

Reconciling the "Two Cultures" Schism

> The clashing point of two subjects, two disciplines, two cultures—of two galaxies, so far as that goes—ought to produce creative chances. In the history of mental activity that has been where some of the break-throughs came. The chances are there now. But they are there, as it were, in a vacuum, because those in the two cultures don't talk to each other.
>
> Charles Percy Snow (1964/1993, p. 16)

Following Heidegger's critique, the enframing of plants entails their removal from cultural contexts that determine their conditions as significantly as biological factors. Since Linnaeus, the technicised plant parallels the larger story of the humanities-sciences standoff. In his 1882 essay, "Science and Literature," presented initially as a lecture to the Senate House in Cambridge, English poet Matthew Arnold (1882/2011, para. 14), a highly influential figure in Victorian England, argued that literature "may mean everything written with letters or printed in a book. Euclid's *Elements* and Newton's *Principia* are thus literature." Arnold envisaged literature as an inclusive term for writing that conveys knowledge, as both *belles-lettres* and technical treatises. In Arnold's view, science and literature need not be the incompatible domains constructed during the Newtonian revolution of natural science, but are rather parts of the education of a well-rounded citizenry. Nearly eighty years later, novelist and research scientist C.P. Snow would return to the theme of conciliation between the arts and sciences. Identifying a growing discord between the "two cultures" of scientists and intellectuals, Snow (1964/1993, p. 61) argued that productive connections could resolve the humanities-sciences estrangement.

In contemporary thought, such epistemological disjunctions are further emphasised by chemist and Nobel Laureate Ilya Prigogine, and philosopher Isabelle Stengers, who argue for a "new alliance" between disciplines. On the view of Prigogine and Stengers (1984, p. xxix), "traditionally science has dealt with universals, humanities with particulars." Concerning temporality, the authors observe a binary "between the atemporal view of classical science and the time-oriented view that prevails in a large part of the social sciences and humanities"

(Prigogine & Stengers, 1984, p. xxviii). Historian and philosopher Isaiah Berlin (1979, p. xxvi) echoes Prigogine and Stengers, identifying several qualitative disciplinary binaries upheld between the sciences and the humanities: "The specific and the unique versus the repetitive and the universal, the concrete versus the abstract, perpetual movement versus rest, the inner versus the outer, quality versus quantity, culture-bound versus timeless principles." Characteristic of the humanities, in Berlin's assessment, are the specific and the concrete (as compared to the abstracted sexualised hierarchies problematised by Schiebinger); perpetual movement and the internal (as compared to Heidegger's time-arresting principle of enframing); and quality and culture-bound principles (as compared to Latour's culture-independent concepts of the circulating reference and synoptic tableau).

The work of French philosopher Michel Serres provides an apotheosis of the vision of the sciences as enculturated and of humanities as scientifically conversant. According to Girard (cited in Harari & Bell, 1982, p. xi), Serres's central aim is to counter "the prevalent notion of the two cultures—scientific and humanistic—between which no communication is possible." Accordingly, Serres (Serres & Latour, 1995, p. 29) observes that "philosophers with a good knowledge of the hard sciences and of the classics—armed with rigor and culture—will never be taken in by folly or ideologies." Envisioning dialogue, Serres (Serres & Latour, 1995, pp. 27-28) calls for greater intellectual exchange towards resolving environmental catastrophes: "The questions fomented since the dawn of time by what we call the humanities help rethink those asked today, about and because of the sciences." Moreover, for Serres, knowledge "transcends academic disciplines and artificial boundaries" (Girard cited in Harari & Bell, 1982, p. xi). Amongst other terms, the unification of disciplines is synonymous with "connectedness," "cross-fertilisation," "cross-breeding" and "mutual enrichment," approaches embodied by Plato, Aristotle, Lucretius, Leibniz and Pascal through a kindred syncretism of knowledge (Serres & Latour, 1995).

According to proponents of rapprochement, the reconciliation has benefits for our relationship to the biosphere. In *The Natural Contract*, Serres (1995, p. 44) uses the "Northwest Passage" to symbolise the convergence between scientific and humanities based epistemologies. The text is stylistically enigmatic and transgressive, eliding categorisation as either a rationalistic treatise or a poetic rumination. A new contract between humanity and the Earth would entail a shift in power structures such that "the natural world will never again be our property, either private or common, but our symbiont" (1995, p. 44). By *symbiont*, Serres means an animate or inanimate participant in a mutually beneficial relationship.

As with Serres, the theme of dialogue between disciplines towards ecological sustainability is adopted by other writers. Cultural theorist Rod Giblett (2011, p. 55) asserts that "greening the humanities and the modern condition is an urgent intellectual and political task whose aim would be to establish an ecologically sustainable relationship with the Earth." The "greening of the humanities" would engage a more ecologically conversant literati and, conversely, scientific professionals who are more sympathetic to perspectives in the humanities. Similarly, Verena Andermatt Conley (1993, p. 77) suggests the need for a green humanities: "Ecology has been studied primarily in areas of biology, meteorology, geography, and demography. Less has been said on the subject in the humanities, where its mention is generally parenthetical." For Serres, Giblett and Conley, greater cohesion between disciplines would produce tangible integrations between plants, people and places.

Towards Transdisciplinary Botanical Thinking

> Science has been about a search for translation, convertibility, mobility of meanings, and universality—which I call reductionism, when one language (guess whose) must be enforced as the standard for all the translations and conversions.
>
> Donna Haraway (1991, p. 187)

The twentieth century brought about scientific constructivism and epistemological anarchism, which situate science within political and cultural contexts and challenge claims towards impenetrable truths (Moran, 2010, pp. 139-140). During this time, the division between the sciences and the humanities, based on a distinction between objective knowledge and subjective interpretations, came under more serious interrogation.

Contemporary attempts to redress the gulf are represented by two related, but discrete, forms of thought: *interdisciplinarity* and *transdisciplinarity.* The works highlighted above argue for rapprochement between the two cultures. They can be characterised as either interdisciplinary or transdisciplinary. Especially when applied to the study of environment, these forms of thought attempt to challenge the distinctions between empirical and embodied knowledges, as well as the prioritisation of scientific truth over qualitative constructivism.

A term first used by social scientists in the mid-1920s, interdisciplinarity is essentially the convergence of disciplines. The concept reflects a larger tendency to confront the epistemological anxiety of Snow's two cultures dilemma in which compartmentalisation constrains integrative knowledge. Endeavouring to address the limitations of specialisation, especially within the academy, and also projecting back to a supposed pre-disciplinary state of unified knowledge, interdisciplinarity refers to the employment of more than a single discipline. Its premise is that the disciplines together form the foundations of interdisciplinarity; individual disciplines maintain discrete identities within theory and practice.

Interdisciplinarity, in which disciplines collaborate to produce integrated knowledge streams, will be distinguished here from transdisciplinarity, which looks towards enquiry-driven research, creating an interconnected knowledge base. The dividing line between interdisciplinarity and transdisciplinarity depends on definition. The two are not mutually exclusive; transdisciplines will always need methods established in disciplines, and disciplines require thought that is transdisciplinary in order to expand their scope. The works of Michel Serres, discussed above, exemplify some of the possibilities of transdisciplinary thinking in environmental terms.

Interdisciplinarity is defined according to the degree of integration between disciplines and the role of the research enquiry itself. Some definitions of interdisciplinarity verge on transdisciplinarity. Moran (2010, p. 14) defines interdisciplinarity as "any form of dialogue or interaction between two or more disciplines." Most fundamental to interdisciplinarity, according to Klein (1990, p. 13), is a "dispersion of discourse" marked by the placing of research activities within a broader conceptual framework. Repko (2008, p. 6) describes the space between disciplines as "contested terrain." Interdisciplinary research cohesively deals with questions or problems that amalgamate multiple disciplines cooperatively, as Repko argues. Soulé and Press (1998, p. 399) stress the reality of engaged formal and informal interactions between disciplines. They assert that interdisciplinarians need to understand the languages of other disciplines before forging cooperative research.

One of the primary theoretical concerns of interdisciplinarity is whether the knowledge produced is the proximation, integration or transcendence of disciplines. Barthes (1977) asserts that interdisciplinarity is more than disciplinary knowledge streams situated side by side or collaboratively producing new epistemological forms towards practical problem-solving; it is the complete dissolution of disciplinary classification:

> It is indeed as though the *interdisciplinarity* which is today held up as a prime value in research cannot be accomplished by the simple confrontation of specialist branches of knowledge. Interdisciplinarity is not the calm of an easy security; it begins *effectively* (as opposed to the mere expression of a pious wish) when the solidarity of the old disciplines breaks down [italics in original]. (155)

The term *interdisciplinarity*, along with the intellectual terrain it interrogates, is itself contested and can be conflated with transdisciplinarity. Repko (2008) identifies three forms of interdisciplinarity: instrumental, conceptual and critical. Instrumental interdisciplinarity seeks to remedy actual, technical problems. Conceptual interdisciplinarity is similarly pragmatic but tends to amplify a critique of disciplinary perspectives through its processes. Critical interdisciplinarity goes beyond problem solving through disciplinary cooperation, and dismantles the boundaries between disciplines towards an essential transformation of knowledge production. In this third sense, critical interdisciplinarity approaches Barthes's requisite "dissolution" and transdisciplinary creation of knowledge forms, not limited by disciplinary borrowing for problem solving in the real world. Soulé and Press (1998, p. 399) argue that "the identity of all disciplines relies in part on a consensus on the body of authoritative works that practitioners consider to be fundamental." Therefore, a discipline is identifiable through its canon; a transdiscipline will be transcanonical or deconstructive of the canon.

At the heart of transdisciplinarity is critical reflexivity on the processes of enquiry. Expanding interdisciplinarity beyond its disciplinary allegiances, the neologism *transdisciplinarity* appeared in the 1970s in the works of psychologist Jean Piaget, sociologist Edgar Morin and astrophysicist Erich Jantsch to indicate the transgression of knowledge boundaries (Nicolescu, 2002). In the nineteenth century, English polymath William Whewell's concept of consilience was a precursor to transdisciplinary thought, and signified the intertangling of knowledge "where disciplines are not juxtaposed additively but integrated into a new synthesis" (Walls, 1995, p. 11). Borrowing from Whewell's call for integration, *Consilience: The Unity of Knowledge* by biologist Edward O. Wilson (1998) adumbrates a contemporary interpretation of synthesis within biological disciplines and between the sciences and the humanities. Wilson (1998, p. 8) defines consilience as "literally a 'jumping together' of knowledge by the linking of facts and fact-based theory across disciplines to create a common groundwork of explanation."

Reflecting the concept of consilience, transdisciplinarity responds to the fragmentation of knowledge by disciplinary strictures and is distinguished from interdisciplinarity, which Nicolescu argues will always remain within disciplinary frameworks. Repko (2008) identifies a critical distinction between interdisciplinarity, which relies on the disciplines for their theories and methods, and transdisciplinarity, where a problem or theme becomes the core focus of research and the disciplines are effectively transcended through diverse methods. Hence, the knowledge forms emerging from transdisciplinary studies are applicable to a broad spectrum of research problems. Repko (2008, p. 15) stresses that within the humanities during the 1990s, transdisciplinarity was frequently referred to as the "critical evaluation of knowledge forms." Central to the transdisciplinary project is a poetics of the world that reconciles the dualisms of the two cultures divide: "If multidisciplinarity and interdisciplinarity reinforce the dialogue between the two cultures, transdisciplinarity permits us to envisage their open unification" (Nicolescu, 2002, p. 100). As such, transdisciplinarity is a contemporary response to the compartmentalisation of knowledge, prefigured by Arnold and Snow.

An example from my research into aesthetic attitudes towards South-West flora may further clarify transdisciplinarity. In researching plant aesthetics, I have queried science, philosophy, historiography, linguistics, ethnography, arts-based research and ecocriticism. My approach draws from ethnographic interviews, poetic enquiry as autoethnography, and the praxis of field walking (see chapter 2). I have come to realise that visual perceptions of flora are omnipresent in popular culture. In this context, transdisciplinarity engenders a constant critical awareness of how disciplinary boundaries might inhibit a networked comprehension of what is being studied. Transdisciplinarity is more than the assemblage of disciplinary methods for problem solving; it is *a priori* a critical approach to expand the bounds of research towards indeterminate patterns rather than delineated outcomes.

Interdisciplinary studies of ecological issues characterise the field of environmental studies, which focuses on human interactions with the environment. However, the question of transdisciplinary environmental knowledge remains open for interpretation and theoretical elaboration. Environmental studies gained popularity in the 1960s as a result of the conservation movement in the United States. Spurred by Aldo Leopold's *A Sand County Almanac* (Leopold, 1949/1987) and Rachel Carson's *Silent Spring* (Carson, 1962/1982) that forewarned of ecological collapse, environmental studies advocated a greater unification of human and ecological concerns. Academic programs responded to the realisation that ecological problems are "fractious, refractory, and expensive" (Soulé & Press, 1998, p. 398) and defy purely scientific or technical approaches. However, the interdisciplinarity of environmental studies tends towards instrumental and conceptual approaches, in which the practical concerns of conservation or policymaking require the strengths of different disciplines. Within environmental studies, the tensions of identity crisis and divergent ideologies aroused by interdisciplinarity have resulted in a great variety of academic programs, variously stressing environmental science; policy and planning; and cultural studies. At the core of the debate are the differing theoretical and methodological stances of the two major fields: social criticism and natural science. Soulé and Press (1998, p. 400) claim that "the second major group—natural scientists—rarely equate intuition (or narrative) and knowledge." Just as environmental problems themselves are fractious, so is the field of environmental studies internally fragmented by the two cultures ideology.

The interdisciplinary study of plants has been mainly confined to the fields of economic botany or ethnobotany. However, the technicisation of these fields limits their transdisciplinary potential. Ethnobotany uses both qualitative and quantitative strategies of anthropology and botany to understand the usage and perception of plants by people. In 1895, the American botanist John William Harshberger proposed "ethnobotany" as the study of plants used by traditional people (Cotton, 1996). Ethnobotany borrows interdisciplinarily from social science and botanical science for researching human-plant interdependencies (Martin, 2004). *Ethnobotany: A Methods Manual* (Martin, 2004, p. 3) enumerates botany, pharmacology, anthropology, ecology, economics, linguistics and conservation science as contributing fields. Ethnobotany strives towards four objectives: documentation of plant knowledge; quantitative assessment of the use and management of plants; experimental evaluation of the benefits resulting from plants; and applied research aiming to maximise the value derived from botanical knowledges (Martin, 2004, p. 3). Economic botany is a specific subset of ethnobotany that stresses the economic benefits of botanical conservation (Martin, 2004, p. 172). Clarke (2008, p. 150) distinguishes between economic botany, focused on industrial uses of plants, and ethnobotany, concerned with indigenous people's interactions with plants. At the centre of economic botany may be the prerogative for local, indigenous foods, fibres and medicines to be transformed into global commodities (Wickens, 2001, p. 55).

Cultural studies, social criticism, literature and philosophy branched off from environmental studies into the environmental or ecological humanities. In these fields, integration between the sciences and humanities is realised outside the dualisms of techno-scientific paradigms. The environmental humanities, as defined by the University of Utah, which in 2007 launched one of the first postgraduate courses dedicated to the emerging field of study, engages "broad-based understanding of social, cultural, ethical, historical, communication, and literary perspectives…with a focus on how these humanities perspectives intersect with and influence public policy, scientific, legal, industrial, and corporate concerns" (Environmental Humanities Graduate Program, 2010, Introductory section, para. 3). The environmental humanities assert that ecological problems have resulted from thinking that posits environment as external to culture. Inherent to the environmental humanities is a critique of classical science's replication of dualistic thinking.

Within Australia, the environmental humanities have taken the more conceptually robust form of the ecological humanities, first theorised by Deborah Bird Rose and Libby Robin. The ecological humanities ameliorate the arts and sciences divide towards ecological thinking by critically addressing two cultures binaries. Ecological issues are "situated across the nature/culture divide" (Rose & Robin, 2004, Introductory section, para. 2). An ontology of connectivity synthesises Aboriginal, embodied and postmodern feminist knowledge, as well as the scientific revisioning championed by researchers, such as Prigogine, who cross-cut the sciences and arts distinction towards connectivity and uncertainty. Griffiths (2007) outlines three techniques of humanities research that enhance the study of environments and ecological issues: scales of space and time; storytelling; and science as subject. The humanities augment the scale of science towards "human-scale geographies" through narrative reflexivity (Griffiths, 2007, Scales of space and time, para. 2). In the ecological humanities, environmental transdisciplinarity is nascent.

Interlude IV: The Western Australian Herbarium

In "When I Heard the Learn'd Astronomer," the nineteenth century poet Walt Whitman (1855/1969, p. 326) found it more life-affirming to experience first-hand the sky that science quantifies with charts and statistics. Poems are like gnamma holes, gathering multiple narratives in a stony crucible and decocting something new from the desirable interfaces between the sciences and the humanities, nature and culture. An herbarium, a collection of preserved botanical specimens, is a resource for plant knowledge, a "circulating reference" (Latour, 1999, p. 38). The discourse of taxonomy engenders ways of knowing plants through ideal examples set out as two-dimensional specimens. Despite its usefulness, an herbarium is nevertheless a place of the dead. The classical scientific culture of plants prescribes rules of engagement to preserve the collections of lifeless matter for the furtherance of botanical knowledge. An herbarium is foremost a place of identification, just as a morgue can be. Yet, in isolating the identity of a plant, the herbarium may conceal the cultural history of a flora, as the second stanza of the poem alludes. Rather than pressed specimens, I am interested in living plants in the field where their sensuous realities pulsate still and are perceivable on their own terms.

The Western Australian Herbarium
Bentley, WA

ambulating between the high-stacked
shelves of pressed botanic specimens,
here it is, the entirety of the plant world
condensed into tightly filed
two-d schemata;

sealed with the drool of Gardner's jealous
lordship, Drummond's imprecise fingers
producing sloppy dispatches, Molloy's love
letters to Mangles in England across swollen
sullen seas;

out of sort, like the survivor wandering
the mortuary, I am unaware of the etiquette
of the dead: living things to be
microwaved, microbes and contaminants,
to be destroyed, then

searched for pistils and stamens
to approximate agreements between
plants, stacked in sheets of
yesterday's news, poised for
the pry of elucidation.

wattles, kangaroo paws, stylidium
sacrificial lambs here to the incessant
need for imprints; day pours through panes
and reminded of Whitman's line from
"When I Heard the Learn'd Astronomer:"

Till rising and gliding out, I wander'd off by myself,

I drift out into the limitless
breathable air again, particles
of light suffusing pores again,
the interior state of being alive

ruptured in the open sky caesura.

Figure 3 Prostrate Rosette in the Wheatbelt of Western Australia. *Image courtesy of the author.*

Poeticising Plant Research: Floral Poetics

Science is often like the grub, which, though it has nestled in the very germ of the fruit, and so perhaps blighted or consumed it, has never truly tasted it.
Thoreau (2000, p. 242)

A poet follows fleeting insight into the natural world, insight that may be unrepeatable and unstructured. Science is thought to embody empirical reason, whereas poetry and the humanities broadly deal with variable states of culture. Such epistemological dichotomies face the ecological transdisciplinarian. Ecocriticism offers a framework encompassing two cultures dialogue towards a poetics of the natural world. Moran (2010, pp. 154-160) describes ecocriticism as a field that melds the concerns of cultural and literary criticism with those of the natural sciences and geography towards ameliorating the conceptual differences between nature and culture. Glotfelty (cited in 2004, p. 3) in the seminal book *The Ecocriticism Reader* defines ecocriticism as "the study of the relationship between literature and the physical environment...ecocriticism takes an earth-centred approach to literary studies." Ecocriticism focuses on the interconnections between cultural forces and natural phenomena, as well as the appropriation of nature and the proliferation of hierarchical power dynamics between non-human and human life.

Yet, ecocriticm may serve literary disciplinarity rather than ecological transdisciplinarity. Beyond ecocriticism's auspices, several hard-to-classify writers evidence a fuller integration of poetics and plants. Borrowing from Berthold (2004, p. 206), "floral poetics" goes beyond ecocriticism towards the intermingling of science, poetry and poetic thinking. A poetics of plants is both literary and scientific. The philosopher and ecologist Henry David Thoreau (1817-1862), the prose writer and conservation biologist Aldo Leopold (1887-1948), and the Western Australian ecologist George Seddon (1927-2007) sought "the scientific ideal and literary temptation" (Serres & Latour, 1995, p. 29). Thoreau, Leopold and Seddon evidence transdisciplinary approaches of generalist curiosity rather than disciplinary closure. Their works exemplify poetic and scientific visions of flora going beyond environmental studies, ethnobotany, economic botany and literary ecocriticism.

Thoreau, in his floristically minded *Faith in a Seed* (1993) and *Wild Fruits* (2000), evidences a Humboldtian view of science and literature. Walls (1995) characterises Thoreau as a paragon of post-disciplinary practice who sought consilience between disciplines through the interrogation of language. As Walls (1995, p. 13) eloquently argues, Thoreau's writings are particularly embodied versions of botany in which the author "celebrates not the crash of metaphysical dualisms but the murmur of multiple voices and actions, not the ecstasy of transcendental disembodiment but embodiment's perilous and bittersweet joys." Thoreau produced a salient transdisciplinary poetics incorporating multi-sensorial experience, over the seasons and grounded within a place: the environs of Concord, Massachusetts (see chapters 3 and 9).

Thoreau's meticulous observation of flora position him as an exemplar of the poet-botanist tradition (see chapter 3). Bradley Dean (2000, p. xi) comments that "the observations he recorded in his journal ranged from the most purely objective and scientific to the aesthetic and highly subjective." Thoreau's interpretations of flora intersect with the botanical knowledge of his day to produce accessible works that simultaneously enlarged the boundaries of botany and oriented his human body. Importantly, Thoreau preferred the "natural" system of botanical classification, developed by Antoine Laurent de Jussieu, over the Linnaean "artificial" system, the former using a broader spectrum of characteristics to define plant groups and the latter focusing on stamen and pistil numbers (Walls, 1995, p. 136).

Thoreau is an exemplar of cultural botany, a transdisciplinarian who invokes literary metaphor, cultural analysis and experiential context in the expansion and critique of science. His botanical oeuvres suggest that the edges between poetry and science, rather than antagonistic or mutually exclusive, can overlap. Thoreau's later works crystallise his achievements as a local plant expert and a writer of poetic prose, ameliorating the fissure between the two cultures. Thoreau's writings evidence the germination of "literary ecology" in North America (Nabhan, 1993, p. xii). His works foreshadow a transdisciplinary space for exchange between poetic and scientific thinking, whereby that which can be tasted, heard, touched or smelled, is not subordinate to that which can be seen.

Perhaps as a reaction to the increasingly technical science of plants, Thoreau's field approach is ostensibly sensuous and body-engaged, with ruminations on the olfactory, audible, gustatory, palpable and ocular qualities of flora. Non-visual experience constitutes a "bodily eye," for Thoreau (1993, p. 26). Olfaction registers the trademark smells of plants, with white pines possessing a "strong spirituous scent, almost rummy, or like molasses hogshead, which would probably be agreeable to some" (1993, p. 39). Hickory forests echo "even in August…the sound of green pignuts falling from time to time" as Thoreau (1993, p. 143) records audible particularities. The sense of touch *reveals* a cranberry plant: "I was obliged with my finger carefully to trace the slender pedicel through the moss to the vine, where I would pluck the whole together, like jewels worn on or set in these sphagnous breasts of the swamp" (Thoreau, 2000, p. 167). Additionally, Thoreau (1993, p. 87) attends to the mixing of the senses; the thistle has silky inner seed capsules guarded by a prickly involucre: "It is a hedge of imbricated, thin, and narrow leaflets of a light brown colour, and beautifully glossy like silk." His "bodily eye" blends visual acumen with multi-sensorial perception of flora over the seasons.

Thoreau's investigations heralded advances in plant ecology. *Faith in a Seed* is concerned with the dispersal mechanisms of seeds. With *Wild Fruits*, it comprises his unfinished project, the "Kalendar," in which he aimed to record all the events of natural history in Concord during a calendar year (Dean, 2000). Thoreau used an inherently seasonal approach to flora. He gathered and articulated sense impressions and deductions over time, rather than fixating on visual instances of form, colour or reproductive isomorphism. Thoreau assembled a life pattern of flora, instead of extracting events from the broader cycle of plants. Through this fusion of careful empirical observation and tonal sensoriality over time, *Faith in a Seed* provided evidence to contradict the prevailing belief in the spontaneous generation of plants, and demonstrates, to the contrary, that the distribution of seeds occurs through a variety of subtle means: by birds, quadrupeds, wind and the actual bursting forth of the seed from its pod.

Along similar lines, twentieth century American biologist and author Aldo Leopold's seminal work on conservation, *A Sand County Almanac*, published first in 1949, outlines a poetic interpretation of science and nature. The work urges science towards a less reductionist interpretation of conservation. Its structure reflects Leopold's integration of poetic and scientific epistemologies. Part I presents a series of essays sequenced according to the twelve calendrical months, while Part II gives a series of geographically organised dirges, elegies, meditations one of and more scientifically grounded proclamations. The book culminates in Part III with a series of analytical essays describing land ethics, wilderness and aesthetics. Berthold (2004, p. 207) observes, "its shifting styles and tones, its unsettling pattern of self-translation and self-transfiguration—is in fact central to Leopold's project of developing a style which would mirror his vision of a transgressive integration of science and poetics."

In Part II, the essay, "Song of the Gavilan," demonstrates that, at the heart of Leopold's poetic science, is the elision of binaries. Leopold distinguishes trenchantly between the song of the river and the instruments of science, which have yet to either disturb or appreciate the river's glissando. The river exists in an idyllic, pre-scientific state in which the denizens of the Gavilan are the original botanists of the river: "Open the crop of a fat little Mearn's quail and you find an herbarium of subsurface foods scratched from the rocky ground you thought barren" (Leopold, 1949/1987, p. 151). Whereas the quail reveals the fecundity of the ecosystem "you thought barren," science interrupts the cadence through "an ironbound taboo which decrees that the construction of instruments is the domain of science, while the detection of harmony is the domain of poets" (Leopold, 1949/1987, p. 153). On the whole, science is preoccupied with the "process of dismemberment." That the health of the river partly depends on the "perception of its music" is not objectively valid (Leopold, 1949/1987, pp. 153-154). Berthold (2004, p. 212) characterises Leopold's acerbic position as a call "upon science to open itself to a metaphysics—a way of seeing beyond or above the characteristics of things as self-enclosed phenomena." For Leopold, seeing is not merely a visual act but begins with the other perceptual faculties that commonly elude scientific practices. The vision of Leopold is one of embodied presence through the senses, in which the distinctions between humans as land managers and nature as managed object, blur indeterminately. Leopold prompts the question, "Who is managing whom?"

In Western Australia, twentieth century essayist and polymath George Seddon's landmark study *Sense of Place* (1972) is an exegesis on the natural history of the Swan River coastal plain in which Perth is situated. *Sense of Place* is one of the earliest works of place writing in the South-West. A notable figure in Western Australian history, Seddon expressed regionally the figure of Hermes courting both "scientific ideal and literary temptation" (Serres & Latour, 1995, p. 29). The text is divided into three sections, The Land; The Plants; and Man. Each details geology; botany; and human ecology, respectively. Seddon asserts that place is defined not only by human activities, but also by its non-human interdependencies. In later works, Seddon continues to connect familiarity with natural history to sense of place in order to "analyse the *genii loci* of our landscapes and celebrate them" (Seddon, 1997, p. 118).

Botanical research in the context of sense of place counters the tendency of classical science to universalise. For Seddon, sense of place develops through multi-sensorial experience over time. To illustrate sense of place, Seddon offers an autobiographical narrative of his arrival in the South-West of Western Australia. He expresses an initial aversion to the plant life: "The country was all wrong… All the plants scratched your legs. The jarrah was a grotesque parody of a tree, gaunt, misshapen, usually with a few dead limbs, fire-blackened trunk, and barely enough leaves to shade a small ant" (1972, pp. xiii-xiv). In *Landprints* (1997, p. 13), twenty-five years later, Seddon arrives upon a lucid language of revelation and a flowering of the senses. He describes a "profusion of creamy spikes" of *Melaleuca huegelii* and *Acacia rostellifera* "wind-sheared into a dense mound which protects the soil and moulds the landscape" whilst *Templetonia retusa* "puts out its brick-red pea flowers in spring." Such passages record transformations in awareness of place and the dynamic of the mediating self in landscape verging on personal, public and natural histories and resting all the while on a continuum between aversion (the country was all wrong) and revelation (the flowers are all right).

Indeed, for Thoreau and Seddon in particular, sense of place builds through sensations, such as working and moving, and sensory memory (see chapters 8 and 13). Within the ecological humanities, place has been theorised as "embodied spatiality" (Rose & Robin, 2004, Connection with nature and place, para. 6). Victoria King (2003, p. 292) notes that the "sense of the haptic (from the Greek *haptesthai*, 'to touch') is an important yet often unacknowledged factor in understanding Aboriginal paintings and culture." Painting becomes "the actual re-creation of…country through the gestures of the body" (King, 2003, p. 294). Place is that which can be felt through the rhythmic motion of digging yams; sense of place emerges through the brushstrokes of a ground painting. Embodied spatiality entails permeability between bodies and the substance of a place: "The country 'gets under the skin' or 'gets into the blood'; people become 'married to' their country" (Rose & Robin, 2004, Connection with nature and place, para. 6).

The contemporary essay collection *Making Sense of Place* (2008) presents an inclusive spectrum of interpretations and definitions of Australian sense of place that bear some similarities to the approaches of Thoreau and Seddon. In his essay "Place Matters," the environmental sociologist Frank Vanclay (2008, p. 3) defines place as "the coming together of the biophysical, social and spiritual worlds. Simply put, place is space that is special to someone. The personal meanings that turn space into 'place' become embedded in people's memories and in community stories." Meaning-making narratives, sensory engagement with landscapes and consciousness of local flora figure into sense of place for many contributors to the collection. For cultural heritage expert Celmara Pocock (2008, p. 77), touch enables immediate and intimate "sensuous knowledge of place" that is being marginalised as popular tourism locales, such as the Great Barrier Reef, are turned into theatres for visual speculation. Additionally, the sound artist Ros Bandt (2008, p. 96) argues that, in Australia, "each region has its own acoustic identity." For Bandt (2008, p. 95), listening attentively to the aural distinctiveness of places invokes the primary traditional act of knowledge-exchange and space-awareness in Australia. Furthermore, the anthropologist Jane Mulcock interviewed plant enthusiasts about the relationship between indigenous Western Australian species and sense of place gestation. Her interviewees express a shared belief that gardening with local flora engenders "sense of identity, sense of place, and feelings of belonging, all of which can be symbolised by their interest in native plants" (Mulcock, 2008, p. 186). Memories of the smells of plants particularly generate "embodied ecological knowledge" of place (Mulcock, 2008, p. 188).

In addition to this Australia-focused collection, other sense of place literature suggests that place is a complex interaction of bodily, cultural, ecological and spiritual variables, rather than rectilinear space only; for such authors, place-making animates space (Casey, 1998; Malpas, 1999, 2006; Relph, 1976). Indeed, returning to Seddon and Thoreau, sense of place comprises embodied experience of plants, animals and landforms. Tyrrell (2005, p. 752) observes that "place and identity are of key importance in Seddon's work...He has strong affinities with local landscapes, as historical interactions of people and land." Seddon's vision of the sciences and the humanities is through the lens of sense of place as embodied spatiality. Seddon's *The Old Country: Australian Landscapes, Plants and People* (2005) provides interpretations of regional botanical science, including banksia ecology, but imbues factual recitations with cultural overtones. Seddon attends to the multi-sensorial experience of the flora as communicated through his poetic prose. The "scents, sights, sounds—all can stir memories" (Seddon, 2005, p. 128). He then recounts an aspect of the ecology of acorn banksias:

> As I write, in the scorching February of a Perth summer, *Banksia prionotes* is in flower along road and rail reserves, and in odd pockets of bushland and park. The inflorescence is at first a creamy white, but as the individual flowers

> open, moving up the cob, their brilliant orange colour is revealed, showing the reason for the popular name, the acorn banksias. (129)

For Seddon, sense of place includes a narrative multi-sensorial aesthetics. Along with the narrative non-fiction writings of Barbara York Main (chapter 11), Seddon's works provide South-West examples from a distinguished scientist who spanned the two cultures in the tradition of Thoreau and Leopold.

Cultural Botany: Bridging Two Cultures, Building on Cultural Ecology

> The mystery of the life of plants is kindred with that of our own lives, and the physiologist must not presume to explain their growth according to mechanical laws, or as he might explain some machinery of his own making.
>
> Thoreau, 1852 (1999, pp. 89-90)

I have maintained that a less fragmented research paradigm into plants is not to be located within environmental studies, ethnobotany, economic botany or in the form of interdisciplinarity where disciplines cooperate but all the while retain their identities and consequently propound divisive ideologies. Cultural botany is a transdisciplinary model that addresses the divide, offering the possibility that plant research attains embodied and poetic character. Such research recognises how poetry and multi-sensoriality infuse the ways in which people interact with plants. Moreover, cultural botany encourages dialogue between the arts and sciences to expand knowledge bases, while embracing techniques of enquiry that foreground the cultural contexts of living flora. As the transdisciplinary study of plants, cultural botany seeks to integrate philosophy, poetry, the visual arts and cultural studies with scientific understandings.

Convergence between poetic language and taxonomic nomenclature, the sciences and humanities, and aesthetics and techniques provides the groundwork for mutually-reinforcing research, rather than the timeworn debates of disciplinary identity. As the term *plant* itself is a product of the scientific vision, humanities-based research will necessarily confront taxonomic ideas. In recognition of the possibility of consilience, cultural botany evokes botanical science, employing its technical terms and acknowledging its limitations, while the science of plants pursues an increasingly poetic and enculturated perspective. Cultural botany strives to reconnect with the diverse knowledge systems of plants that have been subordinated to a universalised model. These include Aboriginal and folk understandings (see chapters 3, 7, 8 and 13).

In addition to the ecological humanities, cultural ecology—the study of the interactions between human societies and landscapes—offers promising precedents from which cultural botany can be advanced. Research into the cultural ecologies of plants affirms the possibility of cultural botany as a transdisciplinary approach. Head (2007, p. 843) proposes the use of "a battery of diverse methodologies" for researching the cultural interstices between plants and people. Hitchings (2003) employed ethnographic methods to understand the perceptions of the materiality of cultivated plants in London public gardens. Hitchings and Jones (2004, p. 8) also used mobile interviews and field observations performed while strolling among living flora. Mobile ethnography introduces taste, smell, touch and sound into floristic research. Head and Atchison (2009, p. 239) detail several studies in which interviewing methods allow people to "talk about or demonstrate everyday embodied interactions with plants" (see chapter 2). As the authors claim, the accounts of corporeality through cultural ecology are more intimate and multi-sensorial than those offered by empirical biogeographic or social science methods (Head & Atchison, 2009).

Embodied research into plants calls for a context that extends cultural ecology. The prevailing models are largely contained within ethnobotany or economic botany. Yet, the limitations of those paradigms highlight the need to synthesise trends in transdisciplinarity, ecocriticism and cultural ecology towards a poetic understanding of plants (see table 1.1). The context of cultural botany brings together the ethnographic and spatial methodologies of the social sciences, the analytic and textual strengths of the humanities, and the taxonomic and ecological understandings of botanical science towards a multi-faceted articulation of the cultural flows between people and plants. Chapter 1 has circumscribed cultural botany, has sketched how it might be positioned in the strata of environmental disciplines, and has provided examples of cultural botanists. Having drawn the circle widely, a specific approach in a specific place is the subject of chapter 2.

Table 1 Ten Qualities of Cultural Botany. Knowledge of plants is situated seamlessly across timeworn disciplinary fissures between the sciences and humanities to foster botanical research that is transdisciplinary, cultural, poetic, imaginative and narrative

- enquiry driven
- critically transdisciplinary
- culturally embedded
- poetically made
- sensorily rich
- temporally positioned
- narratively expressed
- syncretically based
- botanically imaginative
- epistemologically placed

References

Arber, A. (1912/1938). *Herbals: Their origin and evolution* (3 ed.). Cambridge, England: Cambridge University Press.

Arnold, M. (1882/2011). Literature and science Retrieved 4 April, 2011, from http://homes.chass.utoronto.ca/~ian/arnold.htm

Bandt, R. (2008). Place as acoustic space: Hearing Australian identity. In F. Vanclay, M. Higgins & A. Blackshaw (Eds.), *Making sense of place: Exploring concepts and expressions of place through different senses and lenses* (pp. 95-102). Canberra, Australia: National Museum of Australia Press.

Barthes, R. (1977). *Image, music, text* (S. Heath, Trans.). London, England: Fontana Press.

Berg, L. (2008). *Introductory botany: Plants, people, and the environment* (2 ed.). Belmont, CA: Thomson Learning, Inc.

Berlin, I. (1979). *Against the current: Essays in the history of ideas*. London, England: Hogarth Press.

Berthold, D. (2004). Aldo Leopold: In search of a poetic science. *Human Ecology Review*, 11(3), 205-214 Retrieved from http://www.humanecologyreview.org/pastissues/her113/berthold.pdf

Blunt, W. (2004). *Linnaeus: The compleat naturalist*. London, England: Francis Lincoln Limited.

Carson, R. (1962/1982). *Silent spring*. Harmondsworth, England: Penguin in association with Hamilton.

Casey, E. (1998). *The fate of place: A philosophical history*. Berkeley, CA: University of California Press.

Clarke, P. (2007). *Aboriginal people and their plants*. Dural Delivery Centre, Australia: Rosenberg Publishing.

Clarke, P. (2008). *Aboriginal plant collectors: Botanists and Australian Aboriginal people in the nineteenth century*. Dural Delivery Centre, Australia: Rosenberg Publishing

Conley, V. A. (1993). Eco-Subjects. In V. A. Conley (Ed.), *Rethinking technologies* (pp. 77-91). Minneapolis, MN: University of Minnesota Press.

Cotton, C. M. (1996). *Ethnobotany: Principles and applications*. Chichester, England: John Wiley & Sons.

Culpeper, N. (1653/1981). *Culpeper's complete herbal and English physician*. Manchester, England: J. Gleave and Son.

Daw, B., Walley, T., & Keighery, G. (1997). *Bush tucker plants of the South-West*. Kensington, Australia: Department of Environment and Conservation.

Dean, B. P. (2000). Introduction. In B. P. Dean (Ed.), *Wild fruits: Thoreau's rediscovered last manuscript* (pp. ix-xvii). New York, NY: W.W. Norton.

Environmental Humanities Graduate Program. (2010). About the EH graduate program Retrieved April 4, 2011, from http://www.hum.utah.edu/eh/?pageId=1083

Foucault, M. (1966/2002). *The order of things: An archaeology of the human sciences.* London, England: Routledge.

Garrard, G. (2004). *Ecocriticism.* Oxfordshire, England: Routledge.

Giblett, R. (2011). *People and places of nature and culture.* Bristol, England: Intellect.

Glazebrook, T. (2000). *Heidegger's philosophy of science.* New York, NY: Fordham University Press.

Goethe, J. W. V. (1790/2009). *The metamorphosis of plants.* Cambridge, MA: MIT Press.

Goethe, J. W. V. (1828/2010). *Faust* [Google Books]. (G. De Nerval, Trans.) Retrieved from http://books.google.com/

Griffiths, T. (2007). The humanities and an environmentally sustainable Australia. *Australian Humanities Review,* (43) Retrieved from http://www.australianhumanitiesreview.org/archive/Issue-December-2007/EcoHumanities/EcoGriffiths.html

Harari, J., & Bell, D. (1982). Introduction. In M. Serres, J. Harari & D. Bell (Eds.), *Hermes: Literature, science, philosophy* (pp. ix-xl). Baltimore, MD: Johns Hopkins University Press.

Haraway, D. J. (1991). Situated knowledges: The science question in feminism and the privilege of partial perspective. In *Simians, cyborgs, and women: The reinvention of nature* (pp. 183-201). New York, NY: Routledge.

Hassell, E. (1975). *My dusky friends.* Dalkeith, Australia: C.W. Hassell.

Head, L. (2007). Cultural ecology: The problematic human and the terms of engagement. *Progress in Human Geography, 31*(6), 837-846.

Head, L., & Atchison, J. (2009). Cultural ecology: Emerging human-plant geographies. *Progress in Human Geography, 33*(2), 236-245.

Heidegger, M. (1977). The question concerning technology. In D. Krell (Ed.), *Martin Heidegger: Basic writings* (pp. 284-317). New York, NY: Harper & Row.

Hickman, E., & Ryan, J. (*in press*). *Two With Nature: Ellen Hickman and John Ryan.* Fremantle, Australia: Fremantle Press.

Hitchings, R. (2003). People, plants and performance: On actor network theory and the material pleasures of the private garden. *Social & Cultural Geography, 4*(1), 99-113.

Hitchings, R., & Jones, V. (2004). Living with plants and the exploration of botanical encounter within human geographic research practice. *Ethics, Place & Environment, 7*(1 & 2), 3-18.

King, V. (2003). Embodied perceptions: Aboriginal expressions of place. In J. Cameron (Ed.), *Changing places: Re-imagining Australia* (pp. 286-295). Double Bay, Australia: Longueville Books.

Klein, J. T. (1990). *Interdisciplinarity: History, theory, and practice*. Detroit, MI: Wayne State University Press.

Lash, S., & Friedman, J. (1992). Introduction. In S. Lash & J. Friedman (Eds.), *Modernity and identity* (pp. 1-30). Oxford, UK: Blackwell.

Latour, B. (1999). Circulating reference: Sampling the soil in the Amazon Forest. In *Pandora's hope: Essays on the reality of science studies* (pp. 24-79). Cambridge, MA: Harvard University Press.

Leopold, A. (1949/1987). *A Sand County almanac* and *Sketches here and there*. Oxford, England: Oxford University Press.

Malpas, J. E. (1999). *Place and experience: A philosophical topography*. Cambridge, England: Cambridge University Press.

Malpas, J. E. (2006). *Heidegger's topology: Being, place, world*. Cambridge, MA: The MIT Press.

Martin, G. (2004). *Ethnobotany: A methods manual*. London, England: Earthscan.

Meagher, S. J. (1974). The food resources of the Aborigines of the South-west of Western Australia. *WA Museum Records, 3*(1), 14-65.

Miller, D., & Reill, P. (1996). *Visions of empire: Voyages, botany, and representations of nature*. Cambridge, England: Cambridge University Press.

Miller, G. (2009). Preface. In G. Miller (Ed.), *The metamorphosis of plants* (pp. xi-xiv). Cambridge, MA: MIT Press.

Moran, J. (2010). *Interdisciplinarity* (2 ed.). Abington, England: Routledge.

Mulcock, J. (2008). Planting natives: Gardening and belonging to place in Perth, Western Australia. In F. Vanclay, M. Higgins & A. Blackshaw (Eds.), *Making sense of place: Exploring concepts and expressions of place through different senses and lenses* (pp. 183-189). Canberra, Australia: National Museum of Australia Press.

Nabhan, G. (1993). Foreword: Learning the language of fields and forests. In B. Dean (Ed.), *Faith in a seed: The first publication of Thoreau's last manuscript* (pp. xi-xvii). Washington, DC: Island Press.

Nicolescu, B. (2002). *Manifesto of transdisciplinarity* (K.-C. Voss, Trans.). Albany, NY: State University of New York Press.

Pocock, C. (2008). Reaching for the reef: Exploring place through touch. In F. Vanclay, M. Higgins & A. Blackshaw (Eds.), *Making sense of place: Exploring concepts and expressions of place through different senses and lenses* (pp. 77-85). Canberra, Australia: National Museum of Australia Press.

Prigogine, I., & Stengers, I. (1984). *Order out of chaos: Man's new dialogue with nature*. London, England: Shambhala.

Relph, E. (1976). *Place and placelessness*. London, England: Pion.

Repko, A. F. (2008). *Interdisciplinary research: Process and theory*. Los Angeles, CA: Sage.

Rose, D., & Robin, L. (2004). The ecological humanities in action: An invitation. *Australian Humanities Review, 31-32* Retrieved from http://www.australianhumanitiesreview.org/archive/Issue-April-2004/rose.html

Schalow, F., & Denker, A. (2010). *Historical dictionary of Heidegger's philosophy*. Lanham, MD: Scarecrow Press.

Schiebinger, L. (2004). The private lives of plants. In *Nature's body: Gender in the making of modern science* (pp. 11-39). New Brunswick, NJ: Rutgers University Press.

Seddon, G. (1972). *Sense of place: A response to an environment*. Perth, Australia: University of Western Australia Press.

Seddon, G. (1997). *Landprints: Reflections on place and landscape*. Cambridge, England: Cambridge University Press.

Seddon, G. (2005). *The old country: Australian landscapes, plants and people*. Cambridge, England: Cambridge University Press.

Serres, M. (1982). *Hermes: Literature, science, philosophy* (J. Harari & D. Bell, Eds.). Baltimore, MD: Johns Hopkins University Press.

Serres, M. (1995). *The natural contract* (E. MacArthur & W. Paulson, Trans.). Ann Arbor, MI: University of Michigan Press.

Serres, M., & Latour, B. (1995). *Conversations on science, culture, and time* (R. Lapidus, Trans.). Ann Arbor, MI: University of Michigan Press.

Serres, M., & Zournazi, M. (2002). The art of living: A conversation with Michel Serres. In M. Zournazi (Ed.), *Hope: New philosophies for change* (pp. 192-209). New York, NY: Routledge.

Snow, C. P. (1964/1993). *The two cultures*. Cambridge, England: Cambridge University Press.

Soulé, M., & Press, D. (1998). What is environmental studies? *BioScience*, 48(5), 397-405.

Thoreau, H. D. (1993). *Faith in a seed: The dispersion of seeds and other late natural history writings*. Washington, DC: Island Press.

Thoreau, H. D. (1999). *Material faith: Thoreau on science*. (L.D. Walls, Ed.). Boston, MA: Houghton Mifflin Company.

Thoreau, H. D. (2000). *Wild fruits: Thoreau's rediscovered last manuscript*. New York, NY: W.W. Norton.

Tyrrell, I. (2005). placing george seddon. *Environmental History, 10*(4), 752-754.

Vanclay, F. (2008). Place matters. In F. Vanclay, M. Higgins & A. Blackshaw (Eds.), *Making sense of place: Exploring concepts and expressions of place through different senses and lenses* (pp. 3-11). Canberra, Australia: National Museum of Australia Press.

Von Humboldt, A. (1850). *Cosmos: A sketch of a physical description of the universe* (E. C. Otté, Trans. Vol. 2). New York, NY: Harper & Brothers.

Walls, L. D. (1995). *Seeing new worlds: Henry David Thoreau and nineteenth-century natural science*. Madison, WI: University of Wisconsin Press.

Whitman, W. (1855/1969). *Leaves of grass* [Google Books] Retrieved from http://books.google.com

Wickens, G. E. (2001). *Economic botany: Principles and practices*. Dordrecht, Netherlands: Kluwer Academic.

Wilson, E. O. (1998). *Consilience: The unity of knowledge*. New York, NY: Alfred A. Knopf.

Wright, J. (1994). *Judith Wright: Collected poems 1942-1985*. Sydney, Australia: Angus & Robertson.

2

Fields

Abstract: This chapter will specify three methods for researching South-West Australian flora through the lens of cultural botany: poetic enquiry; ethnography; and gestural walking. The research approach grounds, materialises and regionalises a botanical aesthetics. As components of a field-based approach, the methods constitute a specific engagement with indigenous flora, one designed to elicit the dynamics between visual floral aesthetics and embodied floraesthesis; between ocular apprehension and multisensory engagement; and between a metaphysics of being and a poetics of becoming. The three-pronged approach, *botanic field aesthetics*, begins with living plants in their biodiverse habitats, rather than with dried herbaria specimens; cultivated plants in public or private gardens; or popular botanical images.

Introduction to the South-West

The South-West corner of Western Australia (*the South-West*) is a biodiversity hotspot of international significance and one of the most floristically diverse regions in the world (Breeden & Breeden, 2010; Conservation International, 2007, 2008; Corrick & Fuhrer, 2002; Hopper, 1998, 2004). The region is the only globally recognised Australian "hotspot" (Conservation International, 2007). Thirty five percent of its plants are endemic, or found to occur in uncultivated conditions only within the South-West. In the late nineteenth century, botanist Baron von Mueller identified the "botanical province" because of its distinctive flora (Beard, 1979 p. 107). The province holds 80% of the endemic plants of Western Australian (Paczkowska & Chapman, 2000). Extending from Shark Bay in the upper northwest corner to Israelite Bay east of Esperance in the southeast corner, the South-West, including metropolitan Perth, contains plants well-adapted to nutrient-poor soils (Corrick & Fuhrer, 2002).

Constraints of climate and soils have given rise to remarkable species diversity. The relatively flat land exhibits a range of soil types, which give genesis to the unusual diversity of plants (Corrick & Fuhrer, 2002 p. 13). The province has long been isolated from the rest of Australia by the arid limestone of the Nullarbor plains (Hopper, 1993; Seddon, 1972). The biologist and writer Barbara York Main (1967) comments:

> There is no landscape more ancient than this anywhere and, because of its age, it has been able, for aeons, to receive and support a fauna and vegetation, limited in variety and density only by the rigorous requirements set by the relatively barren nature of its soils and hazardous, climactic conditions. (42)

Remarkably varied and venerable plants have evolved through this rare combination of stable climate, geographical isolation and lateritic soils. There are over 8,000 species of indigenous plants, or more than fourteen times the number found in the United Kingdom (Corrick & Fuhrer, 2002; Paczkowska & Chapman, 2000). Four thousand are flowering plants, one-third of which have been scientifically described only in the past three decades (pers. comm., S. Hopper, September 9, 2009).

Interlude V: The Rites of Spring

Wildflower photography and plant conservation mix fluidly in the visually stunning *Wildflower Country* by Stanley and Kaisa Breeden (2010) in which the authors convey an ethos of conservation of indigenous flora augmented by novel photographic techniques. However, photographers are not always "greenies," as "Observing the Rites of Flower Season" attests. The poem retells an encounter with a group of three retired professional photographers in the spring of 2009 in Hopetoun near the Fitzgerald River National Park. I was interested in the visual sensibility skilled image-makers bring to photographic encounters in the field. A technical language of vision surrounds the use of the camera (Crary, 1990, pp. 37, 96). Photographic acts seem ritualistic and purposive, like rites of the spring season. Image-making provides a way for visitors to connect to their surroundings by creating visual products. Wildflower tourism based on images may include the experience of motorised transport. This can create a dispassionate regard for the ecological health of the regions through which image-seekers pass:

> telepathically. next morning, our Queensland driver
> nearly flattens a bobtail sunning on the gravel;
> another is from Kansas. he ends each statement
> kindly with *Do you know what I'm saying man?* (ll. 9-12)

Places like the Fitzgerald River are world-renowned biodiversity sanctuaries, but what else do visitors gain from the experience of the bush besides colourful images of flowers with forgotten names? (to read "The Rites of Spring" in full, see Hickman & Ryan, *in press*).

Figure 4 The Fitzgerald River Coastline. The pounding of surf against rocks is characteristic of the soundscape of this biodiverse place between Albany and Esperance, Western Australia. *Image courtesy of the author.*

Reconfiguring Botanical Field Work

Chapter 1 argued that the fields of environmental studies and ethnobotany perpetuate two cultures divisiveness. Following Latour's position, science claims to de-culturate plants, removing them from specific niches on one side of the globe in order to reconfigure them into predictable abstractions on the other side, far removed from habitats of origin (see chapter 4). The tools of science transport the abstract data of plant names and chemical constituents—as well as the dead and living organic material of specimens and seeds—across space and time beyond contexts of ecology and culture. In contrast, cultural botany comprises a range of perspectives and methodologies towards an embodied appreciation of plants in a place. Cultural botany concerns the artistic, literary, philosophical and cultural practices that engage, invoke, interpret or rely upon living plants. It articulates plants as cultural and poetic phenomena beyond the visual objects of Linnaeanism.

Botanic field aesthetics is a practice of cultural botany at sites in the South-West that brings the humanities, generally, and cultural studies, particularly, to the study of flora. In chapter 2, I delineate three modalities selected for exploring the aesthetics of South-West plants: poetic enquiry, anthoethnography and gestural walking. The approach provides qualitative, humanities-based "data" to support the possibility of an embodied aesthetics of plants (see chapter 13). An integrative Thoreauvian-Heideggerean analytical framework critiques ocularcentrism and emphasises multi-sensoriality through poetic language (see chapters 1, 9 and 10). Through poetry, ethnographic transcripts and gestural walking, the data trace the continuum from floral aesthetics to floraesthesis, from visual apprehension to multi-sensorial sensation, from a metaphysics of being to a poetics of plants in-becoming. As a three-pronged methodology, botanic field aesthetics begins with living plants in their habitats, rather than with dried herbaria specimens, biochemical extracts in a laboratory, cultivated plants in gardens or popular horticultural images. At the end of the chapter, I discuss regionalism and the three selected study sites.

Botanic field aesthetics explores the interworkings of three qualitative techniques. Poetic enquiry extends to the study of flora recent scholarship on the value of poetic practice in the social sciences (Cole & Knowles, 2007; Leggo, 2007; Neilson, 2007; Thomas, 2004). Anthoethnography is the application of ethnography in researching cultures of flora and the forms of language that communicate aesthetic experiences of the natural world. This second branch of the approach explores perceptions of South-West plants through semi-structured interviewing of regional botanists and local experts, as well as participant observation of wildflower tourism (see Clifford, 1986; Geertz, 1988; Head, 2007; Head & Atchison, 2009; Hitchings, 2003; Pratt, 1986). Gestural walking in the botanic field is a participatory aesthetics of flora (see also Edensor, 2001; Mules, 2006, 2008; Solnit, 2001; Thoreau, 1862/2007). In this study, botanic field aesthetics has been organised around three regional sites: Anstey-Keane Damplands, Lesueur National Park and Fitzgerald River National Park. Crucial to the development of botanic field aesthetics is a bodily sense of place: an awareness of a region or locale through sustained physical interaction with its biota (Giblett, 2006; King, 2003; Main, 1967; Rose & Robin, 2004; Seddon, 1972, 2005).

As a practice of cultural botany, botanic field aesthetics involves walking, writing poetry and talking to people about plants. Walking and writing about flora, sometimes poetically, have been integral to the history of South-West botany ever since Willem Vlamingh first collected specimens along the Swan River in 1696 (Marchant et al., 1987, p. 1). In 1699, William Dampier landed the *Roebuck* at Shark Bay where he collected plants and later published drawings (Clarke, 2008, p. 9). Explorers George Grey (1841), John Eyre (1845/1964) and the Gregory brothers (Gregory & Gregory, 1884/1981) traversed some of the most botanically diverse tracts of the state, including present-day Fitzgerald River and Lesueur national parks, although their botanical exploits were limited by the difficulties of field travel (see chapter 6). Between 1829 and 1860, James Drummond (1848) collected a wide variety of plants. Naturalist Ludwig Preiss made collections between 1839 and 1842 (Marchant, et al., 1987). In the mid to later nineteenth century, Georgiana Molloy meticulously pressed specimens of plants near her home in Augusta, many of which are preserved today at the Kew Herbarium in England (Lines, 1994). Botanists and plant collectors like Fraser, Molloy, Lindley and Drummond contributed to early understandings of the flora largely through the emerging international commerce of seeds for greenhouse collections. Early twentieth century biogeographers, such as Charles Gardner (1959/1981), and contemporary conservation botanists, such as Stephen Hopper (1998) and Alex George (2002; 2002), continue the fieldwork tradition.

Gascoigne (1996) and Clarke (2008) suggest that field studies are normatively the domain of taxonomic science and have their origins in the colonising explorations of explorer-botanists. Through a post-colonial and anti-foundationalist framework, botanic field aesthetics is initiated by transdisciplinarity. How do I trace the complex relation between visual appreciation and embodied entanglement? This approach is based in the humanities and supplements field data with readings from a palette of resources: the theories of Thoreau and Heidegger; regional literature (including the sense of place essays of Seddon); Aboriginal understandings; the journals of nineteenth century explorers; natural history writings; poets who write about plants; and the aesthetic narratives of tourists and conservationists. As a language-focused approach, botanic field aesthetics seeks multi-sensorial participation with—rather than visual conquest of—plants.

In considering the question of an aesthetics of flora, I arrive at a question. Should the field approach focus on the qualities of individual plants or the overall plant community as a panoramic scape? Whereas an individual plant is beautiful, handsome or striking, a vista of plants can be picturesque, pleasing, breathtaking or sublime (see chapter 5). The perspectival variations between a plant or its plantscape parallels the difference between the orchid and everlasting effects (see chapter 7). Hence, differing aesthetic vocabularies emerge depending on the character and scale of the plants. The approach considers the relationship between a plant and a plantscape, and the perceptual adjustments and exclusions that happen between discrete plants and the unfolding vista (for example, see "Understanding Parrot Bush," Interlude II). At Fitzgerald River National Park near Esperance, for example, even at the height of the flowering season, the plantscape appears as an olive monochrome, whereas individual plants such as the Qualup Bell, when beheld at a close range, are rich with varied hues of colour, the sound of wind through pendulous flowers and other multi-sensorial affordances. Indeed, there are the smellscapes and soundscapes composed by flora, but any scape implies broad constitutions of sense. As Rodaway (2002, p. 64) maintains, smell evades spatial organisation: "Smells infiltrate or linger, appear or fade, rather than take place or situate themselves as a composition."

Poetic Enquiry into South-West Flora

> Where is the literature which gives expression to Nature? He would be a poet who could impress the winds and streams into his service, to speak for him; who nailed words to their primitive senses.
>
> Thoreau (1862/2007, p. 29)

The first branch of the field approach positions language as the shared medium between scientific discourses and the humanities. Poetic enquiry into South-West flora seeks what Hayles (1990, pp. 176-177) refers to as the "two mingled voices" of literature and science. How does arts-based enquiry differ from scientific investigation of plants? Where do they share commonalities? The use of poetic enquiry connects language to aesthetic values. An aesthesis of sensation develops in conjunction with a poetics that gives corporeal significance to flora. Embodied writing emerges from multi-sensorial approaches to plants. The act of writing itself is enquiry; poetry, therefore, is a creative medium for exploring the complexities between places, plants and human subjects beyond the two cultures categories described in chapter 1.

Botanical nomenclature intrinsically has poetic nuances, or what I call *embedded poetics*. In *Deliciae Naturae* (1773), Linnaeus reveals the poetic nature of his sorting and naming. The Swedish botanist describes the cross-over between science and culture: "The botanical terms correspond to letters, the names of the plants to words and the systems to the grammar" (cited in Sharr, 1978, p. vii). Foucault (1966/2002, p. 176) corroborates the linguistic origins of Linnaeus's taxonomy, asserting that scientific nomenclature "resides in its entirety in the area of language, since it is essentially a concerted use of names and since its ultimate aim is to give things their true denomination." Yet, Linnaean language is more than a technical assemblage of letters, words and grammar. The language incorporates poetic idiom. Hence, nomenclature was learned as a technical system for appreciating the musicality of life. Linnaeus (cited in Sharr, 1978, p. vii) states that "to learn a language requires an acquaintance (to some extent at least) with its letters, words, and grammar. Then only, and not until then, can one enjoy all the beautiful compositions in that language." Circulating in technical treatises, scientific names reiterate poetic perceptions, aesthetic tastes and cultural values. In essence, poetry and science are language forms.

The etymologies of taxonomic names suggest the hidden nodes between poetic language and taxonomy. Scientific names encipher cultural stories. The Golden Kangaroo Paw (*Anigozanthos pulcherrimus*), for instance, only occurs in the South-West. In the nineteenth century, James Drummond first sent a Golden Kangaroo Paw specimen to James Hooker in England. Drummond described it as the loveliest flower of the colony. According to Stephen Hopper (pers. comm., September 9,

2009), Hooker chose the species name, *pulcherrimus* meaning, "most beautiful" in Latin. Despite the culture-independent pretence of science, its language is hand-in-glove with history, poetics and aesthetics. The relationship between taxonomic and poetic enquiry—and between the sciences and the humanities—is not of mutual exclusion but of potential overlay (see chapter 1).

Although a method developed in the social sciences, poetic enquiry is applicable to plant studies. Cole and Knowles (2007, p. 59) describe arts-based research as a "mode and form of qualitative research in the social sciences that is influenced by, but not based in, the arts broadly conceived." Arts-based research strives "to enhance understanding of the human condition through alternative (to conventional) processes and representational forms of inquiry" (Cole & Knowles, 2007, p. 59). Although arts-based enquiry attends to the human condition, it may be extended to the natural world. Prendergast (2009, p. xxxv) defines poetic enquiry as "a form of qualitative research in the social sciences that incorporates poetry in some way as a component of an investigation." Leggo (2007, p. 168) characterises poetic enquiry as "a way of knowing, being, and becoming in the world." Poetry unfolds sense experience and the becoming of bodies. Moreover, poetry expresses "ongoing engagement" with the world beyond the demarcations between the creative arts and social sciences imposed by the academy (Leggo, 2007, p. 168). Neilsen (2007) describes "lyric enquiry" as a research process characterised by song-like outcomes. The use of arts-based enquiry in the social sciences can be relevant to the exploration of alternate—to the scientific paradigm—forms of field-based research into flora.

How do the aesthetic qualities of plants change depending on topographic, environmental or cultural influences? How does an aesthetics of the "scenery cult" (Carlson, 2000, p. 34) return to sensation? Poetic enquiry elicits an embodied aesthetics of plants methodologically through the concept of *poiēsis*, which describes the act of making, producing or becoming over time (see chapter 1). It is the temporal nuance of poiēsis as becoming that I want to emphasise, particularly through the poetic interludes in *Green Sense*. Poetry is a way of apprehending the world; plants, as mutable organisms, are in-being and in-becoming. The scientific work of twentieth century plant morphologists and philosophers Agnes Arber (1950) and Rolf Sattler (1978) affirms that poiēsis as becoming—in dynamic conjunction with stasis as being—characterises the growth of plants. The poet Carl Leggo (2007, p. 166) reminds us importantly that the term *poetry* is derived from the ancient word *poiēsis*. Indeed, poetry is what Heidegger (1982, p. 59) refers to as "an experience with language" (see chapter 9). As a form of enquiry into flora, poetic practice is the undergoing of an experience through the intertwined poiēsis of plants and language. My use of poetic enquiry aims to express that being and becoming are connected to bodily awareness and sensory apprehension. Language traces the dehiscence of seeds, decay of flowers, appearance of barrenness after wildflower season, germination of seeds, irruption of flowers and the appearance of fertility after spring rain.

Poetic enquiry can become what Porteous (1996, p. 244) calls "geoautobiography," or the interrelation between personal history and the land's story. An embodied aesthetics in which the materiality of a place fuses in a geoautobiographical fashion with the body of the researcher emerges in the work of Suzanne Thomas. In *Of Earth and Flesh and Bones and Breath* (2004), Thomas uses poetic enquiry to develop a perceptual approach to coastal island environments. She writes: "I experience island as an inter-subjective, corporeal encounter—my human body moving in relation to natural bodies of island(s); my body and the bodies of islands in relation to one another, and to the immensity of the sea" (Thomas, 2009, p. 128). Poetry of sensation alternates with images of decay towards a process aesthetics positioned in the intertidal zone between land and sea. In the poem "Prima Materia," Thomas (2004, p. 170) writes of the body of a dead seal: "Ripe flesh, rotting skin/ lie transmutable/ carrion, offal, microbe, maggot/ dissolving body returns to earth." Through sight, smell, touch and shifting images, her body is the mediating sensorium, the interface of enquiry. The senses transcend the distance of panoramic sight, and rupture the subject–object rigidity between people and landscape, feeling and intellect. The work of Thomas acknowledges that bodily decay is poiēsis.

As the poetic interludes throughout *Green Sense* aim to demonstrate, poetry is linked to aesthesis. Literally meaning "between play," an interlude is an intervening part between the longer segments of a performance. Unlike an intermission, an interlude entails change in continuity but without the more drastic interruption of a pause. An interlude may be an *entr'acte*, a short play within a play, or an *intermezzo*, a short composition within a larger work. The tone, genre or modality of the interlude might consist of a shift from the larger work, such as the use of comic relief in a tragedy or a musical interlude in a play. Through an embodied narrative, poetic interludes reaffirm my position, not as a disembodied voice, but as a participant in the sensuous worlds of the plants about which I produce knowledge. The interludes are palimpsests in which poetry overlays photographic images to augment visual appeal through sense narratives.

Responding to Porteous's notion of geoautobiography, my approach is "geoautoethnography." It is the exploration of my relation to place through poetic ethnography relating research behaviours to the plants being studied. Poetic enquiry incubates personal memories, cultural histories, quirky anecdotes, taxonomic nomenclatures, metaphorical associations and emotional insights. The interludes include intriguing historical or cultural aspects of the plants I poeticise. As part of botanic field aesthetics, poetic enquiry interweaves subjective experience of plants with its rationalisation. However, as a caveat, place-based sensory embodiment, while a critical component of cultural botany, recognises the difficulties of the transcendent dissolution of the subject-object distinction (Heidegger, 1971) (see chapter 9). Curry (2010, p. 206) characterises such a position as a "viable middle way, grounded in our embodied, imperfect, unstable, liminal nature."

The collection of interludes puts into practice the use of poetic enquiry in the context of South-West cultural botany. By expressing instances of sensation, interludes mark out the complexities between floral aesthetics and aesthesis. Each poem conveys threshold moments between perception of appearances and the experience of sensations. The interlude structure oscillates between personal rumination and theoretical analysis, a style employed by arts-based scholars (Knowles, 2001). In particular, I draw from Carl Leggo (2001, 2004, 2006, 2007) in theorising poetic enquiry, embodiment and flora. Leggo (2007, p. 167) suggests that the ecotones between analytic and creative research are dynamic places, accessible as "textual spaces that invite and create ways of knowing and becoming in the world." An interlude allows a textual space, a pause for rumination, a breath for reflection and the retelling of a moment of contact. Whereas Leggo emphasises the opening of spaces within the logos-privileged social sciences, I propound that poetic enquiry can be similarly extended to studies of nature where the attainment of pithy truths through the objectification of the felt world has produced specialised knowledges. The interludes strive to make tangible the progression to floraesthesis, without nullifying my sensorial experience or my individual position as embodied researcher.

My writing process began with sensation through gesture towards plants. Corporeal responses to the habitat we share for that moment galvanise the poetry. I recorded my impressions in a matrix that indexes the sense features of the plants. I then transmuted the catalogue into verse in the field and over time. Several themes recur, including metaphors that express coterminous and consanguineous relationships between my body and plant bodies. Each poem was finalised months after the field experience, creating a gap of time into which percolates technical facts, namings and lore along with bodily memories (see chapter 8). The poetry aims to express engagements with cultural histories, ecological meanings and sense of place. To this, Porteous (1996) suggests that sense of place occurs narratively through memoirs, essays, fiction and poetry that portray relationships between places and people. However, disjunctions, discontinuities and inversions characterise sense of place as a poetic practice and process (Cameron, 2005).

Interlude VI: Western Quandong

A close relative of sandalwood, Western Quandong (*Santalum acuminatum*) is a small semi-parasitic tree that has been a traditional and contemporary bush food throughout Australia (Isaacs, 1989). Pearman (2005, p. 141) in *The Cultural History of Plants* remarks that the fragrant kernels of the quandong fruit "are generally eaten roasted and have a quite pungent taste...The hard shells have been made into jewelry." "Quandong" alludes to my use of ethnographic engagement with local people to explore cultural knowledge of flora. An informal conversation with Nyoongar heritage guides based at Kodja Place in Kojonup, WA, made me consider ethnography in describing an aesthetics of sensation between people and plants. The fruit of Western Quandong is collected throughout the region. The outer pulp is used for preserves and the inside kernel tastes faintly like macadamia nuts. My memory includes the quandong tree, before its fruiting, against a backdrop of modernity, including sheep trucks, cigarettes and tyre irons:

> pluck—as the double-decker sheep trucks,
> bound for Katanning, tremble the ground;
> now the pulp, tempered to dragon's blood (ll. 7-9)

Originally titled "Quandong Kojonup *Djeran*" to emphasise the mellifluousness of Nyoongar words and the autumn season of *djeran* during which the conversation occurred, the poem suggests that an ecological sense of time emanates from sensorial participation in a place and entrainment to the rhythms of its plants, like the bush plum (to read "Western Quandong" in full, see Hickman & Ryan, *in press*).

Ethnographies Along the South-West Wildflower Trail

The second branch of botanic field aesthetics emphasises collective perceptions of plants through ethnography. The selected methods of participant observation and semi-structured interviewing contribute to the development of an aesthesis of flora (see chapter 13). Ethnographic research conducted during the 2009 and 2010 South-West wildflower tourism seasons suggests that the five senses and the metasenses express various cultural dynamics between plants, people and place. Ethnographic data in the form of interview transcripts, personal observations in the field, photographs of plants and places, poetic interludes drawing from ethnographic encounters, and the desiderata of wildflower tourism allow for comparative understanding of botanical aesthetics in its cultural contexts. All in all, a reading of the language of plant experts and tourists reveals a unifying thread and a monolithic impasse to multi-sensoriality: the visual assignment of value to plants (see chapters 4 and 5).

As a participant, I took part in wildflower tourism through bus tours, wildflower walks on foot or combined with driving, and wildflower shows or celebrations at community centres in the region (see figures 5 and 6). Semi-structured interviewing followed these questions:

- By what names do you know these plants? Do you use scientific, common, colloquial or Nyoongar names?
- How does sensuous experience affect your understanding of plants? Do you smell, taste, touch and listen to plants?
- Would you describe any of your local plants as weird, grotesque, bizarre, strange, beautiful or picturesque?
- To what extent do indigenous plants figure in your sense of Southwest place? In your view, what is especially *South-West* about these species?

Interviews occurred, when possible, as "mobile interviews" conducted while walking with people and the plants being researched (Hitchings & Jones, 2004, p. 8). I used a small, hand-held digital recorder to record the interviews, which were later transcribed. A semi-structured interview deviated typically from the suite of questions, and was tailored to the respondent's expertise, interests, time and

Figure 5 Ethnographies Along the Wildflower Driving Trail, Lesueur-Eneabba. Wildflower touring circuits, including this segment along the Brand Highway, are popular attractions during spring months in Western Australia. *Image courtesy of the author.*

Figure 6 Wildflower Tour from Morawa, WA. Bus tours organised by local wildflower enthusiasts often take tourists into less-frequented areas, such as these everlasting fields near Morawa. *Image courtesy of the author.*

the setting of the interview. Through a Thoreauvian-Heideggerean framework, the transcripts were read as lively texts to locate slippages and convergences between experiences of visuality and corporeality (see chapter 9).

The interviewees were distributed across five types. Briefly, scientific botanists, such as Stephen Hopper, have had formal training and professional experience. Amateur botanists, or local experts, are typically self-taught practitioners of botany who have regional understandings. Whereas scientific botanists are more technically concise about broader biogeographical patterns, amateur botanists are locally focused and broad-based in their knowledge of species. Eclectic botanists draw variously from scientific knowledge, as well as arts-based and embodied knowledges (Thoreau, 1993). Nyoongar botanists invoke Aboriginal worldviews in which sensuous features of plants interdigitate with broader spiritual and ecological meanings. Lastly, wildflower tourists may have complex understandings of plants based on the flowering times of prominent species. These five broad categories of interviewees cross over in actual practice.

Ethnography is a suite of methods used in cultural studies and the social sciences to learn about lived experience and elucidate cultural meanings (Hammersley & Atkinson, 1995). The studies of the five groups produced accounts of flora that are pertinent to broader cultural topographies. Brewer (2000) defines ethnography as fieldwork that requires the participation of the researcher in the social milieu under study:

> Ethnography is the study of people in naturally occurring settings or 'fields' by means of methods which capture their social meanings and ordinary activities, involving the researcher participating directly in the setting, if not also the activities, in order to collect data in a systematic manner but without meaning being imposed on them externally. (10)

Brewer describes participant observation as situating the researcher within settings, activities and rituals. Participant observation differs from ethnographic interviews that use questioning to acquire qualitative data about cultural meanings and practices. Unlike participant observation, ethnographic interviews demarcate settings, times and informants, controlling elements to provide insight that would otherwise be difficult to obtain. Semi-structured interviewing is an effective methodology if meanings are encoded in a practice itself, such as wildflower tourism, and are therefore hard to identify outside of the interview.

Ethnography in cultural research of plants has been historically classed as ethnobotany. Ethnography has a long-standing tradition in ethnobotany for documenting plants in the activities of cultural groups and eliciting life histories in relation to flora (Cotton, 1996; Martin, 2004). Cultural botany, however, is amenable to ethnographic approaches but not constrained to botanical or ethnobotanical logos (see chapter 1). Semi-structured interviews elicit information about views of flora through a conversational format, allowing people to describe their relationships to plants. Participant observation in ethnobotanical studies involves direct participation in the interviewee's activities and behaviours incorporating plants (Martin, 2004). Interviews extend participation in the practice under study. Hammersley and Atkinson (1995, p. 131) note the benefit of using both approaches together: "There are distinct advantages in combining participant observation with interviews; in particular, the data from each can be used to illuminate the other." Gary Martin (2004, p. 96) suggests that "interviews can give us accurate ideas of the ways people describe their lives and their natural surroundings, while participant observation allows us to see how people put their knowledge into practice." Hence, interviews and participant observation complement each other as elements of ethnography.

Reflexive ethnography is the contemporary norm. Rather than a technical account of plants and culture, ethnography engages value perspectives, expressive tendencies and poetic imaginations. The written accounts of ethnography have been upheld as impartial presentations of cultural realities (Pratt, 1986). However, Pratt (1986, p. 28) posits that "it is possible to suggest that ethnographic writing is as trope-governed as any other discursive formation." The presence of authorial voice in ethnographic accounts is argued for extensively by Geertz (1988). An insider approach not only situates a researcher within a culture, in the sense of participant observation. It also permits the examination of textuality through what Geertz (1988) refers to as "I-witnessing," or what I call geoautoethnography. Beyond its social science roots, ethnography may be characterised as an approach to the description of human experience that blends cultural and literary theory (Clifford, 1986). The revaluation of ethnography as a method of cultural botany suggest its utility in the study of plants. Sensory ethnography is discussed in more detail in chapter 8.

Gestural Walking

> The walker in the familiar fields which stretch around my native town sometimes finds himself in another land than is described in their owners' deeds, as it were in some faraway field on the confines of the actual Concord, where her jurisdiction ceases.
>
> Thoreau (1862/2007, p. 3)

As editor of the issue "Peripatetica: The Poetics of Walking" of the e-journal *Landscapes*, I wrote that "'peripatetica' describes the outcomes of walking or its 'productions': a poem, an artwork, a song, a photograph, a story unwritten but recorded in the mind as an oral narrative, a transformation of values or a return to the bodily senses" (Ryan, 2010, p. 1). Certain forms of peripatetic "data" arise from walking. The outcomes of gestural walking include poetry that structurally mimics the process of walking, as well as mobile interviews made interactive because of an outdoor setting. Gestural walking, as I define it, is a *habitus* in the botanic field that amplifies poetic enquiry and ethnography, and provides a basis for experiencing plants in embodied ways (see chapter 12).

Walking is generally thought of as ordinary or perfunctory. As a noun, *pedestrian* refers to one who walks and, as a descriptor, characterises an activity as prosaic. However, places of botanical diversity are commonly preserved as parks where an appreciator is restricted to foot travel. In ascending montane Bluff Knoll and Toolbrunup Peak in the Stirling Range in Western Australia during the Nyoongar season *Djeran*, my heart rate elevated. I experienced sweating, invigorated breathing, occasional cramping in the diaphragm and the cooling wind (see "Three Peaks Triptych," Interlude XI). In contrast, at Anstey-Keane Damplands on the Swan River coastal plain, I became wearied by undulations of sand and pestering flies (see "You Are Known by the Company You Keep," Interlude IX). These sensations constitute a phenology of body responses to place. An aesthetic of flora engendered through gestural walking comprises the varied natural and cultural relations of a plant. Walking from plant to plant, between biotic communities, creates continuity and situates research somatically. Movement on foot summons Georgiana Molloy whose "collecting walks" helped her to learn the flora, connect with knowledgeable Nyoongar people, and overcome frontier isolation (Harper, 2007, p. 8).

Walking puts us in touch with place and disrupts the hierarchies of vision. Rebecca Solnit (2001, p. 12) cogently describes walking as a political act of reclaiming bodily agency: "An apotheosis of speed makes those bodies seem anachronistic or feeble. In this context, walking is a subversive detour, the scenic route through a half-abandoned landscape of ideas and experiences." I conceptualise walking as a dynamic interconnection between being and becoming. The act of walking links people and plants through the corporeal invocation of the senses, both in the moment and over time. In the essay "Walking," Thoreau (1862/2007, p. 9) stresses the relationship between the landscape and his ambulatory body: "There is in fact a sort of harmony discoverable between the capabilities of the landscape within a circle of ten miles' radius, or the limits of an afternoon walk, and the threescore years and ten of human life."

In reading the journals of early explorers who travelled by foot, I noticed a recurrence of picturesque language (see chapter 6). Metaphors that could identify their bodily turmoils tend to be missing. Indeed, such absences may reveal more than their published accounts. Clearly, the walks of the early explorers were not the leisurely perambulations of Thoreau or the Romantic walkers (Edensor, 2001). Starvation, malnutrition, disease, the threat of attack, madness and isolation plagued explorers. Driver (2004) raises questions about the authority of the observations of explorers, and argues that the accuracy of journals was affected by field disturbances: madness, illness, starvation and difficulties with recording information. Yet, how could the body be largely erased? In response to this, I began to consider Thoreau's walking as a *habitus*, as gestural walking in which the senses are open and contact becomes, as Mules (2006, p. 6) puts it, "the openness of the body to the outside, the gesture that makes contact with the world." Thoreau (1862/2007, p. 9) remains sentient when walking as a means of participation: "The thought of some work will run in my head and I am not where my body is—I am out of my senses. In my walks I would fain return to my senses."

Rather than the eyes constructing visual affordances as defences against dangerous environments to be colonised and subdued, bodily experience entails openness towards sensory revelation. Gestural walking involves reaching out to plants, physically through the ears, nose, tongue, mouth and skin, and also through a proximity of vision, putting our eyes closer to the miniscule details that go unregistered by the panoramic juggernaut. Pointing, touching, bending, smelling, picking, tasting, spitting out and throwing away plants are actions that bring us closer. As I practise gestural walking, I become entrained to the rhythms of flora. A phenology of walking connects the seasons to the becoming of plants through phases of flowering, dehiscing, shedding bark and fruiting, as well as interfaces with memory and emotion (see chapters 8 and 11).

Place as Context

At the beginning of 2009, I surveyed sites of botanical diversity in the South-West (see figure 7), scoping out an approach comprising cultural studies and creative writing. In April 2009, I chose two sites of contrasting plant composition, aesthetic character, topography and frequency of visitation: Anstey-Keane Damplands and the Stirling Range National Park. These sites differ dramatically in how the public and policy makers perceive their respective aesthetic values. Anstey-Keane has been called "a seemingly unremarkable parcel of land in Forrestdale" with an abundance of flora species that require sustained contact and attentiveness to appreciate (Giblett & James, 2009). Unlike the coastal plain habitat of Anstey-Keane, the Stirling Range has always held a prominent physical and metaphoric position in the European imagining of the South-West through the mountainous sublime, as well as its significance to Nyoongar people (Keighery & Beard, 1993; Wills & Kinnear, 1993).

At first, I considered including a range of sites, each representing the diversity of the South-West vegetation. Certainly, the aesthetics of the karri forests would differ from the aesthetics of the heathland, just as the Stirling Range contrasted with Anstey-Keane. During the spring wildflower season of 2009, however, I realised the vastness of the biota and decided to limit the scope to three sites with comparable, rather than divergent, character: Anstey-Keane, the Lesueur-Eneabba region in the *kwongan* sandplains between Perth and Geraldton, and the Fitzgerald River National Park between Albany and Esperance on the south coast. Considered barren and sterile by early European explorers, the Lesueur-Eneabba region and Fitzgerald River are places of floristic diversity but historically of low aesthetic and pastoral value (see chapter 6). Similarly, Anstey-Keane Damplands has been characterised as mundane and barren, although it is the second most biodiverse place in the metropolitan area (Giblett & James, 2009). An historical paradox became apparent. Why have some of the most botanically biodiverse places in the South-West been depicted as barren, infertile, mundane and unworthy of protection status? Why, in some instances, do botanical biodiversity and aesthetic perception follow an inverse, rather than parallel, relationship?

Green Sense showcases these three sites (see figure 7). However, passing references are made to other regional places of botanical diversity, such as the Stirling Ranges and the Darling Scarp or the Perth Hills, Western Australia. A regional emphasis contextualises poetic enquiry, anthoethnography and gestural walking in the South-West, rather than other botanically significant places in the

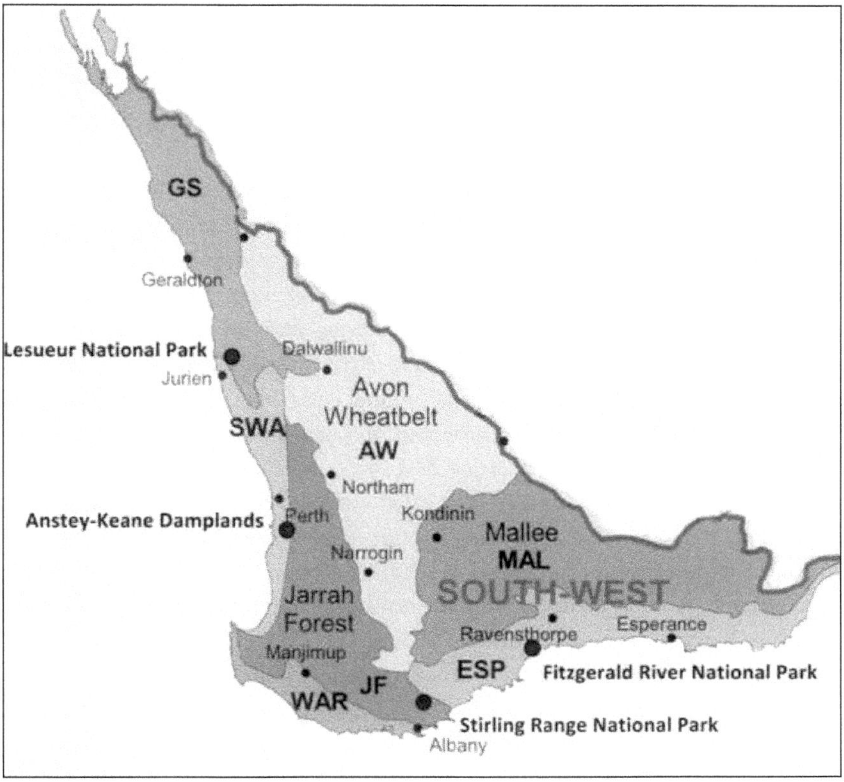

Figure 7 Places of Botanical Diversity in the South-West. The three primary study sites are indicated. The one exploratory site, Stirling Range National Park, is also indicated. *Adapted from* 'Figure 1: Western Australian Biogeographic Regions and Botanical Provinces' (after Thackway and Cresswell, 1995) in *The Western Australian flora: A descriptive catalogue* (inside cover) by Paczkowska & Chapman, 2000, Perth, Australia: Wildflower Society of Western Australia (Inc.), the Western Australian Herbarium, CALM and the Botanic Gardens & Parks Authority.

world such as South Africa (see Cowling, Richardson, & Pierce, 1997). Lesueur National Park lies within the ecological zone known as the Geraldton Sandplains. In 1801, Mount Lesueur was named by the crew of the French corvette *Naturaliste* after Charles-Alexandre Lesueur, a landscape painter who worked onboard with the naturalist François Péron (Hopper, 1990, p. 9). Lesueur-Eneabba has a long history of plant collection since Drummond in the nineteenth century (Hopper, 1990). In the 1950s, Charles Gardner made some of the first proposals for the conservation of land in the Mount Lesueur area. In May 1989, environmental activists initiated a campaign to gazette the Lesueur area and halt the intrusion of mining and the development of a power station (Bailey, *n.d.*, The early campaign, para. 2). Lesueur National Park was gazetted in February 1992 after a thirty-year campaign by conservationists, trade unions and local farmers (Bailey, *n.d.*, Success for the campaigners, para. 2).

Lesueur National Park has over 900 species of plants, or 10% of Western Australia's identified flora, and ranks as one of the three most critical areas for flora conservation in the region along with the Stirling Range and the Fitzgerald River area (Department of Conservation and Land Management, 1995). The park lies at the northern limit of the *kwongan*, a Mediterranean-type shrubland, occurring only in the South-West and exhibiting some of the most complex floristic patterns in Australia (Department of Conservation and Land Management, 1995). The kwongan is a low-growing heath ecosystem prevalent on soils of low fertility, but yielding high plant diversity (Corrick & Fuhrer, 2002, p. 14). *Kwongan*, which comprises 30% of the original flora, denotes Mediterranean-type shrubland ecosystems that are related to maquis and chaparral landscapes from outside Australia (Conservation International, 2007). Plants of the kwongan exhibit characteristic growth habits: compact form, stiff miniscule leaves and small, abundant flowers (Corrick & Fuhrer, 2002, p. 14).

The metropolitan site, Anstey-Keane Damplands, is one of the most botanically significant places on the Swan Coastal Plain and more diverse than Kings Park (Giblett & James, 2009). It lies at the northern tip of the Pinjarra Plains, a system of flat damplands—moist, shallow sinks—including the most suitable soil on the Swan Coastal Plain for pasture and development (Beard, 1979b, p. 27). The five plant communities identified within the damplands are all rare within the metropolitan ambit (Payne, 1993). Keighery (cited in Payne, 1993) suggests that the number of plant species could be 381. Rare and endangered plants include Purdey's Donkey Orchid (*Diuris purdiei*) and 28 other orchid species, as well as sundews (*Drosera occidentalis*). Anstey-Keane Damplands is under threat from road development, off-road vehicle damage, and other pressures of metropolitan expansion (Giblett & James, 2009). According to local conservationist Rod Giblett (pers. comm., September, 2008), aesthetic discourse figures into the protection of Anstey-Keane Damplands. Some local policy makers regard the damplands as monochrome and uninteresting—a place of no consequence.

The third site, Fitzgerald River National Park, is located on the central south coast between Hopetoun on the eastern boundary and Bremer Bay on the western side. The park contains approximately 1,800 "beautiful and bizarre species of flowering plants" (Visit Western Australia Online, 2011, para. 1). The gazetted area comprises 20% of the total flora of Western Australia and 42% of the species of the South-West (Fitzgerald River National Park Advisory Committee, 1987). The Park has three times the number of species found in the United Kingdom and half that of South Australia (Fitzgerald River National Park Advisory Committee, 1987). Sixty-two species are found only within its limits and another 48 are more or less limited to the park, which is unusually biodiverse for a 300,000 hectare land area (Department of Environment and Conservation, *n.d.*) (see chapter 6).

In order to record the being and becoming of plants, I employed botanic field aesthetics at these sites through an awareness of the Nyoongar calendar (see figure 8). Colleen Hayward (2009), a senior Nyoongar spokesperson, explains that the traditional calendar of the South-West recognises six seasons through changes in the flowering, fruiting, seeding and other aspects of plants. The approximate correspondences are *Djeran* (April-May), *Makuru* (June-July), *Djilba* (August-September), *Kambarang* (October-November), *Birok* (December-January) and *Bunuru* (February-March) (Kings Park & Botanic Garden, *n.d.*). Each with

Figure 8 The Traditional Six Season Nyoongar Calendar. Image by Richard Walley (Kings Park & Botanic Garden, n.d.).

distinguishable ecological characteristics, the six seasons are demarcated by regional biota rather than the preset numerical standards of the European calendar. The Nyoongar calendar traces knowledges of local plants that draw from multi-sensoriality and cultural practices. The sensuous features of flora serve as seasonal indicators of the availability of food (Daw, Walley, & Keighery, 1997). Moore (1884/1978) defines the Nyoongar word *yanbart* as ground after the vegetation has been burnt in which the blackened trunks of balga indicate that the land is cared for. The Nyoongar calendar links sense perception to plants and their ecologies. During *Kambarang*, the *Wonil*, or peppermint trees (*Agonis flexuosa*), have a high concentration of aromatic oils that waft in the hot air, imparting a recognisable olfactory quality to the season (Kings Park & Botanic Garden, *n.d.*). Moreover, *Mudja*, or Christmas trees (*Nuytsia floribunda*) flower during *Birok*, heralding the time for migration to the coast. During *Djilba*, *bohn*, or bloodroots (*Haemodorum spicatum*) are harvested for their peppery roots, used as a food seasoning. The calendar is embodied temporality because it encrypts temporal relations between plants and people.

A Practice of Aesthesis

As a transdisciplinary approach, botanic field aesthetics, as I have described it, is engaged with—rather than antagonistic towards—science. I seek complementarity between ways of knowing plants. The field approach brings multi-sensoriality to our interactions with flora and their ecologies (see table 2). By going beyond the "loaded aesthetic appeal of the flower" (Pryor, 2005, para. 9), the field practice underscores temporal, embodied and regionally focused appreciation of flora. My selected methods of poetic enquiry, anthoethnography and gestural walking aim to reveal the possibility of sensuously rich relationships to plants. Although the outcomes could be described as phenomenological, botanic field aesthetics is not intended to be a Goethean phenomenology of plants theorised by such writers as Bockemühl (1981) and others in *In Partnership with Nature* as a specific manner of perceiving morphological formations through time. Through the cultural and arts-based practices of botanic field aesthetics, new knowledge formations are given space to emerge outside the subject-object dynamics of classical science or the visual bias of phenomenological research (see chapters 1 and 5). I admit that the methodology is not new and has precedents in the cultural botanist Henry David Thoreau and others. The nineteenth century English poet John Clare also preferred sensuous interaction with plant life. His poetry reflects conversations with locals and a routine of walking the countryside. In fact, Clare regarded binomialism as "Linnaeus's dark system" (cited in Mahood, 2008, p. 3). As a transdisciplinary methodology, botanic field aesthetics seeks the interstices between flora, human culture and embodiment under the illuminating sun of the South-West.

Table 2 Perception Beyond Images. In addition to their visual appeal, these iconic South-West plant species have distinctive smells, sounds and textures.

Common Name	Linnaean Name	Form	Colour	Smell	Sound	Texture
Grasstree, Blackboy, Balga	*Xanthorrhoea preissii*	round	green black	pungent	swish	sharp
Christmas Tree, Mudja	*Nuytsia floribunda*	irregular	yellow grey	acrid	dsssss	silken
Kangaroo Paw, Kuttych	*Anigozanthos manglesii*	cylindrical	red green	bitter	bob-bob	hirsute
Queen of Sheba	*Thelymitra variegata*	six-petalled	purple yellow	musky	whisper	glossy
Underground Orchid	*Rhizanthella gardneri*	cupular	red white	sweet	thump	sticky

References

Arber, A. (1950). *The natural philosophy of plant form.* Cambridge, England: Cambridge University Press.

Bailey, J. (*n.d.*). Power struggle: Unions' role in a conservation campaign Retrieved April 14, 2011, from http://www98.griffith.edu.au/dspace/bitstream/10072/1574/1/20646.pdf

Beard, J. (1979). Phytogeographic regions. In J. Gentilli (Ed.), *Western landscapes* (pp. 107-121). Nedlands, Australia: University of Western Australia Press.

Bockemühl, J. (1981). *In partnership with nature* (J. Meeks, Trans.). Wyoming, RI: Bio-Dynamic Literature.

Breeden, S., & Breeden, K. (2010). *Wildflower country: Discovering biodiversity in Australia's Southwest.* Fremantle, Australia: Fremantle Press.

Brewer, J. D. (2000). *Ethnography.* Buckingham, England: Open University Press.

Cameron, J. (2005). Place, Goethe and phenomenology: A theoretic journey. *Janus Head,* 8(1), 174-198 Retrieved from http://www.janushead.org/8-1/cameron.pdf

Carlson, A. (2000). *Aesthetics and the environment: The appreciation of nature, art, and architecture*. New York, NY: Routledge.

Clarke, P. (2008). *Aboriginal plant collectors: Botanists and Australian Aboriginal people in the nineteenth century*. Dural Delivery Centre, Australia: Rosenberg Publishing

Clifford, J. (1986). Introduction: Partial truths. In J. Clifford & G. Marcus (Eds.), *Writing culture: The poetics and politics of ethnography* (pp. 1-26). Berkeley, CA: University of California Press.

Cole, A., & Knowles, J. G. (2007). Arts-informed research. In J. G. Knowles & A. Cole (Eds.), *Handbook of the arts in qualitative research: Perspectives, methodologies, examples, and issues* (pp. 55-70). Thousand Oaks, CA: Sage.

Conservation International. (2007). Southwest Australia Retrieved May 5, 2011, from http://www.biodiversityhotspots.org/xp/hotspots/australia/Pages/default.aspx

Conservation International. (2008). Biological diversity in Southwest Australia. K. Caley (Ed.) *Encyclopedia of Earth* Retrieved from http://www.eoearth.org/article/Biological_diversity_in_Southwest_Australia

Corrick, M. G., & Fuhrer, B. A. (2002). *Wildflowers of Southern Western Australia*. Noble Park, Australia: The Five Mile Press.

Cotton, C. M. (1996). *Ethnobotany: Principles and applications*. Chichester, England: John Wiley & Sons.

Cowling, R. M., Richardson, D. M., & Pierce, S. M. (1997). *Vegetation of Southern Africa*. Cambridge, England: Cambridge University Press.

Crary, J. (1990). T*echniques of the observer: On vision and modernity in the nineteenth century*. Cambridge, MA: MIT Press.

Curry, P. (2010). Grizzly man and the spiritual life. *Journal for the Study of Religion, Nature and Culture*, *4*(3), 206-219. doi: 10.1558/jsrnc.v4i3.206

Daw, B., Walley, T., & Keighery, G. (1997). *Bush tucker plants of the South-West*. Kensington, Australia: Department of Environment and Conservation.

Department of Conservation and Land Management. (1995). *Management plan: Lesueur National Park and Coomallo Nature Reserve 1995-2005*. Perth, Australia: Department of Conservation and Land Management.

Department of Environment and Conservation. (n.d.). An online visitor's guide to Western Australia's parks, reserves and other recreation areas: Fitzgerald River National Park Retrieved May 5, 2011, from http://www.naturebase.net/component/option,com_hotproperty/task,view/id,105/Itemid,755/

Driver, F. (2004). Distance and disturbance: Travel, exploration and knowledge in the nineteenth century. *Transactions of the RHS*, *14*, 73-92.

Drummond, J. (1848). Notice of Mr. Drummond's discovery of three remarkable plants in South-West Australia. *Companion to Curtis's Botanical Magazine*, *74*, 1-3. Retrieved from http://books.google.com

Edensor, T. (2001). Walking in the British countryside: Reflexivity, embodied practices and ways to escape. In P. Macnaghten & J. Urry (Eds.), *Bodies of nature* (pp. 59-81). London, England: Sage.

Eyre, E. J. (1845/1964). *Journals of expeditions of discovery into Central Australia, and overland from Adelaide to King George's Sound, in the years 1840-1*. Adelaide, Australia: Libraries Board of South Australia.

Fitzgerald River National Park Advisory Committee. (1987). *An introduction to the Fitzgerald River National Park*. Perth, Australia: The Committee.

Foucault, M. (1966/2002). *The order of things: An archaeology of the human sciences*. London, England: Routledge.

Gardner, C. A. (1959/1981). *Wildflowers of Western Australia* (14 ed.). Perth, Australia: St. George Books.

Gascoigne, J. (1996). The ordering of nature and the ordering of empire. In D. Miller & P. Reill (Eds.), *Visions of empire: Voyages, botany, and representations of nature* (pp. 107-116). Cambridge, England: Cambridge University Press.

Geertz, C. (1988). *Works and lives: The anthropologist as author*. Stanford, CA: Stanford University Press.

George, A. (2002). *The long dry: Bush colours of summer and autumn in South-western Australia*. Kardinya, Australia: Four Gables Press.

George, A. (2002). The south-western Australian flora in Autumn: 2001 presidential address. *Journal of the Royal Society of Western Australia*, *85*, 1-15.

Giblett, R. (2006). *Forrestdale: People and place*. Bassendean, Australia: Access Press.

Giblett, R., & James, D. (2009). Anstey-Keane botanical jewel. *Landscope*, *24*(4), 42-44.

Gregory, A. C., & Gregory, F. T. (1884/1981). *Journals of Australian explorations*. Victoria Park, Australia: Hesperian Press.

Grey, G. (1841). *Journals of two expeditions of discovery in North-West and Western Australia during the years 1837, 38, and 39*. (Vol. 2). London, England: T. and W. Boone.

Hammersley, M., & Atkinson, P. (1995). *Ethnography: Principles in practice* (Second ed.). New York, NY: Routledge.

Harper, M. (2007). *The ways of the bushwalker: On foot in Australia*. Sydney, Australia: University of New South Wales Press.

Hayles, K. (1990). *Chaos bound: Orderly disorder in contemporary literature and science*. Ithaca, NY: Cornell University Press.

Hayward, C. (2009). Engagement and partnership. Paper presented at the CREATEC Special Workshop, Edith Cowan University, Mount Lawley, Australia.

Head, L. (2007). Cultural ecology: The problematic human and the terms of engagement. *Progress in Human Geography*, *31*(6), 837-846.

Head, L., & Atchison, J. (2009). Cultural ecology: Emerging human-plant geographies. *Progress in Human Geography, 33*(2), 236-245.

Heidegger, M. (1971). *Poetry, language, thought* (A. Hofstadter, Trans.). New York, NY: HarperCollins.

Heidegger, M. (1982). *On the way to language* (P. Hertz, Trans.). San Francisco, CA: Harper & Row.

Hickman, E., & Ryan, J. (*in press*). *Two with nature: Ellen Hickman and John Ryan*. Fremantle, Australia: Fremantle Press.

Hitchings, R. (2003). People, plants and performance: On actor network theory and the material pleasures of the private garden. *Social & Cultural Geography, 4*(1), 99-113.

Hitchings, R., & Jones, V. (2004). Living with plants and the exploration of botanical encounter within human geographic research practice. *Ethics, Place & Environment, 7*(1 & 2), 3-18.

Hopper, S. (1990). European Exploration. In A. A. Burbidge, S. Hopper & S. Van Leeuwen (Eds.), *Nature conservation, landscape and recreation values of the Lesueur area: A report to the Environmental Protection Authority from the Department of Conservation and Land Management* (pp. 9-14). Perth, Australia: Environmental Protection Authority.

Hopper, S. (1993). *Kangaroo paws and catspaws: A natural history and field guide*. Como, Australia: Department of Conservation and Land Management.

Hopper, S. (1998). An Australian perspective on plant conservation biology practice. In P. Fiedler & P. Kareiva (Eds.), *Conservation biology for the coming decade* (2 ed., pp. 255-278). New York, NY: Chapman Hall.

Hopper, S. (2004). Southwestern Australia, Cinderella of the world's temperate floristic regions. *Curtis's Botanical Magazine, 21*(2), 132-180.

Isaacs, J. (1989). *Bush food: Aboriginal food and herbal medicine*. Chatswood, Australia: New Holland Publishers.

Keighery, G., & Beard, J. (1993). Plant communities. In C. Thomson, G. Hall & G. Friend (Eds.), *Mountains of mystery: A natural history of the Stirling Range* (pp. 43-54). Como, Australia: Department of Conservation and Land Management.

King, V. (2003). Embodied perceptions: Aboriginal expressions of place. In. In J. Cameron (Ed.), *Changing places: Re-imagining Australia* (pp. 286-295). Double Bay, Australia: Longueville Books.

Kings Park & Botanic Garden. (n.d.). *Aboriginal Life*. West Perth, Australia: Kings Park & Botanic Garden.

Knowles, J. G. (2001). Writing place, wondering pedagogy. In L. Neilsen, A. Cole & J. G. Knowles (Eds.), *The art of writing inquiry* (pp. 89-99). Halifax, Canada: Backalong Books.

Leggo, C. (2001). Research as poetic rumination: Twenty-six ways of listening to light. In L. Neilsen, A. Cole & J. G. Knowles (Eds.), *The art of writing inquiry* (pp. 173-195). Halifax, Canada: Backalong Books.

Leggo, C. (2004). Tangled lines: On autobiography and poetic knowing. In A. Cole, L. Neilsen, J. G. Knowles & T. Luciani (Eds.), *Provoked by art: Theorizing arts-informed research* (Vol. 2, pp. 18-35). Halifax, Canada: Backalong Books.

Leggo, C. (2006). Attending to winter: A poetics of research. In W. Ashton & D. Denton (Eds.), *Spirituality, ethnography, and teaching: Stories from within* (pp. 140-155). New York, NY: Peter Lang.

Leggo, C. (2007). Astonishing silence: Knowing in poetry. In J. G. Knowles & A. Cole (Eds.), *Handbook of the arts in qualitative research: Perspectives, methodologies, examples, and issues* (pp. 165-174). Thousand Oaks, CA: Sage.

Lines, W. (1994). *An all consuming passion: Origins, modernity, and the Australian life of Georgiana Molloy*. St. Leonards, Australia: Allen & Unwin.

Mahood, M. (2008). *The poet as botanist*. Cambridge, England: Cambridge University Press.

Main, B. Y. (1967). *Between wodjil and tor*. Melbourne, Australia: Jacaranda.

Marchant, N. G., Wheeler, J. R., Rye, B. L., Bennett, E. M., Lander, N. S., & Macfarlane, T. D. (1987). *Flora of the Perth region: Part one*. Perth, Australia: Western Australian Herbarium, Department of Agriculture.

Martin, G. (2004). *Ethnobotany: A methods manual*. London, England: Earthscan.

Moore, G. F. (1884/1978). *Diary of ten years eventful life of an early settler in Western Australia*, and also *A descriptive vocabulary of the language of the Aborigines*. Nedlands, Australia: University of Western Australia Press.

Mules, W. (2006). Contact aesthetics and digital arts: At the threshold of the Earth. *Fibreculture, 9* Retrieved from http://nine.fibreculturejournal.org/fcj-058-contact-aesthetics-and-digital-arts-at-the-threshold-of-the-earth/

Mules, W. (2008). Open country: Towards a material environmental aesthetics. *Continuum, 22*(2), 201-212.

Neilson, L. (2007). Lyric enquiry. In J. G. Knowles & A. Cole (Eds.), *Handbook of the arts in qualitative research: Perspectives, methodologies, examples, and issues* (pp. 93-104). Thousand Oaks, CA: Sage.

Paczkowska, G., & Chapman, A. (2000). *The Western Australian flora: A descriptive catalogue*. Perth, Australia: Wildflower Society of Western Australia (Inc.), the Western Australian Herbarium, CALM and the Botanic Gardens & Parks Authority.

Payne, J. (1993). *Wetlands in the city of Armadale: A report to the Water Authority of Western Australia and the Environmental Protection Authority*. Leederville, Australia: Water Authority of Western Australia.

Pearman, G. (2005). Nuts, seeds, and pulses. In G. Prance & M. Nesbitt (Eds.), *The cultural history of plants* (pp. 133-152). New York, NY: Routledge.

Porteous, J. D. (1996). *Environmental aesthetics: Ideas, politics and planning.* London, England: Routledge.

Pratt, M. L. (1986). Fieldwork in common places. In J. Clifford & G. Marcus (Eds.), *Writing culture: The poetics and politics of ethnography* (pp. 27-50). Berkeley, CA: University of California Press.

Prendergast, M. (2009). Introduction: The phenomena of poetry in research. In M. Prendergast, C. Leggo & P. Sameshima (Eds.), *Poetic inquiry: Vibrant voices in the social sciences* (pp. xix-xlii). Rotterdam, Netherlands: Sense Publishers.

Pryor, G. (2005). Black death: Species extinction in WA. *Artlink*, *25*(4) Retrieved from http://www.artlink.com.au/articles.cfm?id=2227

Rodaway, P. (2002). *Sensuous geographies: Body, sense and place.* Abingdon, England: Taylor & Francis.

Rose, D., & Robin, L. (2004). The ecological humanities in action: An invitation. *Australian Humanities Review*, *31-32* Retrieved from http://www.australianhumanitiesreview.org/archive/Issue-April-2004/rose.html

Ryan, J. (2010). Editorial: On the theme of 'peripatetica'. *Landscapes*, *4*(1), 1 Retrieved from http://www.landscapeandlanguagecentre.au.com/Peripatetica/Peripatetica_Ryan_Editorial.pdf

Sattler, R. (1978). *Theoretical plant morphology.* Leiden, The Netherlands: Leiden University Press.

Seddon, G. (1972). *Sense of place: A response to an environment.* Perth, Australia: University of Western Australia Press.

Seddon, G. (2005). *The old country: Australian landscapes, plants and people.* Cambridge, England: Cambridge University Press.

Sharr, F. (1978). *Western Australian plant names and their meanings: A glossary.* Nedlands, Australia: University of Western Australia Press.

Solnit, R. (2001). *Wanderlust: A history of walking.* London, England: Verso.

Thomas, S. (2004). *Of earth and flesh and bones and breath: Landscapes of embodiment and moments of re-enactment.* Halifax, Canada: Backalong Books.

Thomas, S. (2009). Nissopoesis: Visuality and aesthetics in poetic inquiry. In M. Prendergast, C. Leggo & P. Sameshima (Eds.), *Poetic inquiry: Vibrant voices in the social sciences* (pp. 127-132). Rotterdam, The Netherlands: Sense Publishers.

Thoreau, H. D. (1862/2007). *Walking.* LaVergne, TN: Filiquarian Publishing.

Thoreau, H. D. (1993). *Faith in a seed: The dispersion of seeds and other late natural history writings.* Washington, DC: Island Press.

Visit Western Australia Online. (2011). Fitzgerald River National Park Retrieved April 18, 2011, from http://visitwa.com.au/australias-south-west-region/60-fitzgerald-river-national-park-wa.html

Wills, R., & Kinnear, J. (1993). Threats to the Stirling Range. In C. Thomson, G. Hall & G. Friend (Eds.), *Mountains of mystery: A natural history of the Stirling Range* (pp. 135-141). Como, Australia: Department of Conservation and Land Management.

Part II

Botanical Histories

Part II will present an historicised discussion of the aesthetics of South-West flora. Chapter 3, "Leaves of a Tree," will argue that different histories of plants, including Indigenous living heritages, reveal distinctive aesthetic values and practices. Chapter 4, "Values and Evaluations," goes on to provide a detailed textual reading of the English botanist John Lindley's *A Sketch of the Vegetation of the Swan River Colony* published between 1839 and 1840. Visual, scientific and colonial regimes were intertwined during this period. Bearing in mind that these historicised examples are not merely products of the past, chapter 5, "Plants as Objects," will identify the challenges for a theory of embodied appreciation of flora through an analysis of contemporary environmental aesthetics. In particular, models offered by Berleant and Carlson prove insufficient in providing forward-looking groundwork for appreciation of, and engagement with, plants through the bodily senses, as contextualised in Part I.

3

Narratives

Abstract: Science, history, poetry, mythology and personal experience are often thought to contradict one another and are held, commonly, as separate forms of knowledge-making. Like leaves of a tree, however, the botanical works of Henry David Thoreau gather together the diverse stories that give meanings to plants. Drawing from the concept of multiple narrative streams as a method of writing nature and natural history, as inspired by Thoreau, chapter 3 will explore different historical accounts of South-West flora. Botanical sciences, Indigenous spiritualities, nature poetries and colonial histories offer dissimilar though complementary perspectives. The meandering together of narrative streams ensures the perpetuity of non-scientific stories and the potential for cross-pollination between disciplines and multiple ways of knowing flora.

Take their cold seed and set it in the mind,
and its slow root will lengthen deep and deep
till, following, you cling on the last ledge
over the unthinkable, unfathomed edge
beyond which man remembers only sleep.

Judith Wright "The Cycads" (1994, pp. 39-40, ll. 17-21)

Chapter 3 suggests that, as an approach to the writing of plants, it is crucial to consider how stories, including poetic and scientific, augment and amplify each other, rather than create division between human culture and botanical nature (see chapter 1). Importantly for this discussion, the narratives of plants contain distinctive aesthetic modes and values. Through verse instead of science, Judith Wright evokes the primordial character of the cycads, survivors of the age of the dinosaurs and older than the human species. The final line of the poem "beyond which man remembers only sleep" intimates the qualities of adaptation and co-evolution that may distinguish the cycads from recently introduced taxa. Convergent Aboriginal, poetic and scientific stories are etched in the slow-growing cycad fibres. A polyvocal account of the cycads weaves together technical views, cultural histories, Aboriginal understandings and multi-sensorial experiences. The greater cohesion of narratives can result in productive confluences for writers of botanical history, environmental issues and philosophical aesthetics.

Through Thoreau's sense-rich writings, the approach of multiple narrative streams will be outlined. Indeed, this approach could be applied to any account of natural history, from animals and birds to rocks and algae. Barker (2008, p. 483) defines *narrative* as "a sequential account or purported record of events ordered across time into a plot. The concept of narrative refers to the form, pattern or structure by which stories are constructed and told." Although a term in literary theory, *narrative* will be used synonymously with the terms *story*, *account* and *history*. Nyoongar histories, early settler accounts, post-colonial narratives, poetic interpretations and experiential impressions broaden the dominant narrative of science towards storied streams that diverge and converge dynamically. Through narratives, disciplinary accounts that historicise plants, such as those offered by natural history, are networked within diverse knowings (see chapter 1). An orientation towards commingled knowledge streams minimises the potential erasure of certain accounts by dominant narratives (as an example, see Hopper, 2010).

Thoreau's Approach to Multiple Narratives

Despite his criticism of science and empiricism, Thoreau's botanical understandings worked in tandem with his experiential engagement with flora. Thoreau's empiricism led to the development of a novel theory of seed dispersion. His ideas also opened a space for convergence between scientific, sensuous and indigenous modes of interacting with flora (see chapter 1). His works exemplify the seamless integration of narrative streams towards multi-dimensional accounts of plants. Thoreau outlines an approach to flora in *Faith in a Seed* (1993) and *Wild Fruits* (2000), which mark his growing intrigue with scientific botany beginning in the early 1850s. Although his earlier pièce de résistance *Walden* alludes to local species, an interest in flora consumed his later writings, as this journal entry from 1856 indicates:

> I soon found myself observing when plants first blossomed and leafed, and I followed it up early and late, far and near, several years in succession, running to different sides of the town and into the neighboring towns, often between twenty or thirty miles in a day. (Thoreau, 1962, p. 158)

As noted in chapter 2, Thoreau planned to assemble his observations of flowering and leafing into a "Kalendar," modelled after John Evelyn's 1664 *Kalendarium Hortense* or *Gardener's Almanac* (Dean, 2000). The project would be a comprehensive phenology of an archetypal year, setting out all the events of Concord natural history. The work would strive for comprehensiveness and eclecticism spanning scientific botany, Native American ethnobotany, classic Greek and Roman philosophies and first-hand experiences.

Although Thoreau passed away before his ambitions were realised, *Wild Fruits* evidences his methodology, and points to narrative streams as an overarching framework. In his exposition of the strawberry, Thoreau (2000, pp. 10-17) begins with verses composed by the sixteenth century poet Thomas Tusser and a description by the herbalist John Gerard, writing in the pre-Linnaean 1500s. Thoreau (2000, p. 11) reflects upon his observations of wild strawberries, stating "by the thirtieth of May I notice the green fruit." In this evocative passage, he expresses the fragrance of wild strawberries as a quality that evades the visible prominence of its flowers. Strawberries emit:

> An indescribably sweet fragrance, which I cannot trace to any particular source. It is, perchance, that sweet scent of the earth of which the ancients speak. Though I have not detected the flower that emits it, this appears to be its fruit. It is natural that the first fruit which the earth bears should emit and be, as it were, a concentration and embodiment of that vernal fragrance with which the air has lately teemed. Strawberries are the manna found, ere long, where that fragrance has been. Are not the juices of each fruit distilled from the air? (Thoreau, 2000, p. 12)

Thoreau also cites the impressions of explorers to North America, such as eighteenth century Englishman Samuel Hearne's observations of the strawberry in indigenous North American cultures. *Oteagh-minick* in the language of First Nations Canadians of the Churchill River signifies the resemblance of the fruit to a heart, while other names for strawberries, as *Oteimeena* in Cree and *O-da-e-min* in Chippeway, present a signature linking the human heart to the shape of the fruit (Thoreau, 2000, p. 15).

Thoreau read extensively the writings of North American explorers and early ethnographers, citing the American theologian Roger Williams's landmark study of Native American linguistics, *A Key into the Language of America* (1643/1997). On the strawberry, Williams (cited in Thoreau, 2000, p. 16) reveals ethnobotanical interest and what would be called by modern ethnographers a form of participant observation: "The Indians bruise them in a mortar, and mix them with meal, and make strawberry bread...having no other food for many days." Further in his exposition, Thoreau draws from the records of naturalists, such as the missionary George Loskiel, to determine changes to the distribution of strawberries in the eastern United States. Through this vignette, Thoreau exemplifies the use of narrative streams by juxtaposing historical texts, ethnographic accounts, multi-sensorial interaction and astute personal observations. This syncretism augments scientific conjecture, a method he perfects in *Faith in a Seed.*

Echoing his approach to strawberries, the essay "Wild Apples" (Thoreau, 1862/2010) synthesises field science, references to classical writers and the author's physical experiences. Thoreau's reading on wild apples is wide-ranging, including allusions to Tacitus, Palladius and Pliny. Multi-sensoriality mingles with the voices of previous writers towards a cultural botany of the wild apple. Thoreau (1862/2010, p. 25) qualifies some varieties as "acrid and puckery, genuine verjuice," while a particular tree in Concord produces "a peculiarly bitter tang, not perceived till it is three-quarters tasted. It remains on the tongue. As you eat it, it smells exactly like a squash-bug" (Thoreau, 1862/2010, p. 27). The essay celebrates the sensuousness of apples in an age of the increasingly homogenised sizes, shapes, tastes, colours and smells of fruit. Thoreau (1862/2010, p. 28) beseeches the reader to "let your condiments be in the condition of your senses. To appreciate the flavor of these wild apples requires vigorous and healthy senses, papillae firm and erect on the tongue and palate." Although a sensuous naturalist, he relied on emerging botanical science. In the section "The Naming of Them," Thoreau (1862/2010, p. 30) recognises the advantages of standard nomenclature: "I find myself compelled, after all, to give the Latin names of some for the benefit of those who live where English is not spoken, —for they are likely to have a world-wide reputation." As suggested by these essays, Thoreau queried streams of knowledge, guided by his intense curiosity, without rationalising how they might lock together epistemologically. The writing produced offers a confluence of knowledge towards cross-fertilisation between personal, poetic, multi-sensorial, historical, indigenous and scientific ways of knowing.

Interlude VII: Inside a Jarrah Tree, A Black Tunnel Reaching Skyward

Forester John Ednie-Brown (1899), considered the first expert on Australian timber, commented:

> Taken as a whole, there is nothing particularly picturesque about the appearance of a Jarrah tree or forest of these. Indeed, the general effect of the species, en masse, is dull, sombre, and uninteresting to the eye. Except in special spots and localities, the trees are rugged and decidedly inclined to be straggling and branchy. (10)

Before this assessment of the aesthetic value of the forest, Ednie-Brown (1899, p. 10) stated that "Jarrah and Western Australia are almost synonymous words." Clearly, Ednie-Brown brought European perceptual sensibility to Australian forests, one based in appearance and subtended by a managerialist approach to ecology. However, there are multiple narratives—Aboriginal, embodied and poetic—of the jarrah tree (*Eucalyptus marginata*) that come into contact, and at times conflict, with scientific concepts of treeness. In Nyoongar belief, the spirit or *kaarny* of a recently deceased person would be caught and placed in the burned-out trunk of a jarrah to pacify its restlessness (McCabe, 1998, p. 6). For me, the quintessential act of embodiment is physically entering an old tree and feeling the volatile processes of fire, time and age that hollowed out its core, leaving a tunnel to the sky. The final line, intentionally set apart from the rest of the poem, contradicts the common perception held by some recreationalists of the forest as a venue for outdoor sport, an antidote to the city and a reservoir of visual beauty. Although many things, the jarrah forest is also a place that commands spiritual respect (Trigger & Mulcock, 2005). Throughout "Inside a Jarrah Tree," I use bodily metaphor to site myself in relation to the corpus of the jarrah, developing an aesthesis of plants and people in which vegetal and human bodies reach proximity:

> neatly burnt-out innards,
> this tree lives on as skin
> still supple and twisting in pleats
> but where did the heart go, and the breast bone
> and the heavy, unctuous insides? (ll. 1-5)

I am not looking at the tree, as Ednie-Brown did, but looking at the forest from within the tree as an inversion of the picturesque mode of appreciation (chapter 5) (to read "Inside a Jarrah Tree" in full, see Ryan, *in press*).

The Narrative Streams of Zamia Palm

A case study will illustrate the concept of narrative streams applied to the study of South-West flora. The zamia palm (*Macrozamia riedlei*) is a member of the Zamiaceae family of cycads distributed throughout Australia, Africa and warm temperate areas of North and South America. First classified by Charles Gardner, *M. riedlei* is endemic to the lower South-West corner from Hutt River near Perth to Albany. The plant contains macrozamin, a toxin found in most cycads responsible for zamia staggers, a fatal affliction of the nervous systems of animals (Carr & Carr, 1981, p. 18). In the South-West, zamia palms are thought to be found only in iron-rich lateritic soils and as understory plants in jarrah forests. However, other botanists observe that it is common in all soil types throughout Perth (Marchant et al., 1987, p. 57). It includes a short trunk, about three metres high, rigid fronds one to two metres in length with broadly cylindrical or ovoid seeds, reddish brown and fleshy when ripe (Marchant, et al., 1987). Reflecting the condensed structure of most taxonomic descriptions, Paczkowska and Chapman (2000) offer this morphological snapshot:

Macrozamia riedlei (Gaudich.) C.A.Gardner ZAMIA
Cycad, 0.5-3 m high; small, usually trunkless; leaves few, glossy, flat or openly keeled, narrow leaflets; short cones. Fertile plants recorded Sep-Oct. Lateritic soils, jarrah forests. Distribution: SW:ESP, GS, JF, SWA, WAR. (27)

When the Dutch explorer Willem de Vlamingh landed in December 1696 on Rottnest Island, adjacent to the mouth of the Swan River, his party found zamia nuts. Vlamingh reported the initial palatability of the roasted fruits, likening them to "Dutch broad beans, or, when ripe, like hazelnuts," but three hours after consuming them, his crew "began to vomit so violently that there was hardly any distinction between death and us" (Vlamingh, 1985, p. 155). In January 1802, with similar indiscretion provoked by hunger and curiosity, members of the Flinders expedition at Lucky Bay were drawn to eat zamia fruits, but with equally disastrous consequences: "A party of gentlemen were upon the top, eating a fruit not unlike green walnuts in appearance…Mr. Thistle and some others who had eaten liberally were taken sick and remained unwell all the day afterwards" (Flinders, 1814, p. 80). Beaton (cited in Carr & Carr, 1981, p. 17) concludes that nearly every European party, "known for not reading each other's journals and accounts," including Vlamingh in 1696, Grey in 1839 and McDouall Stuart in 1864, suffered cycad poisoning.

Nevertheless, some of the best-preserved records of Aboriginal uses of plants come from the published accounts of explorers and settlers (Grey, 1841a, 1841b; Moore, 1846). Contrary to Beaton's assertion, these records indicate familiarity with the observations of previous explorers and naturalists, and document some positive interactions between explorers and Aboriginal communities. Along the Arrowsmith River north of Perth, George Grey (1841b, p. 61), guided by Kaiber, was alerted to the toxicity of zamia: "Kaiber brought in some nuts of the Zamia tree; they were dry, and therefore in a fit state to eat." The rest of Grey's party indulged impetuously in insufficiently dried fruits, leading to "violent fits of vomiting accompanied by vertigo, and other distressing symptoms" (Grey, 1841b). Grey's party exhibited what is now known as zamia staggers (Carr & Carr, 1981).

Aboriginal cultures throughout Australia, including the Nyoongar of the South-West to whom zamia fruit is known as *by-yu*, have evolved strategies of detoxification, such as roasting, soaking and fermenting, or a combination of techniques, to convert the nut into a staple food. In the South-West, explorers and writers, including James Drummond in 1839, naturalist and Quaker missionary James Backhouse in 1843, HMS Beagle officer John Lort Stokes in 1846 and chronicler J.E. Hammond in 1933, observed the significance of zamia palm in Aboriginal culture (Meagher, 1974, p. 25). Moore (1846) observed Nyoongar use of the cycad fruit and its detoxification:

> This in its natural state is poisonous; but the natives, who are very fond of it, deprive it of its injurious qualities by soaking it in water for a few days, and then burying it in sand, where it is left until nearly dry, and is then fit to eat. They usually roast it, when it possesses a flavour not unlike a mealy chestnut; it is in full season in the month of May. It is almost the only thing at all approaching to a fruit which the country produces. (17)

Western Australian settler and creator of one of the first Nyoongar-to-English dictionaries, Moore (1846 p. 22) lists the term "*djiriji*" for the zamia as containing "a farinaceous matter, which, when prepared, has been used as sago, but is dangerous without preparation." *Gargoin* denotes the pit of the zamia fruit, "edible after being steeped in water or buried in moist earth for a time; but the kernel is considered unwholesome by some persons" (Moore, 1846, p. 28). The complexity of the Nyoongar vocabulary surrounding zamia signifies its cultural importance as a foodstuff. According to Moore's dictionary, the Nyoongar differentiate between coastal species of zamia, such as *kund gor*, and between the outer kernal *d-yundo* and the inner kernel *wi-d* of the nut or *kwinin*.

The poisonous nuts symbolise the progressive understanding of the endemic flora of the South-West by settler society, a process of conciliation that continues with contemporary botanists who forward in the revisioning of botany in developing models for how South-West species adapt to ecological constraints (George, 2002; Hopper, 1998). This brief account of the zamia palm illustrates the potential meeting of multiple narrative streams including poetic, scientific and Aboriginal knowledges, each of which contains sensuous histories. Nyoongar narratives of plants express the edibility and palpability of flora with direct influence on sustenance and longevity, whereas scientific observation represents the apotheosis of visual denomination (see chapter 1). What matters most in post-colonial Australia and elsewhere is the co-existence of narrative streams towards the possibility of dialogue and productive transdisciplinarity.

Nyoongar Conceptualisations of Plants

Before the recent history of colonisation in Australia, there is the cultural richness of 50, 000 years of Nyoongar interaction with plants as food, medicine, tools, ornamentation and totems (Paczkowska & Chapman, 2000). Nyoongar culture and the history of flora are inseparably united because plants have enabled spiritual and material sustenance and, conversely, Aboriginal people have ensured the longevity of plant populations (Hallam, 1975). Joe Northover (1998, p. 40) expresses eloquently the relationship between Nyoongar people and the land: "We don't have Cathedrals or built monuments to celebrate our culture, we have landscape and the very landscape is a reflection on us and we are a reflection on our landscape."

In Nyoongar belief, plants belong to a spiritual landscape. However, an integrated spiritual and material alignment with plants may contradict the imperatives of managerialistic conservation science (for example, see McCabe, 1998).

Through acts of sustenance, Nyoongar peoples have developed complex bodily understandings of flora that deepen the visual understanding of plants. Indeed, the histories of Aboriginal peoples throughout Australia interweave with plant histories (Clarke, 2007). Whereas scientific knowledge relies on universalised structurings that may marginalise embodied experience, Aboriginal knowings are predicated on edibility, palpability, aroma and the elaborate interstices between the senses and eco-cultural meanings. Nyoongar history intertwines plants with the Aboriginal Dreaming, the complex stories and proscriptions that engender acts to ensure the sustained health and productivity of the land. Robert Bropho (1998, p. 31) claims that "all the Dreaming stories are within the roots of that tree, coming from the ground and [the stories and the roots] can never be separated." Bropho (1998, p. 31) aligns human and tree bodies through visceral empathy founded in commiseration: "When I see those photos in the papers of the logs laying there with no limbs on them [I think] that's a body of a Blackman there from the neck down to his ankles and everything's been trimmed...that hurts me." A kindredness between plants and people suggests the shared consequences for both. Ted Wilkes (1998, p. 45) observes that "the trees in the forest in the southwest of Australia have gone through exactly the same thing that Aboriginal people have gone through—annihilation, dispossession."

Traditional Nyoongar interaction with plants exemplifies "embodied spatiality," a term proposed by Rose and Robin (2004) to encapsulate physical connection through acts of sustenance, such as the gathering of plant foods and medicines. Visual features of plants, through the perspective of embodied spatiality, overlay deeper corporeal and cultural resonances. Dorothy Collard (1998, p. 34) reflects on the difference between Nyoongar traditional knowledge and the practices of modern forest management concerning the belief in the restoration of cleared old growth forests: "[The forest] will never, never be the same. [It] might look good, with their eyes but the spirits [are] not there." While an ecosystem may be reconstructed, the soul of the forest is irreplaceable. Furthermore, regarding the consequences of clear-felling old growth forests, Mike Hill (1998, p. 18) alludes to the interdependencies Nyoongar people have maintained with flora and the bodily histories that become endangered when ecosystems are altered.

In the *kwongan*, edible roots, bulbs and tubers have been culturally significant, as elder Ken Colbung (1998, p. 53) observes: "If you ate the food that was around the area, and that was what you had to do is eat the food that was in the region where you went, your...magnetic being was more present." According to Colbung, interactions with wild foods resonate culturally and spiritually as bodily modes of participation in the world. Sense of place becomes palpable and tasted embodied spatiality. The wild yam (*Dioscorea hastifolia*), known as *adjikoh* or *ijjecka*, influenced the degree of sedentism of particular Aboriginal communities (Carr & Carr, 1981, p. 14; Hallam, 1975). Bain (1975, p. 151) suggests that the Nyoongar agricultural practices along the Irwin Valley south of Geraldton were centred around root crops: "The people in the vicinity of the Bowes River lived mainly on *ijjecka* root…it appeared [to settlers] to be a delectable and valuable yam, worth cultivating." Settler Lockier Burges was of the opinion that the variety of edible root crops, like the *ijjecka* growing prolifically and to great sizes, fostered a diet of plant foods amongst the people of the Irwin Valley (Bain, 1975, p. 46).

In the sandplains near the modern suburb of Wanneroo, Colbung (Graham, 1990) demonstrates the edible and medicinal potential of the *bayu* (zamia fruits), *bera* (banksia flower), *boron* (bush red onion), *kojibut* (melaleuca balm), *kollookal* (pig face), *bayini* (wild fig), *mundar* (Christmas Tree) and *balga* (*Xanthorrhea*). Rather than being a sterile place of deprivation, the kwongan nurtures and sustains. The edible and curative influences of species depend on bodily openness to the seasonal life patterns of people and plants. Specific procedures for detoxifying plant parts, such as the fruits of the *by-yu* cycad, have been generated as part of cultural heritage. The bodies of plants intermesh with human bodies towards relationships of mutual benefit (for a classic regional study, see Hallam, 1975).

Concerning another root vegetable, George Fletcher Moore (1884/1978) reported the usage of *konno* or *Platysace cirrosa*:

> I have discovered a bulbous root like a dark-coloured potatoe [*sic*], called by the natives *konno*, which I mean to endeavour to cultivate, and which may be very useful if it succeeds. The taste is something like the meat of a cocoanut [*sic*], or between that and a carrot taste. One specimen is as large as your fist. (301)

In 1842 in the Wongan Hills, the naturalist John Gilbert (cited in Carr & Carr, 1981) reported a harvest festival based on konno: "Their season of meeting in great numbers to dig the edible root called by them *Wargae* is now in full force." One-hundred and twenty-five years later, Meagher (1974, p. 26) observed the collection of *karno* (*konno* in Moore's dictionary) near Mingenew: "The tubers are about half a metre below the ground and are dug up with a digging-stick…These tubers are available throughout the year and, besides being roasted in the ashes, are sometimes eaten raw to quench the thirst." Traditional narratives of flora comprise sensations of hunger, thirst and sickness and are therefore body-engaged accounts.

Similarly, the endemic West Australia Christmas tree, known as *mundar* or *mudjar* in Nyoongar and *Nuytsia floribunda* to scientists, is a conspicuous plant with significant cultural meanings (Hopper, 2010) (see "Seven Names for a Plant," Interlude X). The Christmas tree exhibits a variety of shapes and sizes and flowers around mid-December. Modern science classifies *Nuytsia* as an endemic mistletoe. As a root and rhizome hemi-parasite, it draws nutrients from a number of hosts but also has the ability to photosynthesise food (Paczkowska & Chapman, 2000). The parasitising rootlets coming off the main roots of *Nuytsia* are so tenacious and well-designed that they have been known to burrow into underground utility lines.

In the 1930s, the ethnographer Daisy Bates (1930/1992) noted the associations between the spirit world and the Christmas tree:

> The tree-Moojarr, or Moodurt...was to the Bibbulmun the 'Kaanya Tree', 'the tree of the souls of the newly dead'. From time immemorial the soul of every Bibbulmun rested on the branches of this tree on leaving its mortal body for its heavenly home, Kurannup, the home of the Bibbulmun dead which lay beyond the western sea. (153)

Nuytsia facilitated the passage of souls to the after-world, but as Bates claimed, the tree was feared and avoided because of its power. Other early ethnographers recorded the use of *Nuytsia* as food, water and decoration, suggesting variable spiritual beliefs about the tree (Cunningham, 2005, p. 223). Moore (1846, p. 80) described the Christmas tree as "Mut-yal, s.– Nuytsia floribunda; colonially, cabbage-tree. The only loranthus or parasite that grows by itself. Another anomaly in this land of contradictions. It bears a splendid orange flower." As the world's largest parasitic plant, the Christmas tree epitomised the baffling growth habits of antipodean species. It represented perfectly the departure of Western Australian landscapes from European aesthetic norms (see chapter 4). Even the name *Christmas Tree*, flowering in yellow as it does in the heat of November and December, runs contrary to the image of the evergreen Christmas tree brought indoors from the cold and deep snow of the English countryside.

Colonists reported mixed admiration and disdain for *Nuytsia*. The tree in flower was first recorded by the crew of Dutch explorer Pieter Nuyts's vessel *Gulden Zeepaert* in 1627 (Cunningham, 2005, p. 225). *Nuytsia* was assigned its scientific name in 1831 by Robert Brown and the tree was referred to as "Fire Tree" amongst Swan River colonists (Lindley, 1840, p. xxxix). In the journals of surveyors Alfred Hillman and Septimus Roe, *Nuytsia* indicated infertile country and was described disparagingly as part of the intolerable scrubbiness of the bush (Hopper, 2010). In 1880, peripatetic artist Marianne North painted "Study of the West Australian Flame-tree or Fire-tree," now part of the botanical art collection at Kew Gardens in England. In form, the tree depicted in the painting appears closer to the European elm with its pleasant vase-like symmetry than most Christmas Trees found in the Western Australian wild (to view North's painting, see http://www.kew.org/mng/gallery/764.html).

For some Nyoongar people, the Christmas tree has been considered a sacred plant connected to the souls of the deceased. But the tree has also been a food. The records of settlers and ethnographers point to certain spiritual beliefs and material practices surrounding *Nuytsia*. Writing in the 1880s, Ethel Hassell (1975, p. 26) reported the use of *Nuytsia* root as a candy: "They gave me one of the roots to taste, telling me it was called *mungah*. The outer skin was pale yellow but easily stripped off leaving a most brittle centre tasting very like sugar candy." A ghoulish creature called a *gnolum*, in the form of a very tall, very thin man, enticed boys away by offering them the roots of *mungah* (Hassell, 1975, p. 65). In the 1930s, Bates (1930/1992, p. 153) reported the view of *Nuytsia* as a home for disembodied spirits; the Nyoongar "did not fear the tree; they loved it, but held it sacred for its spiritual memories. The souls of all their forbears had rested on the spirit tree on their way to Kurannup." Nyoongar Elder Noel Nannup explains:

> *Moodjar*, the Christmas Tree, is one of the only flowers late in the [calendar] year when the easterly winds are blowing. The belief was that a spirit of the deceased person sat on the tree until it flowered. Then the spirit moved on to the spirit world, and that would be in conjunction with the easterly winds and fire, so natural fires burned throughout that time of year. The easterly wind would take the spirit out over the sea. (pers. comm., N. Nannup, July 21, 2010)

Noel offers a more complex view of the sacredness of moodjar as integrated to easterly winds and the fire season. According to the Nyoongar calendar, the onset of *birok*, the season encompassing December and January, corresponds to the flowering of mudja along with a host of ecological signs that initiated the movement of people to coastal regions (see chapter 2).

Interlude VIII: Bush Tucker

In the poem "C'mon Pigs of Western Civilization Eat More Grease," the American poet Allen Ginsberg (1994) excoriates, with sarcasm and political edginess, the food choices of Western (and Westernised) people. The poem "Bush Tucker" encourages the obverse sentiment of Ginsberg's work by praising South-West bush foods and asserting that wild-crafted food can also be savoury:

> a slow cook stew of tubers
> for your entrée
> extracted from the ground
> with my bleeding hands (ll. 17-20)

The aestheticised plants that modern tourists come to photograph for their flowering beauty have been food, medicine and fibre to Nyoongar people for thousands of years. George Grey (1841a, p. 12) observed the abundance of wild yam or warren and that the root "is a favourite article of food with the natives." It may be the unseen root below the surface that nourishes rather than the visible parts. Like harvesting vegetables from a garden, the wild-crafting of foods ushers in somatic relationships of contact and sustenance (see chapter 13). A resurgence of popular interest in bush tucker in the last few decades has followed television programs such as "The Bush Tucker Man" with Les Hiddins (Australian Television Information Archive, 2001). In recent years, bush tucker has entered the domain of fine dining and specialty foods. For instance, the small chocolatier, Dark Side Chocolates (2011), in Denmark, WA, crafts chocolates with locally-collected wattle seeds and lemon myrtle. Of course, the irony of the poem is that bush tucker needs to be regulated because gorging freely on the wild foods of national parks has been made illegal, and wild plants would be hard pressed to support the needs of the state's 2.2 million people (to read "Bush Tucker" in full, see Ryan, *in press*).

Figure 9 Sandplain Woody Pear (*Xylomelum angustifolium*), Lesueur National Park. *Image courtesy of the author.*

Plant Narratives and Aesthetics

Springing from Thoreau's expositions, this chapter has presented multiple narrative streams as a way to conceptualise the convergence of the diverse narratives of plants and their aesthetic appreciation. Science has a functional view of plant life aligned to its classificatory prerogatives and taxonomic aesthetics (see chapters 1 and 4). In contrast, traditional Nyoongar stories of plants revolve around corporeality through the seasons towards cultural and spiritual meanings. As suggested by Wright, poetry has the potential to bring metaphor and myth to our understandings of plants (see chapter 10). Narrative hybridity reclaims the sensuous and intellectual histories of interactions between flora and culture. Through narrative streams, the writing of nature and natural history orients towards transdisciplinary knowledges spanning science, poetic thought, Aboriginal traditions and personal experience. That such streams of knowledge meander together in writing of all forms towards creative confluences is more important than their epistemological differences or aesthetic incompatibilities.

References

Australian Television Information Archive. (2001). Bush tucker man Retrieved January 26, 2011, from http://www.australiantelevision.net/btm.html

Bain, M. A. (1975). *Ancient landmarks: A social and economic history of the Victoria District of Western Australia 1839-1894*. Nedlands, Australia: University of Western Australia Press.

Barker, C. (2008). *Cultural studies: Theory and practice* (3 ed.). London, England: Sage.

Bates, D. (1930/1992). *Aboriginal Perth and Bibbulmun biographies and legends*. Victoria Park, Australia: Hesperian Press.

Bropho, R. (1998). Interview 17: Robert Bopho. Swan Valley Aboriginal Community, Circle of Elders. 9:30am, 17:06:98. In T. McCabe (Ed.), *Nyoongar views on logging old growth forests* (pp. 30-32). West Perth, Australia: Wilderness Society.

Carr, D. J., & Carr, S. G. M. (1981). The botany of the first Australians. In D. J. Carr & S. G. M. Carr (Eds.), *People and plants in Australia* (pp. 3-44). Sydney, Australia: Academic Press.

Clarke, P. (2007). *Aboriginal people and their plants*. Dural Delivery Centre, Australia: Rosenberg Publishing.

Colbung, K. (1998). Interview 30: Ken Colbung. Elder. Joondalup. 9:30am. 10:07:98. In T. McCabe (Ed.), *Nyoongar views on logging old growth forests* (pp. 51-56). West Perth, Australia: Wilderness Society.

Collard, D. (1998). Interview 19: Dorothy Collard. Actress and great grandmother. In T. McCabe (Ed.), *Nyoongar views on logging old growth forests* (pp. 32-35). West Perth, Australia: Wilderness Society.

Cunningham, I. (2005). *The land of flowers: An Australian environment on the brink*. Caringbah, Australia: Otford Press.

Dark Side Chocolates. (2011). Dark side chocolates Retrieved January 26, 2011, from http://www.darksidechocolates.com.au/

Dean, B. P. (2000). Introduction. In B. P. Dean (Ed.), *Wild fruits: Thoreau's rediscovered last manuscript* (pp. ix-xvii). New York, NY: W.W. Norton.

Ednie-Brown, J. (1899). *The forests of Western Australia and their development, with plan and illustrations*. Perth, Australia: Perth Printing Works.

Flinders, M. (1814). *A voyage to Terra Australis*. [Google Books] Retrieved from http://books.google.com

George, A. (2002). *The long dry: Bush colours of summer and autumn in South-western Australia*. Kardinya, Australia: Four Gables Press.

Ginsberg, A. (1994). C'mon pigs of western civilization eat more grease. *Literal Latté* Retrieved from http://www.literal-latte.com/1994/11/cmon-pigs-of-western-civilization-eat-more-grease/

Graham, D. (Writer). (1990). *Nyoongar bush tucker with Ken Colbung* [VHS]. The Institute of Applied Aboriginal Studies Mt. Lawley, Australia (Producer).

Grey, G. (1841a). *Journals of two expeditions of discovery in North-West and Western Australia during the years 1837, 38, and 39*. (Vol. 2). London, England: T. and W. Boone.

Grey, G. (1841b). *Journals of two expeditions of discovery in North-west and Western Australia during the years 1837, 38, and 39*. (Vol. 1). London, England: T. and W. Boone.

Hallam, S. J. (1975). *Fire and hearth: A study of Aboriginal usage and European usurpation in south-western Australia*. Canberra, Australia: Australian Institute of Aboriginal Studies.

Hassell, E. (1975). *My dusky friends*. Dalkeith, Australia: C.W. Hassell.

Hill, M. (1998). Interview 7: Mike Hill. RFA Nyoongar Action Group chairperson. Public meeting. In T. McCabe (Ed.), *Nyoongar views on logging old growth forests* (pp. 18-20). West Perth, Australia: Wilderness Society.

Hopper, S. (1998). An Australian perspective on plant conservation biology practice. In P. Fiedler & P. Kareiva (Eds.), *Conservation biology for the coming decade* (2 ed., pp. 255-278). New York, NY: Chapman Hall.

Hopper, S. (2010). Nuytsia floribunda. *Curtis's Botanical Magazine*, *26*(4), 333-368. doi: 10.1111/j.1467-8748.2009.01671.x

Lindley, J. (1840). *A sketch of the vegetation of the Swan River colony*. London, England: James Ridgway.

Marchant, N. G., Wheeler, J. R., Rye, B. L., Bennett, E. M., Lander, N. S., & Macfarlane, T. D. (1987). *Flora of the Perth region: Part one*. Perth, Australia: Western Australian Herbarium, Department of Agriculture.

McCabe, T. (1998). Nyoongar kaarny - Nyoongar spirituality. In T. McCabe (Ed.), *Nyoongar views on logging old growth forests* (p. 6). West Perth, Australia: Wilderness Society.

Meagher, S. J. (1974). The food resources of the Aborigines of the South-west of Western Australia. *WA Museum Records*, *3*(1), 14-65.

Moore, G. F. (1846). *Descriptive vocabulary of the language in common use amongst the Aborigines of Western Australia*. Nedlands, Australia: University of Western Australia Press.

Moore, G. F. (1884/1978). *Diary of ten years eventful life of an early settler in Western Australia*, and also *A descriptive vocabulary of the language of the Aborigines*. Nedlands, Australia: University of Western Australia Press.

Northover, J. (1998). Interview 21: Joe Northover. Police Officer. Collie. 4:30pm. 28:06:98. In T. McCabe (Ed.), *Nyoongar views on logging old growth forests* (pp. 35-37). West Perth, Australia: Wilderness Society.

Paczkowska, G., & Chapman, A. (2000). *The Western Australian flora: A descriptive catalogue*. Perth, Australia: Wildflower Society of Western Australia (Inc.), the Western Australian Herbarium, CALM and the Botanic Gardens & Parks Authority.

Rose, D., & Robin, L. (2004). The ecological humanities in action: An invitation. *Australian Humanities Review*, *31-32* Retrieved from http://www.australianhumanitiesreview.org/archive/Issue-April-2004/rose.html

Ryan, J. (*in press*). John Charles Ryan. In S.-P. Mitchell (Ed.), *Fremantle poets 3: Performance Poets*. Fremantle, Australia: Fremantle Press.

Thoreau, H. D. (1862/2010). *Wild apples*. LaVergne, TN: Filiquarian Publishing.

Thoreau, H. D. (1962). *The journal of Henry David Thoreau in fourteen volumes bound as two: Volumes VIII-XIV (November 1855-1861)*. New York, NY: Dover Publications.

Thoreau, H. D. (1993). *Faith in a seed: The dispersion of seeds and other late natural history writings*. Washington, DC: Island Press.

Thoreau, H. D. (2000). *Wild fruits: Thoreau's rediscovered last manuscript*. New York, NY: W.W. Norton.

Trigger, D., & Mulcock, J. (2005). Forests as spiritually significant places: Nature, culture and "belonging" in Australia. *The Australian Journal of Anthropology*, *16*(3), 306-320.

Vlamingh, W. (1985). *Voyage to the Great South Land* (C. De Heer, Trans.). Sydney, Australia: Royal Australian Historical Society.

Wilkes, T. (1998). Interview 25: Ted Wilkes. Perth Aboriginal Medical Service, Director. 12:55pm. 29:06:98. In T. McCabe (Ed.), *Nyoongar views on logging old growth forests* (pp. 44-45). West Perth, Australia: Wilderness Society.

Williams, R. (1643/1997). *A key into the language of America*. Bedford, MA: Applewood Books.

Wright, J. (1994). *Judith Wright: Collected poems 1942-1985*. Sydney, Australia: Angus & Robertson.

4

Values

Abstract: Why were certain plants privileged as beautiful, while others were considered of no consequence to the aesthetic imagination of colonial botanists, settlers and visitors to the Swan River Colony? How did artistic desirability and scientific prerogative conjoin through the processes of devising names for Swan River taxa? In early botanical documents, how were the immanent senses excluded from—or commoditised through—paradigms of visual beauty? A response is prompted by a reading of the first significant published European account of the flora of Western Australia, John Lindley's *A Sketch of the Vegetation of the Swan River Colony* (Lindley's *Sketch*), issued in three instalments between 1839 and 1840.

Plants and Aesthetics: Reading Lindley's *Sketch*

> What a keep-sake a manual of botany! In which is uttered breathed, man's love of flowers. It is dry as a hortus siccus. Flowers are pressed into the botanist's service.
>
> Thoreau, 1852 (1999, p. 32)

Combining written and illustrated representations, Lindley's *Sketch* is a pivotal document in the history of Australian botany, horticulture and taxonomy. The evolution of plant taxonomy has been well explored by historians of science Thomas (1983) and Farber (2000) who stress social and religious influences. During the nineteenth century, the ordering of plants and animals also corresponded to the establishment of empires, as botanical and colonial explorations were often forged simultaneously (Gascoigne, 1996). However, Lindley's text highlights the underemphasised connections between aesthetic values and the taxonomic systematising of plants. As the basis for structuring the Swan River flora into named patterns, European tenets of beauty abetted Lindley's project by setting out a value system based on colour, symmetry, balance and gracefulness. As science strongly characterises the ways in which people understand plants, evaluations recorded in botanical documents are part of the production of long-enduring cultural perceptions and values. Lindley's *Sketch*, therefore, exemplifies the recursive relationship between evaluations and values. Perceptions of plants instil cultural meanings, which spur visual evaluations, linking human modes of seeing plants to pictorial forms.

Lindley's *Sketch* represents a colonial endeavour to demystify Swan River flora. His publication provides a glimpse into the dissemination of science. Lindley's distanced authorial position on the other side of the globe from the Swan River epitomises some of the core practices of colonial botany (see chapter 1). Through its publication, the vegetation of the colony was ushered into the global stage by a style of value-laden scientific writing. A reading of Lindley's *Sketch* reinforces the assertion by Paul Carter (2010, p. 30) that "however 'scientific' it may purport to be, the language of empiricism remains metaphorical." Hybridising technical nomenclature and European aestheticism, the short treatise appropriated and ordered—through science, naming and ideas of beauty—the newly identified plants of the colony.

For botanists, collectors, artists, writers and visitors, the floral array and faunal menagerie of early Australia and the Swan River Colony confounded taxonomic sentiments. Antipodean species inverted the conventions of flora observed elsewhere on the globe (for example, Clarke, 1876/1993; Seddon, 1972, 1997, 2005). The British botanist and founder of the Linnaean Society, James Edward Smith, published the first book on Australian flora, *A Specimen of the Botany of New Holland.* A colleague of the cosmopolitan botanist Joseph Banks, Smith observed the contradictory characteristics of New Holland plants and the difficulty of invoking European science in the antipodes. Smith (1793) forewarns prospective botanists that the preconceptions of the naturalist trained to see with European eyes will be jarred:

> When a botanist first enters on the investigation of so remote a country as New Holland, he finds himself as it were in a new world. He can scarcely meet with any certain fixed points from whence to draw his analogies; and even those that appear most promising, are frequently in danger of misleading, instead of informing him. Whole tribes of plants, which at first sight seem familiar...prove, on a nearer examination, total strangers, with other configurations, other economy, and other qualities; not only all the species that present themselves are new, but most of the genera, and even natural orders. (9-10)

"Other qualities" were both pleasingly familiar and repugnantly alien. Shifting between curious interest and estranged speculation, Lindley's *Sketch* conveys a comparable tone to Smith's *Specimen.* Although written after decades of investigations by explorer botanists, Lindley's document provides insight into late eighteenth century and early nineteenth century perceptions of New Holland's vegetation (Martin, 1993; Ryan, 1996). Through a hierarchy of naming entwined with judgements of beauty, Lindley's *Sketch* encodes scientific and aesthetic values. It differentiated an unnamed and unknown flora into recognisable classes, bringing the "total strangers" into European proximity as living plants, pressed specimens, seeds, flowers and illustrated images. As a document with popular appeal, Lindley's *Sketch* emphasises taxonomy and aesthetics before the advent of ecology as a discipline.

In this chapter, four themes in Lindley's *Sketch* will be addressed: the character of the vegetation; the horticultural object; the commoditised plant; and the named plant. Lindley sets out to sketch the Swan River environs. The beauty of certain antipodean plants implies the inadequacy of other antipodean species in the imagination of nineteenth century English botanists and horticulturalists. For Lindley, the horticultural object was visibly demarcated, construed as an artefact and evaluated as an object. Disconnected from its Swan River habitat, as well as from Aboriginal networks (see chapter 3), plants were items of aesthetic consumption and objects of visual judgement. Moreover, the document links multi-sensoriality to capitalistic enterprise. Taste and smell, in particular, could be converted into economic terms. Lindley's *Sketch* deploys naming reductionistically as part of the stratification of plants along quasi-social hierarchies with evaluations of beauty the nodes along the chain. All in all, this analysis of Lindley's *Sketch* is meant to identify some of the prevalent historical aesthetic attitudes towards South-West plants.

Words and Images: The Anatomy of a Botanical Document

> According to Linnaeu's [*sic*] classification, I come under the head of the Miscellaneous Botanophilists.
>
> Thoreau, 1852 (1999, p. 35)

British horticulturalist and botanist John Lindley held several highly esteemed positions, including Professor of Botany at London's University College. A prolific author on botanical subjects, Lindley published the journal *Edwards's Botanical Register* (1815-1847) between 1829 and 1847. Established in 1815 by the artist Sydenham Edwards, the *Botanical Register* was an illustrated horticultural magazine. Before leaving editorship to Lindley, Edwards contributed to five volumes of the magazine with the botanist John Bellenden Ker Gawler writing most of the text. Edwards had been a watercolour artist with *Curtis's Botanical Magazine*, which began publishing in 1787 and is now the longest running botanical magazine in the world (Gardham, 2004). Hence, Lindley's *Sketch* arose out of the tradition of

Figure 10 *Laxmannia grandiflora* and Pipe Lily (*Johnsonia hirta*) from Plate 7 of Lindley's *Sketch. Image courtesy of Melbourne University Special Collections.*

combining textual accounts with botanical illustrations. Throughout, there is a cross-current between words and images.

In total, Lindley's *Sketch* named and identified 283 plants, while compiling the classifications of naturalists, explorers and settlers in the Swan River Colony, including botanists von Hügel, Endlicher, Fraser, Brown and Drummond. Adept at identifying the need for new botanical publications, Lindley (1840, p. i) coalesced the "several scattered notices of Swan River plants" into his *Sketch*, giving the most complete and concise portrait of South-West plants then available. Nine coloured lithographs by an unidentified artist, along with four woodcuts, accompany its 58-pages (see table 3). Lindley issued his *Sketch* as an appendix to *Edwards's Botanical Register* (Moyal, 1981, p. 337). Part I, containing pages 1-16 and plates 1-4, was issued on 1 November 1839. Part II, consisting of pages 17-32 and plates 5-7, was published on 1 December 1839. As Part III, the final pages 33-58 and plates 8 and 9 were issued on 1 January 1840.

Before Lindley's *Sketch*, the only significant published western Australian flora was von Hügel and Endlicher's 1837 Latin work, *Enumeratio Plantarum Quas in Novae Hollandiæ ora Austro-Occidentali*. However, unlike *Enumeratio Plantarum*, Lindley's *Sketch* used embellished written descriptions of plants to create broader public appeal. Prior to von Hügel and Endlicher, written works about Swan River plants appeared as abbreviated articles in botanical journals. In 1830, the Colonial Botanist of New South Wales, Charles Fraser, published "Remarks on the Botany, &c. of the Banks of the Swan River" in *Hooker's Botanical Miscellany*. A member of Stirling's 1827 expedition to western Australia, Fraser penned the narrative with compelling visual detail, capturing a sense of awe and intrigue. Fraser (1830, p. 223) precedes Lindley in the use of the aesthetic language to describe flora: "I observed quantities of a species of *Brunonia* growing in great luxuriance on the margin of a salt marsh, its flowers of a brilliant sky-blue." In 1832, Scottish botanist Robert Brown published "General View of the Botany of the Vicinity of Swan River" in the *Journal of the Royal Geographic Society*. Brown compiled the paper from Fraser's 1827 collection, from which Brown received specimens through Alexander Macleay and James Mangles, totalling 140 Swan River species (Brown, 1832).

Unlike many of the naturalists from whom he draws, Lindley never set foot on Swan River soils, declaring that he founded his conclusions on the observations of explorers and botanists including Brown and Fraser, as well as the collections of Mangles supplied by Molloy and Drummond. Whereas some judgements were deduced from living plants in English gardens, others depended on dried examples. The genus *Tribonanthes* includes "plants of no beauty, as far as can be ascertained by their appearance in *the form of dried specimens* [italics added]" (xliv). Lindley (1840, p. ii) acknowledges his secondary sources: "The materials from which the following sketch has been drawn up are the foregoing documents, and

Table 3 A Compendium of Illustrated Plates in Lindley's *Sketch*

#	Original Scientific Name	Common Name	Described By	Renamed
1	Chrysoshöe nitens	Morrison Featherflower	Endlicher	Verticordia nitens
2	Verticordia insignis	Featherflower	Endlicher	-
	Hedaroma latifolium	Lemon-scented Darwinia	Endlicher	Darwinia citriodora
3	Manglesia purpurea	Bottlebrush	Lindley	Beaufortia purpurea
	Calytrix aurea	Starflower	Lindley	-
4	Nuytsia floribunda	Christmas Tree	Labillardière	-
5	Atelandra incana	Silky Hemigenia	Endlicher	Hemigenia incana
	Gastrolobium cordatum	Crisp Roe's Poison	Endlicher	Gastrolobium spectabile
6	Conostylis setosa	White Cottonhead	Lindley	-
	Anigozanthus humilus	Catspaw	Lindley	-
7	Laxmannia grandiflora	-	Lindley	-
	Johnsonia hirta	Pipe Lily	Lindley	Johnsonia pubescens
8	Caladenia longicauda	Long-petalled Caladenia	Lindley	-
	Diuris filifolia	Cat's Face Orchid	Lindley	-
	Thelymitra villosa	Custard Orchid	Lindley	-
9	Borya scirpoidea	-	Lindley	-
	Macdonaldia smithiana	Twisted Sun Orchid	Endlicher	Thelymitra flexuosa
	Macdonaldia antennifera	Vanilla Orchid	Lindley	Thelymitra antennifera

Note: several species have been renamed since colonial classification.

an herbarium of about 1000 species, formed by the communications of Mr. James Drummond, now resident in the Colony, Captain James Mangles..." and others. Most notably, Lindley fails to credit Georgiana Molloy, whose pressed specimens informed his conclusions significantly. Lindley's *Sketch* was based largely on a collection supplied in Molloy's first dispatch of plants to England in 1838 (Moyal, 1981). In a correspondence to Mangles, Lindley applauds Molloy's fine abilities: "That many of the plants are beautiful, you can see for yourself…the seeds we shall be most grateful for" (cited in Moyal, 1981, p. 336). Lindley considered Molloy's seed collections and preserved specimens pivotal to science in the colony. His *Sketch*, therefore, documents more than plants; it records cultural networks, circulating seeds, specimens and ideas in the colonial currents of Australian botany.

Interlude IX: You Are Known by the Company You Keep

With 381 plant species, the Anstey-Keane Damplands south of Perth is the second richest botanical area on the Swan Coastal Plain (Giblett & James, 2009). Through a dream-like encounter with Anstey-Keane, "you are known by the company you keep" relates the biological and sensorial absences that may occur in the experience of a place under pressure. The entangling of sense as synaesthesia intimates the histories of tasting, touching, smelling and engaging with places corporeally that have been relegated to the margins (see chapter 1). Cultural botany situates plants in an ecology of culture and a culture of ecology through intensive interest in the sensations of the body in a botanic field, such as Anstey-Keane, where the imagination and the senses are free to interact with flora. Anna Wallace and Sabina Rahman (2009, para. 2), the editors of the journal *Philament* in which this poem was published, observe that it employs "blank verse to echo the alienation of the self in the modern metropolitan lifestyle," although it is the formlessness of its free verse that refrains from metre or rhyme:

> flora exiled in southern suburbs—
> dying Mangles Kangaroo Paws
> orchid traces, the charred outer
> bark of a Balga: old torpid grower (ll. 17-20)

This poem reflects to me now some of the fragmentation I felt at Anstey-Keane, epitomised by the decaying signs of suburban intrusion and expressed through the disharmonies of free verse (to read "you are known" in full, see Hickman & Ryan, *in press*).

Metaphors and Misnomers: Characterising the Swan River Colony

> But after all where is the flower lore – for the first book & not the last should contain the poetry of flowers.
>
> Thoreau, 1852 (1999, p. 32)

As with the accounts of Brown and Fraser, Lindley's *Sketch* adumbrates the character of the Swan River through the appearance of its flora. Basing opinions on the reports of explorers and visitors, Europeans tended to regard the Australian landscape as not only distant and unknown, but anomalous and inverted (Martin, 1993; Ryan, 1996). The art historian Bernard Smith (1989, p. 242) summarises a predominant early attitude towards Australian landscapes as inversion, or "the belief that the natural productions of Australia were novel creations and that the characteristic features of Australian nature were contrariety and eccentricity." In light of this, it could be said that Lindley's document suffused the consciousness of a northern readership with images of Swan River flora, sorted taxonomically and described compellingly.

Lindley's portrait juxtaposes desirable and displeasing plants. Awe and repulsion intermingle with wonderment and estrangement. A contemporary reader is given a glimpse into his grappling with Swan River plants. Carter (2010) argues that scientific taxonomy enabled the assimilation of unfamiliar Australian plants into a universalised system of translocated specimens:

> Equipped with the artificial system of Linnaeus, novelty ceased to present a problem. Utterly strange forms became type specimens. Less curious plants might be assigned to existing genera. The taxonomy depended on no close examination. (20)

As the historian Keith Thomas (1983, p. 65) plainly states about early eighteenth century taxonomy, "these classificatory schemes represented an ambitious attempt to impose a new form of intellectual order upon the natural world." In the view of Thomas, the hierarchy of plants paralleled the stratification of society into kingdoms, tribes and nations. Lindley's *Sketch* sorts plants into a quasi-social hierarchy, but with judgements of beauty figuring significantly.

Certain families of plants indicate the antipodean nature of the colony. More than other botanical families, the abundant, water-loving carnivorous sundew or *Droseraceae* "evidence…the springy nature of the soil at Swan River" (Lindley, 1840, p. xx). The primeval Proteaeceae, including the wide-spread genera *Banksia* and *Hakea*, distinguishes the landscape: "No where is the many-faced appearance, from which these plants derive their name, more conspicuous than at the Swan River, where they are extremely numerous" (Lindley, 1840, p. xxx). The colony is the "headquarters" of the Haemodoraceae family, including the iconic kangaroo paws "to which the expression *nullibi copiosae* [not widespread] recently applied to it, is no longer applicable; for at the Swan River they seem to form about one-fiftieth of the species" (Lindley, 1840, p. xliii). Written and illustrated images evoked Swan River place in the consciousness of audiences across the seas.

Fraser's "Remarks on the Botany" also relates narratively the appearance of the colony, including a comparative geography between western and eastern Australia. Fraser emphasises the beauty of plants. For example, he speaks of "two other species of the same genus [*Metrosideros*], but of less beauty" (Fraser, 1830, p. 228). The colony varies between barrenness and luxuriousness. Unlike Lindley, Fraser's comments came from a surveyor's perspective, though his assumptions about agricultural suitability proved controversial and later inaccurate (Russo & Schmitt, 1987).

Specific hues of the colour green compelled Fraser to report optimistically about the fertility of the soil (see chapter 6). Describing the view from Pelican Point near present-day Bunbury, Fraser (1830) writes:

> The contrast between the dark blue of the distant mountains and the vivid green of the surrounding forests is such as must in a peculiar manner strike the attention of a person long accustomed to the monotonous brown of the vegetation of Port Jackson. (226)

He attributed the colour to underground water: "I was much astonished at the beautiful dark green and vigorous appearance of the trees, considering that the season had been evidently unusually dry" (Fraser, 1830, p. 224). Brown offered a scientific explanation for the lustreless quality of the forests, a subject of much speculation amongst colonial botanists and explorers. Cutaneous glands on each side of the leaf were thought to account for "that want of lustre which is so remarkable in the forests of New Holland" (1832, p. 21). For Fraser and Brown, form and colour underpinned their evaluations. In particular, Fraser's effusiveness about the prospective settlement could have reflected poetic license granted by Stirling to amplify the perceived suitability of the region for British rule (Russo & Schmitt, 1987).

Like Fraser and Lindley, Brown also attempted a sketch of the Swan River vegetation. The Proteaceae are "the most striking, as well as the most extensive... tribe which, from its general dispersion, and the remarkable forms of its numerous genera and species, includes many of the chief peculiarities of the vegetation of New Holland" (Brown, 1832, p. 19). The "chief peculiarities" include the zamia palm (*Macrozamia riedlei*) and the Christmas tree (*Nuytsia floribunda*) (see chapter 3). The more conspicuous, though anomalous, plants "which greatly contribute to give a character to the landscape" include *bullanock* (*Kingia australis*) (1832, p. 17). In 1801, Brown made the first documented collection of bullanock. In 1825, he presented his findings to the Linnaean Society. A year later, he published "Character and Description of Kingia." In contemporary taxonomy, *Kingia* is assigned to the Dasypogonaceae, a small family presently unrecognised by the worldwide Angiosperm Phylogeny Group (Paczkowska & Chapman, 2000).

Springing from these precedents, Lindley regards the singularity and eccentricity of Swan River plants. Australian botanists struggled with the challenging of norms by endemic taxa that were specially adapted to low soil nutriment, high solar exposure and unusual geological stability (see Breeden & Breeden, 2010; Seddon, 2005). The inflorescence of *Calothamnus* "is so peculiar as to deserve to become the subject of special enquiry" (Lindley, 1840, p. ix). Falling outside the scope of taxonomy, other species are strange and befuddling: "*D. Quadrilatera* has leaves which look more like objects prepared to puzzle a geometrician than any thing already known in the vegetable kingdom" (Lindley, 1840, p. xv). Expectations are turned upside-down. *Jacksonia densiflora* are "very strange looking plants, with branches so like leaves that they would certainly be taken for them by an incautious observer" (Lindley, 1840, pp. xii-xiii). The leaves of *Hakeas* and *Grevilleas* are "so varied and peculiar that a young Botanist might be excused for mistaking them for ferns" (Lindley, 1840, p. xxx), whereas *Anadenia* "might be mistaken for a Holly, when not in flower" (Lindley, 1840, p. xxxi). These species contradicted conventions formulated elsewhere.

Lindley employs a figurative strategy, one used throughout colonial history by Banks and others and preserved in common names for plants (Carter, 2010, p. 29). Old World analogues, such as small heath-like flowers, brought the foreignness of colonial plants into familiar corridors. Comparisons with European analogues recur in the English names of Swan River species, including conspicuous examples of catachresis: sheoak (*Allocasuarina fraseriana*), holly-leaved dryandra (*Banksia sessilis* or parrot bush) and Swan River mahogany (*Eucalyptus marginata* or jarrah). Well-worn metaphors were expository devices, invoking the comfortable language of a European readership. The difficulty of envisioning the flora necessitated rhetorical tropes. Cyclogyne "calls to mind the European species of *Onobrychis*" (Lindley, 1840, p. xvi) and a species of sundew bears "bright scarlet bulbs the size of the largest kind of hazel nut" (Lindley, 1840, p. xx). *Synaphea* is "a strange oak-leaved plant…whose leaves seem as if they were intended to be larger, but starved into hard dry lobes" (Lindley, 1840, p. xxxii). Lindley's *Sketch* calls upon linguistic turns to bring the colony's strange configurations and economies into the familiar topographies of European natural history.

Vision and Beauty: The Horticultural Object

> I suspect that the child plucks its first flower with an insight into its beauty & significance which the subsequent botanist never retains.
>
> Thoreau, 1852 (1999, p. 33)

The esteem of colonial plants in the European imagination depended on their aesthetic appeal. Whereas some plants were horticultural objects, others were objects of no beauty, and therefore of no potential use to gardens. The nine lithographs executed by an unnamed artist illustrate aesthetic objects (see figures 10 and 11). Lindley's *Sketch* exemplifies the meaning of objectification as "object-making." Allied to taxonomy, botanical illustration captures aesthetic features and fuses art and science (Hewson, 1999). Through images and words, Lindley's union of art and taxonomy privileges sight. The visual construction of value is predicated on distance between the appreciator and the "horticultural object" (see chapter 5). Although careful scrutiny and tactile dissection of plants are part of botanical practice, the resulting depiction isolates a specimen in time and space.

The European quest for the "horticultural object" idealised or rejected plants according to pleasing criteria. Select species reflected compositional ideals of garden beauty developed outside Australia. The horticultural object was circumscribed as an object of art. Carter (2010) critiques incisively the taxonomic object of science, which in Lindley's terms, is an aestheticised object:

> Each object, found, translated into a scientific fact and detached from its historical and geographical surroundings, becomes a complete world in itself. It loses all power to signify beyond itself, to suggest lines of development or the subtler influences of climate, ground and aspect. In short, its ecology, its existence in a given, living space is lost in the moment of scientific discovery. (22)

For example, a depiction of *Laxmannia grandiflora* and Pipe Lily (*Johnsonia pubescens*) displays anatomised organs like footnotes (see figure 10). Single idealised species against a white backdrop with reproductive parts flayed open constitute botanical portraiture, disconnecting the plant from its seasonality and habitat. The exposed and dissected specimens of illustration parallel human anatomical illustration, as embodied by the *écorché* renderings of Vesalius and da Vinci (see, for example, Saunders & O'Malley, 1982).

Lindley's *Sketch* disseminated information about the potential to cultivate aesthetically-worthy plants in English gardens. It estimated the suitability of taxa along two-dimensional lines. The text is both horticultural in emphasis and commercial in character. It seeks to prevent "double names" through the universalisation of nomenclature for colonial plants. Lindley (1840, p. i) created standardised references for his readership to determine "whether particular species are worth possession, either for the sake of their *beauty* or *singularity* [emphasis added]." He states:

> The frequent arrival of seeds from this Colony, the excellent state in which they are received, and the facility with which further supplies can be procured, appear to render some Botanical account of this remarkable country a desirable appendage to a work which, like the Botanical Register, forms an

original record of new plants introduced, or worthy of introduction, to our Gardens. (1840, p. i)

By "Gardens," Lindley intimates decorative plantings, valued for their beauty, in contrast to Aboriginal gardens, valued for their sustenance (see chapter 3). The use of "singularity" implies the cultivated plant's uniqueness, but also underlies the fixing of a specimen as a single, translocated and appropriated object. Lindley's *Sketch* offers scant reference to climatic or habitat differences between the Earth's hemispheres that would affect propagation and cultivation. Rather than purely ecological or horticultural, Lindley's *Sketch* is aesthetic. He focuses on the beauty, prettiness, strikingness, colour and brilliance of Swan River plants. As stressed above the phrase, "beauty or singularity" signifies the transoceanic imagination that underpins the work. Beauty is accorded through form and colour, a theme also found in Brown and Fraser.

In order to disseminate information about the commercial viability of Swan River plants, Lindley strives towards popular appeal. As a resource for nineteenth century markets, his *Sketch* capitalises on the impact of images rendered in words. To promote Swan River plants, aesthetic language compels feelings of attraction through insinuations of beauty or strikingness. As with Lindley, Fraser's descriptions of the Swan River flora carry aesthetic overtones, possibly to increase the influx of settlers to the colony. Appealing through sight and drawing extensively upon the capacity of the mind to visualise places, plants are depicted as objects to forward colonial prerogatives. Beautiful plants attract settlers, seduced by the promise of fertile land, and seed purchasers hunting for exotic beauty. The Myrtle family, for instance, is composed of "bushes, with small heath-like leaves, and white, yellow, or purple flowers, of great brilliancy" and forming "a most striking object in the vegetation" (1840, p. v). Lindley constructs the flora as objects of art, with select plants described as isolated jewels in the monolithic rough of the bush. *Thomasia* species "bear fine showy flowers and deserve a place in a conservatory" (Lindley, 1840, p. xix). *Corethrostylis bracteata* bears "a profusion of graceful forked racemes of crimson flowers…and is one of the most beautiful plants of the Colony" (Lindley, 1840, p. xix). Numerous evaluations of beauty are inflected.

But while some species are horticultural objects, others are not suitable as European plantings. Such species fall outside beauty speak. Lindley (1840, p. xxvi) refers to the Goodeniaceae in dismissive terms with most of the family "not at all suited to the objects of cultivators." With the exception of the fine royal blue *Leschenaultias* and the indigo *Dampieras*, "all the other species, and there are many, are by no means beautiful objects" (Lindley, 1840, p. xxvii). Similarly, to further affirm the hierarchy of aesthetics, in reference to the genus *Stenopetalum*, Lindley (1840, p. xxxviii) writes that "none of them [are] of any Horticultural interest" and the *Stackhousias* encountered are "species of no beauty." The genera *Roea* and

Dichosema are "two others of little beauty…neither of which possess the slightest interest for horticultural purposes" (Lindley, 1840, p. xvi).

The invocation of beauty inversed continues from the beginning to the end of Lindley's *Sketch*. Attractive features are visual, even though taste and smell might be pleasing. Species of *Rhodanthe* are "annuals of no beauty" (p. xxiii) and *Cylindrosorus* and *Myriocephalus* "are of no beauty" (p. xxiv). Regarding the *Leucopogon*, "none of them deserve particular record" (p. xxv). Grasses and sedges "have no connection with Horticultural objects" (p. lviii). While objects of no beauty fail to meet visual criteria, the same species might be aromatic or tactile. In Lindley's terms, the abnegation of multi-sensoriality seems prerequisite for an aesthetics of beauty. The Myrtle family is "not in general so handsome as those already mentioned" even though "the fragrance of which exceeded any thing" (Lindley, 1840, p. x). A plant is pleasing only if it appears so. Two species of the genus *Lyperanthus* "have no pretensions to beauty, but have a very singular appearance with their dingy sad-coloured flowers, and are very fragrant" (Lindley, 1840, p. liii). Lindley excludes smell, placing colour along a hierarchy of emotion: singular, sad-coloured, aromatic but not beautiful flowers.

Botany and Commercialisation: Multi-Sensoriality in Economic Terms

Whereas species with "no pretensions to beauty" are excluded from gardens, fragrant and edible plants could be exploited. The reader is exposed to Lindley's speculation about the prospective economic importance of certain aromatic or palatable plants. Lindley weighs the potential for multi-sensorial qualities to be capitalised upon for purposes of empire. Instead of a hierarchy of beauty, his *Sketch* in some passages reveals a taxonomy of economics. Indeed, the commodification of South-West plants is found throughout historical booster literature. Ogle (1977) and Groser (1927) published accounts to attract colonists to the west coast. The conceptualisation of plants as potential exports or resources marks the beginning of colonial history, as exemplified by the sandalwood industry (Loneragan, 1990; Russo & Schmitt, 1987). The smell, taste, sound or sensation of a plant may be etched into its nomenclature. The history of plant sensuousness is one of its imbrication into economic networks. The fragrant leaves and half-ripe fruits of *Hedaroma* "might be worth collecting for the use of the perfumer; and if so they would furnish a new and most agreeable article of luxury to Europe" (Lindley, 1840, p. vii). Commodified as exports, plant foods like the *Drosera* "may be easily obtained for the purpose of exportation, and may assist the poorer settlers in turning to account the produce of their land" (Lindley, 1840, p. xxi). Certain species of sundews possess "scarlet roots, not unlike in shape and size to tulip bulbs…they may be considered the bread of the natives who live near the coast" (Lindley, 1840, p. xxi).

The requisitioning of plants for enterprise characterises colonial emigration literature, in which rich descriptions of landscapes persuade prospective emigrants to the burgeoning state. Historian James Belich (2009, pp. 153-154) identifies a correlation between certain types of published material and the practices of settlement, arguing that "formal settlerism manifested itself in 'booster literature' or 'emigration literature': books, pamphlets, newspapers and journal articles, lectures, and advertisements [that] almost monopolized published information about emigration destinations." Ocularcentric attitudes towards plants amplified the appeal of the booster literature of the Swan River colony. Published in 1839 by Nathaniel Ogle, *The Colony of Western Australia* enticed "emigrants," "capitalists" and "the younger branches of the higher classes" to the nascent colony. To serve the colonial agenda, Ogle's book relies on a visual vocabulary of plants with the speaker situated distantly: "The earth is enamelled with flowers; at some seasons the meadows have been compared, when at a distance, to surfaces of golden chintz" (Ogle, 1977, p. 260). Hence, plant appearances augment the persuasiveness of boosterism.

Imagistic language worked hand-in-hand with utilitarian attitudes towards flora as commodities. Appearances are folded into an aesthetics of appropriation. Visitor Fred Davis, cited in Thomas Groser's 1927 *The Lure of the Golden West*, evidences the drawing power of "pictorial statements" (Ivins cited in Jay, 1993, p. 69). Fred Davis (cited in Groser, 1927, p. 245) depicts a drive alongside the Blackwood River in Western Australia as "what is indeed a charming picture. How a landscape painter would revel in these lovely spots!" A concatenation begins with an aesthetic evaluation and terminates in the broad-scale conversion of the wild to the pastoral. The drive is interspersed with cleared areas and "beautiful as [the forest] undoubtedly is in its natural state, yet, from a utilitarian point of view, it will be more beautiful still when all under cultivation" (Davis cited in Groser, 1927, p. 245). This statement underscores the problematic contradiction of sense knowledge gained exclusively through the eyes bringing to rise, as an imperative, the conversion of an idle-looking scene into productive, techno-industrialised land.

By no means should Lindley's work be classed as boosterism. Yet, it does employ booster strategies. A comparable web of aestheticism and settlerism tied to commercial prerogatives is evident in his *Sketch*. Lindley ponders other ways to capitalise upon Swan River plants. Such values emerged through non-visual pleasure, particularly smell and taste. The sundew family "appear likely to be in some cases of commercial value as dyer's plants" (p. xxi). The bulbs possess "a deep scarlet powder secreted by the scales of the bulb…more like the colour obtained from Archil than any thing else to which I can compare it" (p. xxi). Archil, a violet dye from lichens of the Canary Islands, acts as a trope overlaying the Swan River flora with pre-existing commercial implications. *Petrophila* yields "so brilliant a yellow colour that it is worth examination as a probable source of dyeing material" with the species *P. brevofolia* possessing "dyeing properties that deserve investigation" (p. xxxv). The orchid *Glossodia brunonis* has "large roots, enveloped in numerous coarse skins, and as sweet as a chesnut [*sic*], even when dried; they would certainly afford a delicate article of food" (p. li). Lindley's evaluation of the flora carries values of commodification. Most importantly, the colonial agenda disregarded existing Aboriginal networks of sustenance that depended on plants (see chapter 3).

Colonisation and Classification: Naming Plants

> Botanies instead of being the poetry are the prose of flowers. I do not mean to underrate Linnaeus' admirable nomenclature much of which is itself poetry.
>
> Thoreau, 1852 (1999, p. 33)

Scientifically systematic and artistically concerned, Lindley's publication centres around classification. His *Sketch* shows the value-laden process of name-granting, a practice engaged in by European botanists classifying antipodean taxa. Lindley (1847, p. i) standardised plant names and strove for uniform terminology because "half a dozen names were proposed in different places to express the same idea." As Lindley (1847, p. i) lamented, botany was suffering a crisis of communication: "The language of Botany is marvellously in want of reformation." Thomas (1983, p. 66) and Schiebinger (2004) have observed that the standardisation of names entails stratification into social hierarchies of families, classes and orders. Taxonomic hierarchies in Lindley's *Sketch* are connected to aesthetic values.

The naming of a plant is the ultimate point of scientific reference, but names are also cultural productions. Nomenclature enciphers cultural values, reiterated in every utterance and re-utterance of names. Before the arrival of colonists, Swan River plants were denoted by Aboriginal names. The pre-colonial Christmas tree (*Nuytsia floribunda*) invoked neither the Christian holiday nor the Dutch explorer Nuyts, but rather a web of lifeways inherent to *mudja*. The overlayering of settler terminology onto indigenous signifiers is a nuance of colonisation and empire. Arranged in stratigraphic layers, the plant world becomes a palimpsest of nomenclatures. On the power asymmetries of naming, Skene (1997) comments that:

> The early settlers overlaid the place names of the indigenous people with their own; the botanists were likewise guilty of erasing Aboriginal plant names and imposing a different way of relating to the flora of Western Australia. (11)

In taxonomy, name-granting is the defining practice, whereas Aboriginal names are endemic to the Swan River or the broader region, like the plants themselves. In an Aboriginal sense, names are coterminous with the plants and their habitats. In taxonomic terms, names govern the universal expression of information, while generally minimising the conveyance of local character. However, botanical naming is rich with aesthetic and poetic meanings, suggesting that, despite claims to universality, science is cultural (see chapters 1 and 2).

Often names for South-West plants commemorate significant male figures in the history of botany, despite the fact that, in colonial times, women collectors were central to the development of plant taxonomy. *Manglesia* was "named by Endlicher in compliment to Captain James Mangles…to whose exertions the country owes the greater part of the plants as yet introduced from this colony into our gardens" (Lindley, 1840, p. xxxvii). Skene (1997, p. 2) avers that "in eighteenth and nineteenth century Australia, exploration and the concomitant act of naming was an overwhelmingly masculine activity: there seems to be little 'space' for women in this founding practice of cultural definition." Like the names of mountains, deserts, bodies of waters and localities, botanical names often perpetuate pioneer myths as honorifics for explorers and figures in colonisation, impressing a gendered convergence of values, histories and ideologies upon flora, place and people. The newly identified genus *Loudonia*, for instance, is "a tribute to the eminent services rendered to Horticultural Botany by John Claudius Loudon, Esq. author of the Arboretum Britannicum, and of many other valuable works well known in every part of the civilized world" (Lindley, 1840, p. xlii). Named *Hedaroma latifolium* by Lindley for its rich fragrance (but now known as *Darwinia citriodora*), the naming of Lemon-Scented Darwinia associates the male figure Darwin perpetually with the fragrance of the plant.

Names are the outputs of systems, either artificial in the Linnaean sense or natural. Lindley preferred the natural system advocated by de Jussieu. Well aware of the challenge of Australian flora, botanist Robert Brown viewed Linnaean taxonomy as a hindrance to botany in the antipodes. Instead of the Linnaean method, which relied on stamens and pistils, Brown adopted the natural system, founded on the overall physiology and anatomy of plant parts (Moyal, 1986, p. 20). Brown's preference for the natural system is reflected in the illustrations of botanical artist Ferdinand Bauer, each carefully depicting the structures of seeds, roots and flowers. Similarly, Lindley's preference for non-Linnaean taxonomy is apparent in the illustrated plates depicting the various anatomised parts of plants (see figures 10 and 11). In both systems, names bring plants into conceptual and physical proximity.

The aestheticism of Lindley's *Sketch* undergirds scientific universalisation. His *Sketch* is a technical output and a cultural production. Constrained by the mechanical, reductionist and poetically sober nature of technical enquiry, the

Figure 11 Twisted Sun Orchid (*Macdonaldia smithiana*), Vanilla Orchid (*M. antennifera*) and *Borya scirpoidea* (Plate 9). *Image courtesy of Melbourne University Special Collections.*

treatise expresses names and judgements of beauty, but disengages endemic knowledges, with Lindley himself positioned at a great distance. In the 1850s, Thoreau (1999, p. 35) expressed discontent with the artificiality of all taxonomies: "By the artificial system we learn the names of plants, by the natural their relations to one another; but still it remains to learn their relation to man. The poet does more for us in this department." The botanical names and terminologies which Lindley sought to reform and homogenise "approach ever nearer to the dryness of an algebraic formula" (Thoreau, 1999, p. 92).

Thoreau argues that poetic thought has been removed from plants through the mechanisation of language (see chapter 9). Thoreau (1999) goes on to say that:

> The artificial system has been very properly called the dictionary – and the natural method, the grammar of the science of botany, by botanists themselves – But are we to have nothing but grammars and dictionaries in this literature? Are there no works written in the language of the flowers? (33)

Beauty is not merely visual speculation but an experience integrated to human culture and corporeality:

> The natural system may tell us the value of a plant in medicine or the arts or for food. But neither it nor the Linnaean to any great extent tell us its chief value & significance to man, which in any measure accounts for its beauty. (Thoreau, 1999, p. 32)

For Thoreau, nomenclature lacks interconnection that emerges complexly out of indigenous grammars and "the language of the flowers." As such, beauty is not gestated solely through visuality, but also through contact with the earth.

Interlude X: Seven Names for a Plant

"Seven Names for a Plant" comments on the multiple naming histories for the fascinating West Australian Christmas tree (*Nuytsia floribunda*) (see figure 12) (see chapter 3). In the 1600s, the crew of Gulden Zeepaard named the tree after Pieter Nuyts, the highest ranking figure on the ship. The orange-yellow blossom of Nuytsia could have been one of the earliest features of the Western Australian coastline sighted by European explorers. Plant names encode the history of colonisation of the land by settlers who brought Linnaean science and European aesthetic tastes. Nyoongar names signify different conceptualisations and interdependencies with plants (see chapter 3). Lindley (1840, p. xxxix) recorded that "such is the abundance of the orange-coloured blossoms, that the Colonists at King George's Sound compare it to a tree on fire; hence it has gained the name of 'Fire Tree'." Ednie-Brown (1899, p. 28) references the tree "not because of any value attached to its timber but simply on account of the picturesque appearance of its flowers." Of the awe held by some Nyoongar people towards the Christmas Tree, colonial ethnographer Daisy Bates (1930/1992) observed the respect of the local people towards the tree:

> No living Bibbulmun ever sheltered or rested beneath the shade of the tree of souls; no flower or bud or leaf of the tree was ever touched by a child or adult; no game that took shelter beneath it was ever disturbed. (153)

The eighth line of the poem refers to the Nyoongar belief that deceased souls rest on mudja branches on their way west to the afterlife (Bates 1930/1992):

> ghost bush,
> way station of those passed,
> glissading spirits to the sea
> branches pose ghost-like buds (ll. 5-8)

The italicised Nyoongar and Latinate words in the poem emphasise the intersection of scientific and Aboriginal terminologies. The final lines suggest the role of poetics in opening up other intriguing convergences between plants and language (to read "Seven Names" in full, see Hickman & Ryan, *in press*).

Figure 12 The West Australian Christmas Tree (*Nuytsia floribunda*), Lesueur National Park. *Image courtesy of the author.*

Absences and Aesthetics

As an important document in the cultural history of South-West flora, Lindley's *Sketch* shows that aesthetic judgments may be inherent in scientific documents of this period. Rather than situated in regional awareness and local understandings, his *Sketch* is globally interested. In its focus on identifying horticultural objects, the document records the transhemispherical exchange between the colonies and Europe through plant material, such as seeds and dried flowers, but also as values towards plants. Prioritising visual evaluations and visually derived values, Lindley's work reveals the aesthetic leanings of taxonomy and the hierarchical arrangement of life forms based on beauty. To this end, the document evidences the core impetus of taxonomic botany: to forge standardised groupings. Whereas local knowledge inhabits regions, taxonomy universalises, not only through nomenclature, but also through aesthetic judgements. Most glaringly for a contemporary reader, Lindley's *Sketch* absents the strata of ecological relations embedding species within habitats, a reality now being considered by botanists, such as Stephen Hopper (2004), in researching Western Australian flora.

Based heavily in form and colour, an aesthetic focus on plants often conflicts with values of conservation and ecosystem integrity. The botanophilist of the twenty first century might engage an aesthetics of flora, rooted in multi-sensoriality and Aboriginal heritages. As contemporary ecology continues to reveal, landscapes and plants that look idyllic but idle, unproductive, unattractive or repulsive—as objects of beauty or no beauty—are biologically critical to a host of organisms, including human beings. Extending the notion of a counter-aesthetic, or a conservation aesthetic (Giblett, 1996, 2011), a post-colonial aesthetics of plants would articulate the bodily senses. The crowning achievement of Lindley's *Sketch*, a visual aesthetics of Swan River plants, hence would become a corporeal aesthesis of flora (see chapter 13). It is on this final point that Thoreau's cues become salient.

References

Bates, D. (1930/1992). *Aboriginal Perth and Bibbulmun biographies and legends*. Victoria Park, Australia: Hesperian Press.

Belich, J. (2009). *Replenishing the Earth: The settler revolution and the rise of the Anglo-world, 1783-1939*. New York, NY: Oxford University Press.

Breeden, S., & Breeden, K. (2010). *Wildflower country: Discovering biodiversity in Australia's Southwest*. Fremantle, Australia: Fremantle Press.

Brown, R. (1832). General view of the botany of the vicinity of Swan River. *Journal of the Royal Geographic Society*, 17-21 Retrieved from http://books.google.com

Carter, P. (2010). *The road to Botany Bay: An exploration of landscape and history.* Minneapolis, MN: University of Minnesota Press.

Clarke, M. (1876/1993). Preface to Adam Lindsey Gordon's Sea spray and smoke drift. In M. Ackland (Ed.), *The penguin book of 19th century Australian literature* (pp. 43-46). Ringwood, Australia: Penguin Books.

Ednie-Brown, J. (1899). *The forests of Western Australia and their development, with plan and illustrations*. Perth, Australia: Perth Printing Works.

Farber, P. (2000). *Finding order in nature: The naturalist tradition from Linnaeus to E.O. Wilson*. Baltimore, MD: The Johns Hopkins University Press.

Fraser, C. (1830). Remarks on the botany, &c. of the banks of Swan River, Isle of Buache, Baie Geographie, and Cape Naturaliste. *Botanical miscellany, 1*, 221-236 Retrieved from http://books.google.com

Gardham, J. (2004). Curtis's botanical magazine Retrieved 7 April, 2011, from http://special.lib.gla.ac.uk/exhibns/month/oct2004.html

Gascoigne, J. (1996). The ordering of nature and the ordering of empire. In D. Miller & P. Reill (Eds.), *Visions of empire: Voyages, botany, and representations of nature* (pp. 107-116). Cambridge, England: Cambridge University Press.

Giblett, R. (1996). *Postmodern wetlands: Culture, history, ecology*. Edinburgh, Scotland: Edinburgh University Press.

Giblett, R. (2011). *People and places of nature and culture*. Bristol, England: Intellect.

Giblett, R., & James, D. (2009). Anstey-Keane botanical jewel. *Landscope, 24*(4), 42-44.

Groser, T. (1927). *The lure of the golden West: Experiences and adventures in a bush brotherhood of Western Australia: Early problems and conquests: All about group settlements of the West: The land of sunshine and opportunity*. London, England: Alexander-Ouseley.

Hewson, H. (1999). *Australia: 300 years of botanical illustration*. Collingwood, Australia: CSIRO Publishing.

Hickman, E., & Ryan, J. (*in press*). *Two with nature: Ellen Hickman and John Ryan*. Fremantle, Australia: Fremantle Press.

Hopper, S. (2004). Southwestern Australia, Cinderella of the world's temperate floristic regions. *Curtis's Botanical Magazine, 21*(2), 132-180.

Jay, M. (1993). *Downcast eyes: The denigration of vision in twentieth-century French thought*. Berkeley, CA: University of California Press.

Lindley, J. (1840). *A sketch of the vegetation of the Swan River colony*. London, England: James Ridgway.

Lindley, J. (1847). Glossary of technical terms. In *The elements of botany, structural and physiological* (pp. i-civ). London, England: Bradbury & Evans.

Loneragan, O. (1990). *Historical review of sandalwood (Santalum spicatum) research in Western Australia*. Como, Australia: Department of Conservation and Land Management.

Martin, S. (1993). *A new land: European perceptions of Australia 1788-1850*. St Leonards, Australia: Allen & Unwin.

Moyal, A. (1981). Collectors and illustrators: Woman botanists of the nineteenth century. In D. J. Carr & S. G. M. Carr (Eds.), *People and plants in Australia* (pp. 333-356). Sydney, Australia: Academic Press.

Moyal, A. (1986). *A bright & savage land: The science of a new continent - Australia - where all things were "queer and opposite"*. Ringwood, Australia: Penguin.

Ogle, N. (1977). *The colony of Western Australia: A manual for emigrants 1839*. Sydney, Australia: John Ferguson.

Paczkowska, G., & Chapman, A. (2000). *The Western Australian flora: A descriptive catalogue*. Perth, Australia: Wildflower Society of Western Australia (Inc.), the Western Australian Herbarium, CALM and the Botanic Gardens & Parks Authority.

Russo, G., & Schmitt, H. (1987). *Swan River mania: A colonial history of Western Australia*. Perth, Australia: Lynwood Enterprises.

Ryan, S. (1996). *The cartographic eye: How explorers saw Australia*. Cambridge, England: Cambridge University Press.

Saunders, J., & O'Malley, C. (1982). *The anatomical drawings of Andreas Vesalius*. New York, NY: Bonanza Books.

Schiebinger, L. (2004). The private lives of plants. In *Nature's body: Gender in the making of modern science* (pp. 11-39). New Brunswick, NJ: Rutgers University Press.

Seddon, G. (1972). *Sense of place: A response to an environment*. Perth, Australia: University of Western Australia Press.

Seddon, G. (1997). *Landprints: Reflections on place and landscape*. Cambridge, England: Cambridge University Press.

Seddon, G. (2005). *The old country: Australian landscapes, plants and people*. Cambridge, England: Cambridge University Press.

Skene, J. (1997). The power of naming: Women botanical collectors and the contested space of 19th century botany. In J. Gregory (Ed.), *Studies in Western Australian History: Historical Traces* (Vol. 17, pp. 1-12). Nedlands, Australia: University of Western Australia.

Smith, B. (1989). *European vision and the South Pacific* (2 ed.). Oxford, England: Oxford University Press.

Smith, J. E. (1793). *A specimen of the botany of New Holland*. London, England: J. Sowerby.

Thomas, K. (1983). *Man and the natural world: Changing attitudes in England 1500-1800*. New York, NY: Penguin Books.

Thoreau, H. D. (1999). *Material faith: Thoreau on science.* (L.D. Walls, Ed.). Boston, MA: Houghton Mifflin Company.

Wallace, A., & Rahman, S. (2009). Editorial. *Philament,* (December) Retrieved from http://sydney.edu.au/arts/publications/philament/issue15_pdfs/Editorial.pdf

5

Aesthetics

Abstract: Chapter 5 will present the conceptual challenges to a model of botanical aesthetics based in embodied human interaction with plants through smell, taste, touch, sound and sight. The chapter will argue that the science of aesthetics is deterministically visual. Drawing from theories of landscape aesthetics of Carlson and Berleant, four primary obstacles to an embodied aesthetics will be detailed: plants as objects of sight; plants as objects of art; plants as objects of disinterestedness; and plants as objects of scientific discourse. A multi-sensorial aesthetics of flora requires physical proximity and degrees of entanglement between the appreciator and the appreciated plant that raise important philosophical questions about aesthetic experience of the natural world.

Chapter 4 argued that Lindley's *Sketch* hybridises science and aesthetics. An aesthetics, based in sight, fastens the value of botanical habitats to appearances. The kind of visual regime suggested by Lindley factors into some contemporary conservation research. For instance, the Lesueur management plan (Western Australia) includes a study of landscape character that identifies high, moderate and low scenic—or aesthetic—value. Vegetation distinguished by colours and patterns is privileged. Perspectival contrasts identify exemplary plants that stand out in the otherwise undifferentiated bush. Focal point specimens offer high scenic quality, and include the "windshaped, gnarled or dwarfed vegetation unusual in form, colour or texture" (Department of Conservation and Land Management, 1995, p. 60). Of moderate and low value are plants without distinguishable scenic characteristics of texture, form, colour and structural variation (Department of Conservation and Land Management, 1995, p. 60).

Chapter 5 will theorise an aesthetics of flora in which our embodied selves contact living plants. The discussion will acknowledge a continuum between wild, semi-cultivated and domestic plants. The contexts of plants range from national parks, garden plantings, wildflower shows and the interior of a home. Unlike a flower in a vase, however, wild plants form habitats where there is little rectilinear orderliness. Moving shorelines and scarps define the perceptual field in which a plant is situated. The scene is not composed but *decomposing* before our eyes through the processes of weathering and other temporal progressions. Depending on rainfall, the prominent colours of flowers might be missing. Eyes gaze expectantly in search of affections of sight, but disorientation may follow. Notions of beauty might be disrupted—possibly redefined—because of the mutability of places. In contrast, domestic plants are typically arranged according to fixed patterns, shapes, colours, textures, scales and specimen numbers. Such plantings form a portrait, each flower a constituent brushstroke framed by manicured borders. In the same way, wildflower shows at churches or community centres display plants harmoniously to accentuate their artefactual beauty (see chapter 7).

The chapter begins with the complexities of Western environmental aesthetics, and then arrives at the limitations of botanical aesthetics. An aesthetics of flora, as an extension of environmental aesthetics, is inhibited by significant objectifications: (a) plants as objects of sight; (b) plants as objects of art; (c) plants as objects of disinterestedness; and (d) plants as objects of science. To begin with the first point, a visual regime problematises an aesthetics involving tasting, smelling, hearing and touching plants through rhythms of seeding, flowering, fruiting, emerging and dying. In Western aesthetics, multi-sensoriality has been relegated to the bottom rungs in the hierarchy of the senses (see chapter 13). With an emphasis on distance rather than proximity, environmental aesthetics has yet to account sufficiently for diminutive bodies and intimate gestures of contact. Regarding the second point, environmental aesthetics conflates art and plants. Despite the forward-thinking philosophies of Allen Carlson and Arnold Berleant, the diction of artistic objects presides over the appreciation of nature (see chapter 4). Disinterestedness, rather than engagement, is an underlying ethos, as the third point suggests. Concerning the fourth point, the ocularcentrism of science aligns to aesthetics. Perception orders nature with reason, rather than engaging nature through sense immanence.

In outlining these limitations, a variety of aesthetic theories will be analysed. However, the focus will be on Berleant's "engagement" model (1993, p. 25; 2005) and Carlson's "natural environmental" model (2000). Carlson and Berleant are conscious of the culture of visuality that constructs plants as the desiderata of tourism. These two aestheticians theorise disinterestedness, beauty, sublimity, the picturesque, perception, cognition, detachment, visualism and embodiment with respect to landscape appreciation. However, neither model offers a framework for exploring visceral multi-sensoriality. After identifying these shortcomings, the works of Thoreau and Heidegger will be returned to in theorising botanical aesthetics through the concept of aesthesis.

Plants as Objects of Sight: From Visuality to Sensation

The literature of aesthetics includes works from philosophy (Berleant, 2005; Carlson, 2000; Hepburn, 1966/2004), cultural theory (Eagleton, 1990; Giblett, 2011), psychology and the social sciences (Averill, Stanat, & More, 1998), anthropology (Shelton, 1994), the visual and literary arts (Schroeder, 2010), architecture, regional planning and urban design (Porteous, 1996), and advertising and marketing (Baisya & Das, 2008). There is a remarkable variety of definitions of aesthetics and aesthetic experience. Yet, most disciplines concur that sensory perception and the experience of pleasure are connected to aesthetics. As pleasurable perception of the world, aesthetics concerns sensory apprehension that is desirable. Averill, Stanat and More (1998, p. 154) assert that "most theoreticians and laypersons agree that aesthetic experiences are in some way pleasurable or enjoyable." Aesthetics, as such, does not concern everyday experience, but rarefied moments of pleasure.

As the analysis of beauty, aesthetics deals with the qualities that constitute an enjoyable experience. Carlson (2000, p. xvii) defines aesthetics as "the area of philosophy that concerns our appreciation of things as they affect our senses, and especially as they affect them in a pleasing way." Not all experience is aesthetically pleasing. Some experiences affect us by impressing, moving, inspiring, invigorating or captivating us, whereas some will do so by disgusting, horrifying or repulsing us. The analysis of aesthetic experience is the core aim of philosophical aesthetics. Environmental aesthetics comprises natural and cultural experiences of pleasure. Landscape aesthetics further describes the perception and appreciation of places: deserts, forests, heathlands or oceans. Carlson (2000, p. xx) taxonomises between environmental aesthetics as including various theories of built and natural environments, and landscape aesthetics as applicable to natural places: "Within the genus of environmental aesthetics fall a number of different species, such as the aesthetics of nature, landscape aesthetics, the aesthetics of cityscapes and urban design, and perhaps the aesthetics of architecture." Landscape aesthetics thus refers to the philosophical analysis of the sensory perception of the natural world. For this discussion the term *aesthetics of nature*, rather than landscape aesthetics, will be used. The concept of *landscape* invokes the picturesque as physiographic space rather than multi-dimensional interactive place (Cosgrove, 1998) (see chapter 10).

In Western aesthetics, pleasurable experience engages vision mostly (Berleant, 2005; Heidegger, 1938/2009; Jay, 1993). To consider something an object of aesthetics is to assess it as optically beautiful, and hence desirable, by virtue of its surface features. Jay (1993, p. 9) comments that "the distinct historical manifestations of visual experience in all its possible modes" have given rise to the pre-eminence of sight and the attendant marginalisation of other senses. An image-focused culture can pictorialise the world in the mind only insofar as it can construct word images. Visualism has a close functional association with language, entailing morphologies of visual pleasure (Jay, 1993, p. 9) (see chapter 9). Canons of beauty, derived through language, exclude the less quantifiable bodily senses. As specialists, aesthetic appreciators demarcate object appropriateness; some objects will be beautiful, whereas others deserve no special consideration (see chapter 4). Aesthetic taste implies the attractiveness of an artistic object as judged by an aesthetic subject; the object thus becomes aestheticised. A caption beneath the photo of a mountainscape refers to the limitations of images in expressing landscapes, summed up by Porteous (1996, p. 40) as "visual aesthetics only; other senses elude representation."

Whereas aesthetic perception has its locus in a perspectival point *out there*, corporeal appreciation has its hexis in bodily sensation *in here*. The framing of an aesthetic object implies distance between bodies (see chapter 4). Aesthetics is a neologism first proposed in 1750 by Alexander Baumgarten in *Aesthetica* (Bychkov & Sheppard, 2010, p. xi). The term derives from the ancient Greek *aesthesis* for open sensory perception. Since Baumgarten, aesthetics has evolved into a narrowing of the experience of perception, a closing off of an object (including a landscape) from its network. Closed sensory perception denominates, essentialises and resists nature as bodily affective. As the exploration of "perceptual complexity," aesthesis has been eroded from aesthetics by the privileged role of the "aesthetic subject" (Singer, 2003, pp. 14-15). Whereas aesthesis denotes broad apprehension and bodily sensation, aesthetics has come to mean the pseudo-scientific apprehension of "a sensible image of perfection," in Baumgarten's terms (Caygill, 1995, p. 53). Moreover, Caygill (1995, pp. 53-54) stresses that Kantian aesthetics extracts pure forms from sensation: "Some aspects of Kant's [*Critique of Judgment*] are anticipated by Plato in *Timaeus* when he relates *aisthesis* to pleasure and pain, but this was no part of Baumgarten or Kant's concern." In contradistinction to aesthetics, *open multi-sensoriality* would explore the "diffuse, inchoate, transient, and emotional" senses (Tuan, 1982, p. 117).

Aesthetics reinforces the idea of an unknown place as foreign and threatening, as alienated from human inhabitants and conceptualised into an aggregation of objects rather than a complex network of sustenance (see chapter 3). In order to minimise harm, a potentially dangerous landscape or plant could be approached through the safe distance of sight. However, for all of its practical benefits, seeing might inhibit participating. Porteous (1996, p. 31) notes that *perception* almost always refers to the intake of visual stimuli. He differentiates between autocentric (subject-centred) and allocentric (object-centred) senses. Vision is allocentric and is associated with cognition, detachment, distance and maturation. Autocentric senses (hearing, smell, touch and taste) are physical, primitive, emotive, immediate, proximal and intrinsic to children (Porteous, 1996, p. 31). Visual appreciation requires a percept, whereas the autocentric senses reduce—or even eliminate—appreciative distance. In consuming wattles, for example, the seed can be physically incorporated through mastication, swallowing and digestion.

The predilection for sight is acknowledged by Berleant (2005, p. 3) who opens his most recent volume on environment and aesthetics with an apology: "Aesthetic perception is usually described in visual terms: we are given not an aesthetic of *experience* but an aesthetic of *appearance* [emphasis added]." Sight affirms cognition; perception becomes intellectual and the appreciator, a rational subject. Berleant (2005, p. 3) further argues that Western aesthetic thought privileges "sight, along with the other distance receptor, hearing, [as] the only senses traditionally admitted as legitimately aesthetic." Through the surface qualities of landscapes, aesthetics has been a "triumph of a pure visuality, concerned solely with formal optical questions" (Jay, 1993, p. 160). Whereas vision allows the analysis of landscape qualities, the smellscape, for example, is "an enveloping, unstructured, often directionless space" (Porteous, 1996, p. 36). Vision aids an experience of perceptual uniformity through recognition of form, balance and symmetry that enable systematic appreciation and knowledge. Smell and taste are too messy and entangled for a perceptual science.

Aesthetics has been defined as the "science of sensory knowledge," requiring visual parameters (Berleant, 2005, p. 3). Aesthetic judgements are made objective and communicable through the power of reason to mediate pleasure. Yet, contemporary aestheticians have endeavoured to redeem the narrowing of aesthetics into a science by emphasising the appreciator's discerning abilities. In response to the perennial question of what constitutes aesthetics, Stolnitz (1998, p. 83) remarks that "any object at all can be apprehended aesthetically, i.e. no object is inherently unaesthetic" and that one needs only to adopt an "aesthetic attitude." The aesthetic attitude broadens appreciation. Aesthetics is restricted neither to specific occasions nor to privileged environments. Something is aesthetic insofar as it is judged so by a discerning subject in a certain context.

However, the aesthetic attitude privileges the separative function of visual perception, reinforcing the dualisms of object and subject, consciousness and matter, thought to have reached an apotheosis in the seventeenth century philosophy of Descartes (see chapter 1). Stolnitz posits an axis between the appreciator and the natural world through the unconstrained power of the human subject to determine perceptual merit. Inherent to the positioning of an aesthetic subject is the percept and the accompanying problems of disembodiment, sensory distance and visual speculation. Constructed by the gaze, a percept is separate. What happens when the object is absorbed into the viscera, collapsing the distance required to maintain an aesthetic subject and an object of art? I appreciate quandong fruits by subversively eating them. If perception transgresses the allocentric senses as sensation, has the aesthetic attitude been foiled through the (*dis*)integration of the object?

Figure 13 Sign Near the Summit of Bluff Knoll. An element of physical danger is part of the mountain sublime of the Stirling Range National Park. *Image courtesy of the author.*

Interlude XI: Three Peaks Triptych

The Stirling Range National Park consists of about 1,500 plant species, or about one-third of the flora in the South-West, including 87 endemic species dispersed throughout five main botanical communities (Keighery & Beard, 1993). The geoautobiographical "Three Peaks Triptych" stresses the disjunctions and disorientations that occur in a new place and the tendency to associate unfamiliar plants with the flora of the "default country" (Arthur, 2003). Through staccato verse, the poem suggests the grappling for sense of place as the *poiēsis* (*something in-the-making*) and *aesthesis* (sensation) of physical interaction with an unfamiliar flora. So, to this end, I walked up Mount Trio, Toolbrunup Peak and Bluff Knoll, three prominent mountains in the Stirling Range to interact with plants during the early months of my residency in the South-West. Montane landscapes are classically associated with sublime exaltation stimulated by the fear of heights in tandem with the expansiveness of vistas (see figure 13) (see Abraham, 2006). In 1835, John Septimus Roe, a surveyor, exclaimed that "the Stirling Range burst on our view in great magnificence as we rounded the crest…The whole extent of the conical summits were spread before us" (cited in Thomson et al, 1993, p. iii). The arrangement of experience into triptych symmetry is foiled by the pangs of the body ascending through the lived spaces of the Stirling Range. The poem also shows the generation of a syncretic vocabulary of place, drawing together Nyoongar, European settler and scientific dictions, as well as sensation as the substance of aesthetics.

Three Peaks Triptych

i. *la montagne a les yeux de nombreux*

bluff knoll watches
each and every finger flick
every twitch of the eye
bala mial   bula mial
his eyes   many eyes
watch the alluvium
of north-stretching
Bremer Basin: sentinel
of the sedimentary.

I flick bread crust flakes
into a bivouac of plastinated
shrubs, into a bevy of
slope-hungry bushes,
        mial   bala mial
the staccato of autos
duodenum of road   car lot
a vestigial tissue
bursting into the pink

inflamed irradiance
of bottle brushes' late
blooming so, measuring
its own metes and bounds,
so registering its own
cadastre between grazing land
and mountain, limina transected

by an impetuous bitumen
tube. it goes as such
and should be readily noted:
the choir of darwinia eunuchs
of the autumnal light
is castrated
let them grow   let them

sing, eternally nodding
against the white
organs of
granite.

ii. *il ya scandale dans l'air*

quiet on the spire
of toolbrunup peak
& a wedge-tail catching thermals
catching djeran
windy season before winter
sandalwood   sweat   a sweet
fiery distillation time-lapsed
between the desert and here.

western shield   ten-eighty
harboured by poison pea
gastrolobe to woylie
shielding thee
        shielding thee

on this solitary crag
on this scree-strewn slant
scent carried on the
old limbic corridors
of the brain, fire-oil
fused in the fury and fragrance
of the hemi-parasite's flesh:

the scandal of *santalum*.

plants await sleep too
a reprieve from
the permanent view to the
porongorups   positioned
for receiving light, holding
rubble at the chute-top
the cold settles in
lowering sun
      a time to go.

iii. *plantes comme les plantes - une convocation anonymes*

mt trio in the morning
sore calves and a calling
of several unknown birds;
north, tires reeling, supersonic
spinning of wheels, I shift
from rock to rock, mountains
irrupt out of the grazing
land like boils

on the back of the sheep plain,
plant consumes
word   cells diffract
into asexual new tongues,
a nouveau lingua belly rising
as adipose ripples under shallow sea—
Hume peak holds the
western-most corner

clover-like   triangulate plants
sweetfern-like toothed
& hemp-like, bay laurel-like,
sprout from   the stem
(highly unusual)
ephedra-like whorls of spikes.

huddled in below gust-line
we talk our trade: animalia-plantae.
have history, make ranges,
brood, surveil, we are emblems
(there are guidebooks to us)
beside cairns up here,
we duck the wind and
the aster-like bursts of angst.

before names,
chthonic associations,
presence
now defined:
an anonymous convocation
of palm-like fingers holding
a coarse line of air.

Plants as Objects of Art: From Artefacts to Habitats

The previous section connected aesthetics to the creation of a perceptual object, posing the problem of distance between the appreciator and that which is being appreciated. The challenge faced by landscape aesthetics is that art has been the framework for the enjoyment of the natural world. But how do we appreciate plants outside of the model of object appreciation? Visual apprehension serves the evaluation of human-made objects, but sight alone limits the appreciation of living, sensuous flora. Carlson (2000) suggests that aesthetic appreciation is made untenable when a plant or animal is separated from the human appreciator as an object of perception. The emphasis on visual experience in the form of scenery, focal point specimens or wildflower imagery narrows experience. An embodied aesthetics would not revolve around sight alone. As chapters 4 and 6 relate, visual aesthetics may even degrade the cultural value of flora when plants contradict aesthetic ideas generated in other places.

The fuller disambiguation of the appreciation of art from the appreciation of nature is considered here. The artefactualisation of the natural world into a series of well-designed objects is a core issue faced by aestheticians, as well as conservationists as the Lesueur plan reveals. Plants *in situ* are not dead scenes, but living entities capable of alterity and becoming. A multi-sensorial aesthetics disengages the appreciation of plants from the appreciation of visual artefacts. In this respect, the "nonaesthetic model of nature appreciation" argues nihilistically "that aesthetic appreciation is paradigmatically appreciation of works of art" (Carlson, 2000, p. 8). Aesthetics, as the theory maintains, is the science of perceiving art, not nature. The non-aesthetic model addresses the core complication of distance. Carlson (2000, p. 9) holds that the appreciation of art and nature are different experiences, the former manufactured through the creative intent of a designer and the latter generated through ecology: "To aesthetically appreciate the natural world, we do not need to actually make it, as we make works of art; nor do we need to conceptualize it in artistic categories."

In aesthetic philosophy, the theorising of the difference between the appreciation of art and the appreciation of nature has occurred concertedly in the last forty-five years. Hepburn's essay "Contemporary Aesthetics and the Neglect of Natural Beauty" (1966/2004) argues for the role of experience of living places in contrast to the appreciation of static objects. Hepburn (1966/2004) affirms that artistic pieces differ to living environments. Since a forest surrounds an appreciator, disengaged viewership collapses. Immersive sensation encourages participation that brings the distant tree, stone, herb or bird into proximity. The appreciating subject is incorporated into the object of appreciation through moving, eating, tasting, smelling, touching and feeling. For example, the kino sap of the marri tree might be likened to blood rather than paint (see "Blood," Interlude XIX).

In Nyoongar culture, kino has been used as a topical antiseptic for burns, stings and other wounds (Moore, 1884/1978). The visual appreciation of marri for its flowering beauty differs considerably from embodied networks of sustenance. In Hepburn's view, unlike works of art, natural objects are without frames that delimit perception with defined edges. Hepburn stresses that, in contrast to fixed artworks, natural environments must be approached through change and multi-sensoriality.

Since Hepburn, contemporary eco-aesthetics has tried to distinguish between the appreciation of artistic objects and natural environments, considering differences of immersion and viewership. The question of whether there should be one aesthetic model encompassing both art and nature or separate accounts for each is taken up by Berleant (1993, p. 228), who states that "few would deny the possibility of obtaining aesthetic satisfaction from both works of art and from nature." The "contemplative" model of aesthetic experience balkanises an object of art from its surroundings in order to project visual focus on the object. According to Berleant (2005, p. 4), the contemplative model has been axiomatic in modern aesthetics and rests upon an attitude of disinterestedness, which Stolnitz (1998, p. 80) defines as "disinterested and sympathetic attention to and contemplation of any object of awareness whatever, for its own sake alone." Berleant (2005, p. 5) regards disinterestedness as a "doctrine of separation and distance" that creates an objectified space for ordering everything animate and inanimate within it. Liberated from utiltarian biases, a disinterested attitude rarefies aesthetics.

In *Art as Experience*, Dewey (1934) departs from the contemplative model to argue for experience of the living world that reduces perceptual distance. Rather than a model of viewership, an "active" model or an aesthetics of pragmatism, as suggested by Dewey (1934, p. 3), envelops appreciation of art objects in experience: "When artistic objects are separated from both conditions of origin and operation in experience, a wall is built around them that renders almost opaque their general significance." Dewey (1934, p. 25) argues that the interaction between the object of art, the appreciator and the environment reinstates "the union of sense, need, impulse and action characteristic of the live creature." An organism participates in an environment that "is not exclusively visual but rather somatic: the body energizes space" (Berleant, 2005, p. 7). However, whereas plants have aesthetic qualities, they lack the agency of an artist. Instead, life forms bear ecological agency through habitat relations: their networks of acting and being acted upon. Carlson (2000) evidences a concern over the distinction between the appreciation of art and nature. Whereas an object of art is dimensionally fixed, typically immobile and sensorily limited, "in our aesthetic appreciation of the world at large we are confronted by, if not intimately and totally engulfed in, something that forces itself upon all our senses" (Carlson, 2000, p. xviii). Carlson intimates that the world at large is temporally and spatially unbounded, defined by rhythm and flux.

Despite somatic emphasis, the pragmatics of Dewey (1934, p. 25) valorises art as "the living and concrete proof that [humanity] is capable of restoring consciously" the union of the senses through experience. The *live creature* is an analogue for the object of art. However, an inanimate creature of art still differs intrinsically from a living creature. An experience of Michelangelo's *David* might engage gallery conditions, the mood of the appreciator, the journey of arriving, knowledge of the work, and the viewer's anticipation. The statue is positioned in a gallery that, with the materiality of the statue itself, creates a scope of aesthetic possibilities. Similarly, Andy Goldsworthy's *Neuberger Cairn* (2001), a stone obelisk, has a setting: an outdoor environment. Yet both *David* and *Neuberger Cairn* differ from plants; a statue is inorganic stone or dead organic wood, whereas a plant is alive. An appreciator can neither exchange life nutriment with a statue nor receive a response from the object. Plants are not static objects temporally fixed into position. As ever-changing, self-determining and responsive organisms, certain plants might not fit canons of visual beauty. A vocabulary derived from the appreciation of art objects constrains aesthetic appreciation of plants.

The installations of Australian artist Gregory Pryor are physically immersive, including sound and renderings, to convey the urgency of species decline (for example, see Marchant, 2005). However, a visual depiction differs to the field experience of plants through ecological phases of variability and sensuality. Berleant (2005, p. 108) distinguishes between living creatures and objects or renderings by rejecting "the ethnocentric assumptions of modern Western aesthetics that restrict art and the aesthetic to carefully circumscribed objects and occasions of museums, galleries, and concert halls." As Berleant (2005, p. 152) argues, aesthetic appreciation is broader than the viewership model; it extends to environments: "The occasions on which aesthetic appreciation can develop are unlimited and can involve any objects whatsoever… aesthetic involvement need not be rare or restricted." An immersive gallery is evocative, yet mediated, whereas living plants in botanic fields engage networked experience. In the field, an appreciator may enter into a wide circle of possibilities between bodies. Two very different aesthetic experiences occur; neither the gallery space nor the botanical field substitutes adequately for the other. Indeed, the skill of the artist and the particularities of the exhibition space can accentuate aesthetic qualities of a plant that go undetected by the senses in the field. For instance, Pryor's renderings elicit congruities between plant shapes through an image tableau.

Carlson and Berleant respond to the challenge of aesthetics to distinguish between objects of art and living organisms. An embodied aesthetics would respond to the overwhelming cultural tendency to construct plants as artistic objects. An aesthetics of flora would need to address the scenery cult (Carlson, 2000, p. 32) of picturesque detachment and formulaic beauty in which certain species might be labelled inconsequential scrub (see chapter 6). Carlson (2000, p. 32) goes on to describe the scenery cult as linked closely to landscape paintings and postcards in which "form-constituting elements, resultant overall form, and formal qualities such as balance, proportion, and organization are of considerable importance." An aesthetics of plants would respond to the appreciation of plants as objects of art, as well as the taxonomic perspective in which a plant is represented as somehow interchangeable with its representation, its two-dimensional rendering.

Plants as Objects of Disinterestedness: From Detachment to Engagement

As the previous section argued, aesthetic disinterestedness parallels the project of objective knowledge of nature. This section returns to the role of disinterestedness in the categories of the sublime, beautiful and picturesque. Through the perceptual gulf rendered by disinterestedness, an environment is made to feel safe, whereas in close proximity it can be hostile or repulsive. A threatening plant endangers the body directly through noxious smell, bitter taste or stinging sensation (see chapter 2). Through visual safety, this potential danger or discomfort is averted. Whereas the autocentric senses bring plants into proximity, vision rarefies experience distantly , creating a perspective. Disinterested appreciation is unhinged when we eat, drink, smell, taste, touch, consume and wear wild plants. In his disputatious account of aesthetics, Dixon (1995, p. 80) maintains that the term is a corruption of aesthesis for "sensible." The invocation of aesthetic taste splits perception and cognition, the senses and reason. Sense has come to denote rationality. In Dixon's view, the ancient notion of sensibility, in contrast, embeds reason in sense experience. In other words, aesthetics before Baumgarten marked the attainment of knowledge through all senses. Dixon (1995, p. 80) maintains that aesthetics, as it has been theorised, "signals a great philosophical crime, an expulsion of the carnal body from the garden of knowledge."

Contemporary aestheticians return multi-sensoriality to aesthetics through critiques of disinterestedness in the sublime, picturesque and beautiful. The beautiful is the appreciation of human-scale objects or small-scale objects such as flowers. The sublime involves places or objects of formless magnitude, height and volume. The sublime is predicated on disinterestedness to create a chasm between the appreciator and the appreciated that is then bridged by reason. In the eighteenth century, Edmund Burke (1756/1990) propounded aesthetic ideas in *Philosophical Enquiry into the Origin of our Ideas of the Sublime and Beautiful.* He associates the sublime with terror and awe as sensations arising from the exhaustion of vision. Kant (1798/1974) distinguishes between the beautiful as pleasure evoked through well-formed objects, and the sublime as pleasure and pain experienced through the immensity of formless objects. Expanding Burke's concept, Kant deepened the sublime. Firstly, sublime experience depends on the ability of reason to comprehend size or force where that formlessness eludes the senses. Secondly, the distance between the appreciator and the sublime object converts the initial experience of terror into comfortable pleasure. The Kantian notion of the sublime depends on the intellectual transformation of terror into pleasure through rationalising the experience of unbounded objects. As Kant argues, where vision fails to frame the object as beautiful, reason modulates the sensory overload. In this regard, Kant (1790/2008, p. 71) distinguishes between the mathematical and dynamical sublime in order to argue for reason's precedence. The former is invoked if the subject is overwhelmed by physical magnitude, whereas the latter occurs if one is overwhelmed by force, but both forms entail mastery of nature through reason.

A sublime experience of flora might occur in an old-growth eucalypt forest or in the open *kwongan* in Western Australia. In both situations, scale and size can affect the appreciator with terror, disorientation and anxiety. The sublime figures into the engagement and natural environmental models. For Carlson (2000), the sublime is characterised by disinterestedness, which excludes a range of economic, moral and bodily investments. Carlson (2000) connects disinterestedness to the sublime and to scientific objectification, both of which can physically subdue. In contrast, Berleant (1993, p. 234) asserts that the sublime marks the movement of the aesthetics of nature away from the appreciation of artistic objects through "the capacity of the natural world to act on so monumental a scale as to exceed our powers of framing and control, and to produce in their place a sense of overwhelming magnitude and awe." For Berleant, the sublime signifies the impossibility of disinterested contemplation through the "experience of continuity, assimilation, and engagement" (235). For Carlson, disinterested contemplation is central to the sublime, whereas, for Berleant, engagement weakens sublime disinterestedness.

As another mode of aesthetics, the picturesque theorised by William Gilpin (1786) describes the painterly qualities of a scene. The term *picturesque*, meaning "picture-like," aims to distil land into artistic ideals governed by landscape painting and poetry (Carlson, 2000, p. 4). The beautiful depends on smoothness and symmetry for creating the experience of pleasure, whereas the sublime entails oscillations between discomfort and pleasure as moderated by reason. The picturesque, however, can be defined by the contrasts of roughness, irregularity and abrupt variations against the smoothness of background elements, such as a placid lake (Brady, 2003, p. 40). The picturesque is based in the empiricist thinking of eighteenth century British aestheticians who found that landscapes that resemble works of art are most appealing (Carlson, 2000). Desirable works of nature are composed pictorially. The picturesque mode is inherent to most forms of nature tourism and its affiliated desiderata of photos, postcards, brochures and web pages (Andrews, 1989; Urry, 2002). The picturesque is also identified with panoramic prospects, scenic outlooks and other distanced ways of appreciating—and indeed formulating—landscapes.

The picturesque, sublime and beautiful, as Giblett (2011, p. 64) argues, are the three philosophically valid aesthetic modes. A sublime experience of flora would entail the fearful, yet pleasurable, sensation of being engulfed in the eucalypt forest or isolated on the kwongan. The sublime exceeds the senses, leaving reason to accept our position within enormity or at the foot of monumentality. Giblett (2011, pp. 71-72) observes the perceptual differences between these three aesthetic modes: "The sublime is evoked by looking up at monumental, formless, phallic and patriarchal objects...the beautiful involves looking down at small, well-formed, feminine objects lying before one; the picturesque entails looking out across feminised pleasing prospects." A discrete natural object might become an artistic object, whereas a prospect is framed pictorially as the picturesque. In the Fitzgerald River National Park in Western Australia, sublimity characterises the expansiveness of the prospects and the sea's tumult. The drooping red Qualup Bell is beautiful because of its well-formed flower (see "Qualup Bell," Interlude XXIX).

The uncanny confounds the beautiful, picturesque and sublime by injecting the resonance of that which lingers and the unseen perceptibility of smell, aftertaste, and drifting, unlocalised odour (Giblett, 1996, 2011). Whereas the beautiful and picturesque are pleasing and the sublime entails physical unease converted into mental mastery, the uncanny engages immanent smell. The odour of a rotting flower evokes the uncanny, especially if the source of the smell is not identifiable. The uncanny might entail the displeasurable experience of ugliness (Sibley, 2001). Such experience might transmute into pleasurable or beautiful experience as plants undergo changes over seasons or lifespans. A beautiful Qualup Bell will have an uncanny post-flowering phase in which it decays back to the Earth. Enrapturing perfume decomposes to pungent detritus, which engenders the possibility of the flower again. Alterity points to the fluidity of aesthetic experience, shifting between aesthetic modes comprising changes through seasons (see chapter 13).

The uncanny is significant to Berleant's engagement model. The beautiful, sublime and picturesque occur through disinterestedness in order to achieve separation. However, the uncanny engages smell and confounds sight. Passing the rotting Qualup Bell, we search for the source of the nauseating stench. Smell eludes visual depiction. Berleant (1993, p. 237) calls, not for disinterested contemplation, but for an aesthetics of engagement, "a sensory immersion in the natural world that reaches the still uncommon experience of unity." Participatory engagement counterbalances the beautiful, picturesque and sublime orders. Berleant broadens the aesthetic character of natural places beyond that which is visually validated: "The attraction of a spreading patch of bunchberry or a stand of wild columbine on the forest floor [lies in its] colour, shape, poignant simplicity, delicacy and, as much as any anything, its gratuitousness and profusion." According to Carlson (2000, p. 6), the engagement model emphasises "the contextual dimensions of nature and our multi-sensory experience of it…in an attempt to obliterate traditional dichotomies such as subject and object, and ultimately to reduce to as small a degree as possible the distance between ourselves and nature." To some extent, engagement involves bodily integration between the appreciated and the appreciator.

Despite his emphasis on multi-sensoriality, Berleant is unclear about how engagement would invoke the many senses. The attraction of a spreading patch of bunchberry lies further in its promise of edibility, the subtle fragrance of flowers, or the sounds of ice flakes on leaves. Berleant (2005, p. 241) goes on to advocate "an aesthetics of engagement [that] encompasses both art and nature," aligning nature and art again. Objects of art cannot be tasted, smelled, heard or sometimes cannot even be touched, although contemporary participatory art may redefine viewership through an interest in multi-sensoriality (Almenberg, 2010). By amalgamating the appreciation of nature and art, Berleant marginalises the immanence of living creatures in distinction to the mediated experience of artistic objects. Art aesthetics is the experience of inanimate objects, whereas nature aesthetics is the appreciation of life. Berleant's engagement risks constructing a plant as an object of art in which visual pleasures supersede embodied connectivity. Hence, Berleant's model provides uncertain ground for a corporeal aesthetics of plants.

Plants as Objects of Scientific Discourse: From Reason to Embodiment

Emphasising multi-sensorial experience, Berleant's engagement model responds to the "dichotomising metaphysics" subtending the aesthetics of art and nature (2005, p. 152). This section considers science's role in the appreciation of flora. The epistemological tension between cognitive and non-cognitive appreciation is another issue confronted by landscape aesthetics. The debate stems back to Kant (1798/1974) who trifurcated perception into the senses, reason and imagination. Through the principle of *sensus communis*, Kant argued that recollection or imagination are not appropriate foundations for an aesthetics because they do not allow for a standard of communicability between appreciators (Brady, 2003). Hence, Kantian aesthetics aligns sense experience to reason. Challenging scientific precedence, some environmental aestheticians argue for non-cognitive appreciation (Hepburn, 1966/2004).

The cognitive debate reinforces destabilising Cartesian fragmentations between knowledge and intuition, corporeality and reason. In Berleant's model, aesthetic experience occurs integratively and syncretically. For instance, the ecology of the Qualup Bell, along with its colour and the tactile memory of its stiff leaves, the stories of Nyoongar and colonial European histories that surround it, and imaginative recollection of a trip to the Fitzgerald River create collectively a form of syncretic appreciation (see chapter 3). Scientific classification has a troubled relationship to multi-sensoriality. The episodic movement of experience involves splicings, interruptions and punctuations between the cognitive and non-cognitive. Ideally, science complements non-cognitive appreciation towards multiple narrative streams (see chapter 3).

The relationship between scientific knowledge of plants and their appreciation is emphasised by Carlson. However, Carlson's "natural environmental" model parallels the Kantian sublime in which the validity of sensation is subordinated to reason. For Carlson (2000, p. 9), science is sufficient for aesthetics because rational knowledge dispels inappropriate judgements: "When we cast the conceptual net of common-sense and scientific understanding over nature we do enough to it to make possible its aesthetic appreciation." Casting a conceptual net involves capturing, which subjugates by conceptualising. Carlson (2000) is concerned with the distinction between the "designed" object of art and the "ordered" object of nature and the particular narrative that informs the natural world. In order to distinguish the aesthetics of nature from art, cognitive theorists argue that natural science comprises the appropriate basis for appreciating nature (Carlson, 1993, 2000). The "natural environmental" model asserts that the natural sciences enable a distinction between the designed object of art and the ordered object of nature. Whereas the designed object can exist on its own, an ordered object, such as a rock or plant, does not stand apart from its story of creation (Carlson, 1993, 2000). Based in science, the model is inconclusive about multi-sensoriality and non-cognitive knowledges for fear of distorting the "objects for what they are and...the properties they have" (Carlson, 1993, p. 219).

The "natural environmental" model requires the probity of aesthetic experience and the narrative of science. For instance, the appreciation of Patterson's curse, an aggressive purple-flowered exotic plant in Western Australia, would be constrained by the understanding of its growth habits and its tendency to displace indigenous plants. Similarly, Carlson (2000, pp. xix-xx) approaches an aesthetics of wildflowers with the scientific awareness of the diminutive plants of an alpine meadow. Understanding that the meadow survives in an intense high altitude climate, the appreciator is less inclined "to overlook miniature wild flowers." This aspect of the "natural environmental" model is especially germane to the aesthetics of South-West plants, which often contradict expectations of form, colour and behaviour. Botanical knowledge of plants may augment their appreciation, instilling admiration that stimulates physical engagement with plants over time. Knowing the colonial history of a plant such as the zamia palm can entice me into the field to locate the plant, even if it is out of flower. I then attune to the species, taking note of each time I encounter it and perhaps planning occasions to visit it, an approach perfected by Thoreau (see chapter 1).

However, the main limitation of the "natural environmental" model is its basis in science rather than the body. Science relies on the visual denomination of living forms through disinterested detachment, despite arguments for the corporeal production of scientific knowledge (Driver, 2004). Sensation is traditionally excluded from scientific discourse for its subjective basis. Recalling the scepticism of Thoreau (2000, p. 242) observing science as the grub "nestled in the very germ of the fruit," a multi-sensorial aesthetics would reinstate sensation. Thoreau's healthy scepticism is a model here: "Let us not underrate the value of a fact; it will one day flower in a truth" (cited in Walls, 1995, p. 40). Indeed, Thoreau's experience of plants transgresses distanced art appreciation and scientific knowledge through corporeal acts in bodily space. Closely related to Thoreau's position is Heidegger's challenging of objectivism through *Ge-stell* or enframement, linking visual apprehension to technological domination (Heidegger, 1977). Chapter 9 will return to the positions of Heidegger and Thoreau in relation to language.

Interlude XII: Katoomba Incantation

"Katoomba Incantation" is a three-part poem that recounts a cross-continental train trip on the Indian Pacific railway from Perth to Katoomba, NSW, Australia through spaces of memory, sound and ecology. Viewed from a train, the pre-eminent geomorphological feature of the terrain is the flat and dry Nullarbor Plain from Kalgoorlie, WA, to South Australia. Train travel induces a soporific of spectatorship, gazing out upon a landscape from behind a window as the machine slices through it with quintessential linearity. Indeed, a section of the Trans-Australian Railway from Kalgoorlie to Port Augusta is the longest stretch of straight rail track in the world. Passing references to plants occur in the poem as ravished casuarina, the fractalised mulga bush of the Nullarbor and a delinquent suburban gum tree in the suburb Wembley Downs near Perth. But, due to the rapidity of travel, plants are fleeting objects:

> in the sunken green dimples
> of treed land between tin roofs,
> obstreperous white cockies
> ravish seeds of casuarina
> agitating the backstreet silence
> above The Gully (lines 10-15)

The poem prefigures the discussion of wildflower tourism (see chapter 7) in which the spectatorship mode of appreciating plants can be augmented by the bodily engagement of gestural walking (see chapter 12). Travelling in a train simply does not afford the opportunity to get down on hands and knees and smell a wild plant. Instead, experiences on trains and other mechanised forms of transport aid in the perception of the landscape as a succession of quickly moving images and place names (to read "Katoomba Incantation" in full, see Ryan, 2010).

Figure 14 Cook, South Australia. Established in 1917, Cook serves as a railway station along the Trans-Australian Railway. *Image courtesy of the author.*

Theorising a Corporeal Aesthetics of Plants

This chapter has critiqued the limitations of eco-aesthetics for developing a theory of embodiment with respect to plants. As argued, the shortcomings of the theories of Berleant and Carlson in particular might replicate traditional aesthetic concepts of plants as objects of vision, art, disinterestedness and scientific discourse. Visual aesthetic theories position the experience of flora outside the appreciator towards a distanced locus. Vision is implicated with disembodied apprehension, coterminous as it is with scientific knowledge production. According to these theories, the point of contact between the appreciator and the appreciated occurs cognitively through the recognition of form, colour, symmetry, exposure, vista and other qualities beyond the intimate spheres of bodies. But what if an aesthetics of flora situates the nexus of perception in the tissues and fibres, in the sensation between bodies? This aesthetics would hold the body as a plenum of contact with plant life through the senses (see chapter 13). As will be theorised, an embodied aesthetics of flora—a *floraesthesis*—would close the perceptual gulf between the appreciator and the plant, bringing bodies closer to one another rather than holding them at bay.

References

Abraham, T. (2006). *Mountains so sublime: Nineteenth-century British travellers and the lure of the Rocky Mountain West.* Calgary, Canada: University of Calgary Press.

Almenberg, G. (2010). *Notes on participatory art.* Central Milton Keynes, England: AuthorHouse.

Andrews, M. (1989). *The search for the picturesque: Landscape aesthetics and tourism in Britain, 1760-1800.* Aldershot, England: Scolar.

Arthur, J. M. (2003). *The default country: A lexical cartography of twentieth century Australia.* Sydney, Australia: UNSW Press.

Averill, J., Stanat, P., & More, T. (1998). Aesthetics and the environment. *Review of General Psychology,* 2(2), 153-174.

Baisya, R. K., & Das, G. G. (2008). *Aesthetics in marketing.* New Delhi, India: SAGE.

Berleant, A. (1993). The aesthetics of art and nature. In S. Kemal & I. Gaskell (Eds.), *Landscape, natural beauty and the arts* (pp. 228-243). Cambridge, England: Cambridge University Press.

Berleant, A. (2005). *Aesthetics and environment: Variations on a theme.* Burlington, VT: Ashgate Publishing.

Brady, E. (2003). *Aesthetics of the natural environment.* Tuscaloosa, AL: University of Alabama Press.

Burke, E. (1756/1990). *A philosophical enquiry into the origin of our ideas of the sublime and beautiful* (5 ed.). Oxford, England: Oxford University Press.

Bychkov, O., & Sheppard, A. (2010). Introduction. In O. Bychkov & A. Sheppard (Eds.), *Greek and Roman aesthetics* (pp. xi-xxx). Cambridge, England: Cambridge University Press.

Carlson, A. (1993). Appreciating art and appreciating nature. In S. Kemal & I. Gaskell (Eds.), *Landscape, natural beauty and the arts* (pp. 199-227). Cambridge, England: Cambridge University Press.

Carlson, A. (2000). *Aesthetics and the environment: The appreciation of nature, art, and architecture.* New York, NY: Routledge.

Caygill, H. (1995). *A Kant dictionary.* Malden, MA: Blackwell Publishing.

Cosgrove, D. (1998). *Social formation and symbolic landscape.* Madison, WI: University of Wisconsin Press.

Department of Conservation and Land Management. (1995). *Management plan: Lesueur National Park and Coomallo Nature Reserve 1995-2005.* Perth, Australia: Department of Conservation and Land Management.

Dewey, J. (1934). *Art as experience.* New York, NY: Capricorn Books.

Dixon, R. (1995). *The Baumgarten corruption: From sense to nonsense in art and philosophy.* East Haven, CT: Pluto Press.

Eagleton, T. (1990). *The ideology of the aesthetic.* Cambridge, England: Basil Blackwell.

Giblett, R. (1996). *Postmodern wetlands: Culture, history, ecology.* Edinburgh, Scotland: Edinburgh University Press.

Giblett, R. (2011). *People and places of nature and culture.* Bristol, England: Intellect.

Gilpin, W. (1786). *Observations relative chiefly to picturesque beauty, made in the year 1772* [Google Books] Retrieved from http://books.google.com/

Heidegger, M. (1938/2009). The age of the world picture (J. Veith, Trans.). In G. Figal (Ed.), *The Heidegger reader* (pp. 207-223). Bloomington, IN: Indiana University Press.

Heidegger, M. (1977). The question concerning technology. In D. Krell (Ed.), *Martin Heidegger: Basic writings* (pp. 284-317). New York, NY: Harper & Row.

Hepburn, R. (1966/2004). Contemporary aesthetics and the neglect of natural beauty. In A. Carlson & A. Berleant (Eds.), *The aesthetics of natural environments* (pp. 43-62). Ontario, Canada: Broadview Press.

Jay, M. (1993). *Downcast eyes: The denigration of vision in twentieth-century French thought.* Berkeley, CA: University of California Press.

Kant, I. (1790/2008). *The critique of judgement* [Google Books] (J.C. Meredith, Trans.) Retrieved from http://books.google.com

Kant, I. (1798/1974). *Anthropology from a pragmatic point of view.* The Hague, Netherlands: Nijhoff.

Keighery, G., & Beard, J. (1993). Plant communities. In C. Thomson, G. Hall & G. Friend (Eds.), *Mountains of mystery: A natural history of the Stirling Range* (pp. 43-54). Como, Australia: Department of Conservation and Land Management.

Marchant, N. G. (2005). Documenting the diverse Western Australian flora. In *Black solander, Gregory Pryor.* Perth, Australia: Perth Institute of Contemporary Arts.

Moore, G. F. (1884/1978). *Diary of ten years eventful life of an early settler in Western Australia,* and also *A descriptive vocabulary of the language of the Aborigines.* Nedlands, Australia: University of Western Australia Press.

Porteous, J. D. (1996). *Environmental aesthetics: Ideas, politics and planning.* London, England: Routledge.

Ryan, J. (2010). Katoomba Incantation. *Philament,* December (Wallace, A., & Rahman, Eds.) Retrieved from http://sydney.edu.au/arts/publications/philament/issue16_pdfs/RYAN.pdf

Schroeder, S. (2010). *Philosophy of literature.* Chichester, England: John Wiley & Sons.

Shelton, A. (1994). Predicates of aesthetic judgement: Ontology and value in Huichol material representations. In J. Coote & A. Shelton (Eds.), *Anthropology, art and aesthetics* (pp. 209-244). Oxford, England: Clarendon Press.

Sibley, F. (2001). Some notes on ugliness. In J. Benson, B. Redfern & J. R. Cox (Eds.), *Approach to aesthetics: Collected papers on philosophical aesthetics* (pp. 190-206). Oxford, England: Oxford University Press.

Singer, A. (2003). *Aesthetic reason: Artworks and the deliberative ethos.* University Park, PA: Pennsylvania State University Press.

Stolnitz, J. (1998). The aesthetic attitude. In C. Korsmeyer (Ed.), *Aesthetics: The big questions* (pp. 78-83). Malden, MA: Blackwell Publishing.

Thoreau, H. D. (2000). *Wild fruits: Thoreau's rediscovered last manuscript.* New York, NY: W.W. Norton.

Tuan, Y.-f. (1982). *Segmented worlds and self: Group life and individual consciousness.* Minneapolis, MN: University of Minnesota.

Urry, J. (2002). *The tourist gaze* (2 ed.). London, England: Sage.

Walls, L. D. (1995). *Seeing new worlds: Henry David Thoreau and nineteenth-century natural science.* Madison, WI: University of Wisconsin Press.

Part III

Botanical Cultures

Part III will develop the notion of South-West cultures of flora through readings of ethnographic, literary and scientific material. Chapter 6 transitions from Part II's historicised discussion to a contemporary cultural and literary contextualisation of plant aesthetics. Chapter 6 will present the notion of green tropism as the privileging of greenness, evident throughout some historical depictions of South-West flora. As A.D. Hope, Wallace Stegner and Alex George argue, a cultural reckoning—specifically with the South-West biota and more broadly with dry landscapes or *xeriscapes*—has involved a reconfiguration of green semantics and a reconsideration of aesthetics. Chapter 7 will further elaborate on anthoethnography, the approach proposed in chapter 2, beginning with an historical treatment of South-West wildflower tourism and then turning to readings of 2009-2010 interview transcripts from tourism proprietors and participants. Further bringing to light the nodes between culture and flora through ethnography, chapter 8 will theorise botanical memory as a form of environmental memory comprising personal and collective recollection of plant life. Emotional responses of mourning and celebration, as well as somatic memories of contact with flora, constitute varieties of botanical memory.

6

Green

Abstract: The colour green has been symbolically linked to fertility and productivity as a metonym for nature. In popular use, greenness is a trope for the environmental and sustainability movements, as well as political eco-consciousness. American author Wallace Stegner, however, remarks on the difficulties of green speak, considering the perceptual inversions that occur in dry landscapes. To become appreciative of arid country and attain xeri-consciousness, he asserts that "you have to get over the color green." Chapter 6 will explore greenness as it occurs in depictions—and cultural expectations put upon—South-West flora. Drawing from A.D. Hope's poem "Australia" and the journals of early European explorers to the South-West, the prevalence of green semantics will be noted, along with the consequences of green speak for botanical appreciation. Getting over the colour green in the South-West has required the responses of writers and scientists, such as Barbara York Main, George Seddon and Alex George, all of whom reconfigure the perception of green through an expanded descriptive vocabulary of indigenous flora and towards regional aesthetic appreciation.

The force that through the green fuse drives the flower
Drives my green age; that blasts the roots of trees
Is my destroyer.
And I am dumb to tell the crooked rose
My youth is bent by the same wintry fever.

Dylan Thomas (1952, ll. 1-5)

Green Speak

Why are green landscapes considered beautiful? Why does the colour green evoke particular associations with fertility and growth? Every back yard gardener knows that green leaves are healthy, whereas yellow leaves are withering and in danger of dying. When a plant turns yellow, a sensible green thumb waters it urgently to quash its colour change and restore the foliage to a healthy greenness. Green indicates the growth, vitality and fertility of landscapes and plants, but brown and yellow—the dun colours—tend to symbolise decay, senescence and barrenness. In climates of intense seasonal heat and dryness, such as the South-West, the predictability of greenness as a perceptual indicator becomes disjointed or inverted as the environment adapts to severely dry conditions by altering the colour and size of its foliage (for example, see Seddon, 2005).

Green tropism may be defined as a cultural leaning towards greenness, just as phototropism, in biological terms, is the sidling of a plant towards light. American author Wallace Stegner (1992b) in his essay "Thoughts in a Dry Land" from 1972 argues that ideas of greenness formulated in temperate climates can be inverted in dry habitats. In order to appreciate the character of the American West, "you have to get over the color green" (Stegner, 1992b, p. 54). He suggests that arid landscapes require an adjustment of perceptual faculties, and that green symbolics forged in other climates might not apply to arid habitats. Nature is neither "red in tooth and claw" (Tennyson, 1850/2009) nor "green in leaf and branch" (Giblett, 2011, p. 256), but often brown and yellow in bark and foliage. Moreover, a deficiency of green in the visual domain is not a lack of value, nor is dunness the dying exhortation of a barren land. Drought-prone places, or xeriscapes, vacillate between brown, yellow and orange interspersed with seasonal irruptions of greenness.

Here is a bold contention: green is one of the most complex descriptors in the English language. As a noun and adjective, green has been connected symbolically to vegetation, nature and fertility. In particular, since the 1970s, the colour has become a synecdoche for environmentalism and land consciousness (for example, see Herndl & Brown, 1996), ecological politics and social change (see Reich, 1970) and the revisioning of science towards a sacred conceptualisation of the natural world (Sheldrake, 1991). In popular terms, *greening* is used as a trope for the modification of corporate, institutional or political practices towards minimising environmental impacts and energy consumption (for example, see Sarni, 2010). Greening as a progression towards sustainability has become an increasingly accepted, though contested, turn-of-phrase (Herndl & Brown, 1996). However, the colour green and its implicit associations are more complex than common symbolic associations: vegetation, nature, fertility and environmentalism. Green speak can have destructive consequences when unrealistic expectations gestated in other climates are imposed on dry places.

This chapter addresses the complexities of green tropism in relation to the perceptions of the botanical biodiversity and ecological value of arid places. After presenting the symbolic aspects of green with focus on the relevance of the colour to ideas of beauty and fertility in nature, I give a reading of A.D. Hope's "Australia." The poem satirises greenness as a value system that has been forced upon Australian landscapes. In the South-West, green values have factored into written representations of botanically important subregions. The region is often fully alive, vegetated and even spectacularly biodiverse by international standards, but often not green (Seddon, 1972, 2005). Historical representations of Lesueur National Park, one of the most significant botanical areas in the South-West, evidence overtones of green tropism. Returning to Stegner's call to "get over the color green," the chapter concludes by intimating that the appreciation of arid landscapes begins with an expanded vocabulary, an interrogation of language. Botanist Alex George (2002), zoologist Barbara York Main (1967) and ecologist George Seddon (2005) all suggest the need for new linguistic techniques to describe an aesthetics of xeric places.

The Positive-Negative Nature of Green

The word *green* is derived from the Old English term *growan* meaning "to grow" (Harper, 2010). In addition to growth, green has various symbolic nuances. Table 4 distributes green symbolism across five categories—bodies, cultures, emotions, environments and politics—to show its diversity of meanings. Inherently dichotomous, green signifies both the appearance of verdure and the advance of decomposition, as well as all meanings on the spectrum between growth and decay. De Vries (1974, pp. 226-227) lists nearly thirty variants of green symbolism, including positive associations of growth and fertility—"earthly, tangibly growing things, *vegetation*"—but also negative connotations of poison and jealousy. Since de Vries's dictionary was published nearly thirty five years ago, new symbolic dimensions for green have been forwarded. Especially since the environmental justice movements of the 1970s, green as vegetative growth—or in Stegner's (1992b, p. 53) terms, green as the term for "universal chlorophyll"—has been advanced politically and socially to encompass environmental consciousness and the eco-politics of the Green party, the Greens, or even *greenies* with ecological sympathies (for example, see Burchell, 2002). Green signifies positive reconstruction of abused landscapes in the form of environmental remediation, or mine-site revegetation, as in the greening of the landscape. However, on an anti-ecological note, green conjures the American greenback as a metonym for Western mass consumption and the exploitation of natural resources. The malevolent twin of green remediation is the practice of greenwashing in which *green sheen* companies deceptively advertise the environmental integrity of their products (for a review, see Firestein, 2009, pp. 196-235).

Greenness has signified clarity of vision, virtue, piety and power throughout the history of cultures. In Antiquity, green represented harmony, balance and nature. The colour embodied the Greek idea that vision depends on variations between light and dark. During the classical era, green, as an intermediate between white and black, was regarded highly for its pleasing and soothing qualities. Aristotle considered green a particularly restful colour between the white of water and the black of earth (Gage, 1993, p. 13). Echoing Aristotle, medieval French Scholastic philosopher William of Auvergne (cited in Gage, 1993, p. 82) considered green to be beautiful because it "lies between the white which dilates the eye and the black which contracts it." In the thirteenth century, Pope Innocent III sanctioned the colour green as a liturgical colour for its balanced, pleasing qualities (Gage, 1993, p. 82). Green New Zealand jade, also known as greenstone or *pounamu*, is significant in Maori culture as a symbolic colour for virtue and power (Keane, 2009). Jade in Chinese traditions represents virtue, imperial power and piety; the Jade Buddha for Universal Peace, exhibited in Perth in 2009, is carved from a rare piece of translucent jade found in Canada (Buddhist Council of Western Australia, 2009).

Table 4 Dichotomous Colouration. Green is a complex descriptor comprising a positive-negative symbolic structure.

Bodies	the productive body fertility vegetation chlorophyll	the toxic body sickness pus and bile chlorosis
Cultures	ancient nature-aligned green man	inexperienced unproven green horn
Emotions	neutrality harmony virtue	envy monstrosity jealousy
Environments	spring growth remediation biodiversity	noxious proliferation green washing green wood
Politics	environmental movement ecopolitics	capitalism the greenback

While green is meaningful as a distinguished colour of power and balance, it is also linked animistically to nature. The archetype of the Green Man—the carving of a foliated head in medieval churches—is an historically recurring motif associating green to the natural world in European traditions (Basford, 1978). In his exposition of archetypes and the collective unconscious, Jung refers to the Russian fairytale "King of the Forest" as expressive of the relationship of the unconscious mind to green symbols. A peasant in the forest encounters Och, a little wrinkled old man whose "green beard hung down to his knees" (cited in Jung, 1959, p. 222). The peasant's son accompanies Och to "a green hut…In the hut everything was green: the walls were green and the benches, Och's wife was green and the children were green…and the little water-women who waited on him were as green as rue" (cited in Jung, 1959, p. 222). In Jung's analysis, the Green Man symbolises vegetation and woodland numen. The conducting of the peasant into a green underworld represents the capacity for conscious human transformation in nature and unconscious transmutation of the human into the vegetative.

In different cultures, green is indistinguishable from nature, growth, fertility and the spring season. Embodied in the archetype of the Green Man, the colour green, as Gage (1993, p. 258) observes, has "an indissoluble association with nature" and the fertility of the plant world. In ancient Egypt, green was used extensively in fertility rites (De Vries, 1974, p. 226). Moore (1846, p. 78) defines the Nyoongar word *won-gin* as "living; also green, when applied to leaves or wood." In ancient Chinese five phases medicine or *wu xing*, green is associated with the wood element, which further corresponds to the emotion of anger, the organs liver and gall bladder, and spring growth (Kaptchuk, 2000, p. 439). In Europe, by the thirteenth century, the term *sinople* came to stand for the vibrant green of nature in contradistinction to the rudimentary green of paints and dyes (Gage, 1993, p. 82).

The colour green has a complex symbolic history. However, its associations with growth in nature are the most contradictory, for while green indicates fertility and increase, it can also evoke decay and toxicity (see table 4). For instance, Hippocrates coined the word *chlorosis* to denote the green colour of the skin during sickness associated with excess bile in the Greek humoural system. Eventually, green skin came to symbolise love sickness and jealousy (King, 2004). Paradoxically, however, the root of *chlorosis* can also mean flourishing, in the shared sense that chlorophyll underlies the greenness of plants. Folklorist John Hutchings (1997, p. 55) maintains the "positive-negative nature of symbolic green" and argues that "all common symbolic meanings are derived either from the green growth that occurs in springtime or from the green mould of decay." Hutchings (1997, p. 55) recognises the symbolic dichotomies of the colour, indicating that which is both beneficial and destructive to nature: "Not only is green seen as natural and good for the environment (as in 'green' farming), but it is also used to symbolise noxious growth (green for poisonous packaging and for atomic power stations)."

Green is a dichotomous descriptor, a fluid perceptual category of simultaneous contradictions. Though not referring specifically to green, Freud (1997, p. 94) in "The Antithetical Meaning of Primal Words" identifies the tendency of older languages for "combining contraries into a unity or for representing them as one and the same thing." According to Freud's reading of classic philologists, words can combine contrary meanings to highlight the contrasts that define each through the juxtaposition of their opposite meanings. Thus, flourishing encompasses a recollection of sickness, growth a comparison to decay, and neutrality a reference to jealousy when expressed as *green*. Freud's discussion of antithetical meanings suggests the relationality of language, expressing concepts through chiaroscuro between opposed meanings. Green is only green in relation to dun; fertility is only fertile with respect to sterility.

Interlude XIII: Balga Jarrahdale

The image of balga (*Xanthorrhea preissii*) pressed against the Western Australian sky presents a lucid visual contrast between aquamarine blue, luxuriant green and charred black hues. Its colloquial labels "grasstree" and "blackboy" entwine the apprehension of colour with the exertion of naming (see chapter 4). "Balga Jarrahdale" alludes to the multiple and often contradictory terms for balga, several of which are no longer in circulation. Although not a grass, balga has been known as grasstree because of the appearance of its foliage. Additionally, the name blackboy reflects the colonial-era notion of its trunk resembling the distant appearance of an Nyoongar person in the bush. Numerous parts of the balga have been consumed traditionally. Its leaves have been used for thatching or torches, dead flower stems for fire and spear-making, living flower stalks soaked in water to produce a fermented drink, and resin applied as an adhesive for constructing implements. The dark black trunk of balga contains highly nourishing bardi grubs prized by Nyoongar hunters and gatherers. George Fletcher Moore (1884/1978) notes the "fragrant, aromatic flavour" of the bardi:

> The presence of these grubs in Xanthorea [sic] is thus ascertained: if the top of one of these trees is observed to be dead and it contain any Bardi, a few sharp kicks given to it with the foot will cause it to crack and shake, when it is pushed over and the grub extracted, by breaking the tree to pieces with a hammer. (5)

The last couplet of the poem alludes to this embodied cultural history of balga: "covenant in its roots/ dark shelter in its limn" (ll. 20-21). But the poem is also an audio-centric commentary on the futilities and asymmetries of visual speculation in the bush. Sensation emerges as a kind of curious cross-checking between the senses. Is what I hear what I see? Can I taste and smell the specificities of place in this moment beyond visual prompts, arrays of colour and the imaginative distractions?

Balga
Jarrahdale, WA

beyond the slender jarrah stalks
parsing the lazuline into snaggy bits
is the roar, like an indomitable river
hammering rocks on a violent descent
out of some snowy silence:

the boom of terrestrial pilots
    on burning bitumen.
before me, a song flies by, pure song,
only song, of a bodiless fluttering katydid;
between the metronomic clicking,
the bush breathes out an off-beat whistle
and for all I know, below:
    Alph, the sacred river runs
    through a chasm in Xanadu.
somewhere a siren sings sweetly
for a grasstree, but here, neither tree,
nor grass, nor lily—not a black boy
but a balga—this two-trunked
oldster splits into heaven, offering:
    covenant in its roots
    dark shelter in its limn.

Green Fertility, Dun Sterility: South-West Australian Contexts

In the South-West, green has been associated with fertility and dunness with sterility in some historic depictions of the land. As cued by Stegner (1992b), a perceptual inversion occurs when biodiverse places lack the lush greenness of tropical or temperate climes. The disjunction between biodiversity and greenness—and the correlation of dun colours to barrenness and sterility—is evident in the writings of Charles Darwin (Armstrong, 1985), Captain Stirling (1827/2009) and the Gregory brothers (1884/1981). A barren landscape is considered infertile, incapable of producing offspring, and, in the colonial agenda, unable to support agricultural enterprise. In Western Australia, with increased attention to flora through local and international scientific focus, however, a greater appreciation of indigenous plants has emerged through a regional environmental ethos (Hopper, 2004). South-West flora has come to symbolise successful adaptation to the climatic constraints confronted by societies in xeriscapes.

Some early European explorers and settlers to Western Australia were shocked by the drab monochromes of the vegetation and the arid conditions of the landscape. Rose Saulces de Freycinet, onboard the French vessel *Uranie* in September 1818 with her husband Louis Claude de Saulces de Freycinet, was repulsed by her first stark and unwelcoming glimpses of Shark Bay, at the north-western tip of the South-West Botanical Province: "It is without a single regret that I left that hell on earth, the west coast of New Holland...My courage forsook me utterly, and I could see nothing but horror about me" (cited in Seddon, 2005, p. 34). Approaching the tropical islands of Simão and Timor, she was relieved of her horror and eased into the verdant scene before her: "Imagine our satisfaction at seeing the lovely vegetation of these islands. Our eyes were pleasantly rested by this greenery after the sand dunes and the dry or stunted shrubs of New Holland" (cited in Seddon, 2005, p. 34). Rose de Freycinet intimates a view espoused by Aristotle and others of green as a harmonious, pleasing and restful colour. Whereas the greenness of the tropical islands welcomed, the dun coloured earth of Western Australia repulsed and threatened. Nineteenth century American geologist Clarence Dutton (cited in Stegner, 1992b) in the canyon country of the American West, a similarly arid and stark environment, notes the shock that occurs to sensibilities acculturated in lusher places:

> The lover of nature, whose perceptions have been trained in the Alps, in Italy, Germany, or New England, in the Appalachians or Cordilleras, in Scotland or Colorado, would enter this strange region with a shock, and dwell there for a time with a sense of oppression, and perhaps with horror. Whatsoever things he had learned to regard as beautiful and noble he would seldom or never see, and whatsoever he might see would appear to him as anything but beautiful and noble. (53)

Dutton (cited in Stegner, 1992b, p. 54) goes on to identify a transformation of perception, in which "some day he would suddenly become conscious that outlines which at first seemed harsh and trivial have grace and meaning." Perhaps Rose de Freycinet was not immersed long enough in the West Australian setting to reconfigure her European-borne modes of perceiving beauty and flora.

On a related note, Captain James Stirling (1827/2009) after a turbulent passage around Cape Leeuwin, south of Perth, reported the bleak character of the landscape:

> The first appearance of the Coast we were now to explore presented nothing attractive; the monotony of its outline and the dusky hue of the meagre vegetation it supported at once accounted for the sterile and hopeless character attributed by early navigators to this region. (21)

Stirling associates the "dusky hue of the meagre vegetation" with sterility. The terrain near Cape Leeuwin "presented nothing attractive" and appeared monotonous in outline. Stirling shared the perceptual legacy of the early navigators who concurred that South-West Australia was hopeless for agriculture. The coastline lacked beauty and potential because of its deficient greenness, the sinople vividness also observed as absent by de Freycinet. The dusky hue of the vegetation reflected the dusky hue of the Aboriginal inhabitants who lived in sustaining relationship to plant foods, medicines and fibres (for an example of the use of the descriptor, see Hassell, 1975). Aboriginal inhabitants and indigenous plants become the metonym of each other with the same pejorative connotations of monotony, sterility, hopelessness and lack of attractiveness attaching to both. Though appearing sterile and hopeless to explorers such as Stirling, the South-West has been the home of ancient human societies and unusually well-adapted taxa (see chapters 2 and 3).

Echoing the attitude of his predecessor Stirling, in 1836 young Charles Darwin was confronted by the aridity and apparent sterility of Western Australia as signified to him by the appearance of the flora. Darwin (cited in Armstrong, 1985, p. 35) observed that "the general bright green colour of the brushwood & other plants viewed from a distance seems to bespeak fertility; a single walk will, however, dispel any such illusion." For Darwin's young eyes, the fertility of the landscape—signified externally by the shade of its greenness as an aesthetic quality—was an illusion dispelled upon closer inspection. More exactly, the standard of fertility invoked was contingent on the potential of the land to support European crops. Bearing these criteria in mind, the eventual proponent of evolutionary theory dismissed the prospective fertility of the land: "I do not think this country can ever rise to be a second North America. The sterile aspect of the land, at once proclaims that Agriculture will never succeed" (cited in Armstrong, 1985, p. 35). With Darwin's northern-hemispherically trained eyes, an environment that was not tropically or temperately green was barren and infertile, incapable of yielding to the offspring of diasporic Europeans and the displaced crops of other climates.

Lesueur National Park lacks the compelling greenness suited to the picturesque mode of appreciation (see chapters 2 and 5). In 1801, Nicolas Baudin, the post captain who kept a detailed journal aboard the vessel *Naturaliste*, estimated that the barrenness of the Lesueur landscape meant little value for natural history: "As this coast appears to be of no interest for navigation and even less for Natural History, I did not think it necessary to stop there" (Baudin, 1974, p. 200). Explorers who travelled physically through the environment depicted the sandplains of Lesueur in even less flattering terms. In 1837, on his southward traverse of the *kwongan*, Grey (1841) referred to the "waste and barren plains" (p. 59), "arid and barren in the extreme" (p. 66) and of a "bare, sterile, and barren nature" (p. 118). Approaching Mount Lesueur from the south, English-born explorer A.C. Gregory (1884/1981) wrote that "the country traversed almost wholly worthless sand and scrub" (p. 29) and that "the hills produced little besides coarse scrub...the land [on the banks of the Hill River] was very scrubby and indifferent" (p. 29).

The apparent barrenness and inhospitability of present-day botanical reserves like Lesueur thwarted agricultural development. The kwongan was not only infertile to settler needs, but was also resistant to stock grazing, which prevented its conversion to pastoral land. Gregory and pastoralist Lockier Burges concluded that the Lesueur area should be avoided. Its poisonous plants were "so abundant on that part of the range of hills on which Mr. Drummond has taken up his licence near Mt. Lesueur that stock cannot be driven through it with safety" (cited in Hopper, 1990, p. 11). Early European settlers, such as Gregory, found the Mt. Lesueur area "different, daunting…dangerous, and unpredictable" (Stegner, 1992a, p. 57). Inaccessibility hindered colonial incursions and allowed the original plant life to remain intact, though presently only accounting for a meagre 3% of the original heath lands (pers. comm., A. Tinker, August 29, 2009).

Why has the concept of fertility been defined exclusively through the criteria of human enterprise with explorers and settlers providing the seed to produce new growth? The South-West coastline gazed upon by Darwin and Gregory would later prove to be some of the most botanically biodiverse in the world. But why did it appear sterile and worthless in Darwin's opinion? To modern botanical science, places such as Lesueur are extraordinary repositories of floral diversity, dun-coloured, barren, sterile and infertile though they appear. Bailey (*n.d.*, The Mt Lesueur region, para. 2) comments that "like much of the ancient WA landscape, Mt. Lesueur's treasures are subtle. The scruffy heath consists of great diversity." As a critical area for conservation in the South-West, Lesueur National Park contains 10% of Western Australia's flora (Department of Conservation and Land Management, 1995) (see chapter 2). As suggested by Stegner, the paradox of perception is that scruffy-looking lands of dun colours can be richer in botanical life than jungles. Perceptual preferences trained in "the Alps, in Italy, Germany, or New England, in the Appalachians or Cordilleras, in Scotland or Colorado" (Dutton cited in Stegner, 1992b, p. 53) necessitate a retraining in the regional aesthetics of dry country.

Hope and Stegner: Unravelling Greenness

Australian poet A.D. Hope and American writer Wallace Stegner offer literary examples of the shift in perception required to appreciate the dry landscapes of Australia and the American West respectively. Both authors point to an aesthetics of xeriscapes through a revisioning of the traditional attachment of greenness to fertility. Although markedly different literary voices—the satirical verse of Hope and the earnest prose of Stegner—both authors grapple equally with green tropism as it has been pressed upon dry lands. Published in 1938, Hope's poem "Australia" satirises contradictory views of Australian plants, both as antithetical to the familiar leafy green plants of the Old World and strangely intriguing symbols of nationalism and intellectual freedom in their own right. By the conclusion of the poem, the "drab green and desolate grey" (Hope, 1977, l. 1) of the flora instead comes to represent the inherent strength and potential of Australia to depart from European stereotypes. Hope suggests that dunness, rather than sinople greenness, signifies potential. The poem's dramatic tension lies in its jostling of the truisms of greenness and fertility and its satirising of the feminine body trope that connects dryness to sterility and infertility. Additionally, Hope pushes the edges of green tropism by connecting the colour green to the stifling intellectual corridors of the Old World, evoking instead arid fields of brown and yellow, along with the biblical allusion "from the deserts the prophets come" (l. 24).

"Australia" has been read as the pejorative, Eurocentric prattle of a cloistered academic writing about a nation and a landscape with which he has lost touch. Considering the satirical slant of much of his writing (McLeod, 1980), however, a closer reading shows Hope's crafted iconoclasm. The poem begins with "a nation of trees, drab green and desolate grey/ In the field uniform of modern wars" (ll. 1-2). Australia as a nation is already bound interchangeably to its plant life and, more implicitly, the campaigns to clear trees, shrubs and scrub, and to conserve them. The trees conjure images of army-issued green. The colour represents militancy against the trees as impediments to pastoral expansion. Additionally, one of Hope's more controversial iconoclasms in the poem is his description of the nation as a post-menopausal woman incapable of bearing progeny. Rather than in its vigorous prime of life, Australia is "the last of lands, the emptiest," (l. 6) and the drab quality of colour signifies a declining fertility, barrenness, senescence and dryness. The nation is "a woman beyond her change of life, a breast/ Still tender but within the womb is dry" (ll. 7-8). The ensuing three stanzas seem to constitute a diatribe against the nation as Hope regurgitates the stock biases against Australia of the 1930s: "the river of her immense stupidity" (l. 12), "monotonous tribes" (l. 13), "the dying earth" (l. 16), "and her five cities, like five teeming sores" (l. 17), and "a vast parasite robber-state/ Where second-hand Europeans pullulate" (ll. 18-19).

The sixth stanza presents the volta, in which Hope suggests a values-based change in which a hopeful feeling for the nation and the potential of the land to support an emerging population supersedes the cynicism of the earlier five stanzas. The "lush jungle of modern thought" (l. 22) refers to the sinople European countryside. Hope proffers the view that the intellectual and cultural pioneers still come like prophets from the xeriscapes, not from the thickets of Old World ideas and the claustrophobic quarters of European traditions. The poet turns "gladly home" (l. 21) to "the Arabian desert of the human mind" (l. 23) away from "the learned doubt, the chatter of cultured apes" (l. 27). The sarcasm of the earlier stanzas is the voice of the learned doubt of the Old World. Australia offers new life and ways of perceiving, not from clutching to indoctrinated modes of seeing, but from reappraising the arid lands where "no green hills dare" (l. 25).

Through the shroud of excoriating images and green assumptions, Hope sees the potential of Australia. Similarly, Stegner (1992b, p. 54) calls for an aesthetics of arid landscapes in which the deficiency of greenness is not, by default, sterility: "You have to get over the color green; you have to quit associating beauty with gardens and lawns; you have to get used to an inhuman scale; you have to understand geological time." In his essay "Thoughts in a Dry Land," Stegner argues for a revisioning of aesthetics that is suited to the arid expanses, phantasmagoric landforms and harsh light of the American West. Culturally engrained modes of seeing the world engender predictably unsympathetic responses to dry landscapes: "To eyes trained on universal chlorophyll, gold or brown hills may look repulsive" (p. 53). Rather than moisture, aridity shapes the character of the xeric landscape and "exposes the pigmentation of the raw earth and limits, almost eliminates, the color of chlorophyll" (p. 46). Indicating the presence of plant chlorophyll, greenness is not a stable visual quality of dry places.

Stegner's assertions relate to the ways we view landscapes, and the decisions about how places should be used based on how they appear. Stegner goes on to say that perceptions trained elsewhere "have had to be modified. That means we have had to learn to quit depending on perceptual habit. Our first and hardest adaptation was to learn all over again how to see" (p. 52). In "Living Dry," Stegner (1992a) further points to the relation between the mistreatment of arid landscapes and habits of perception:

> [The American West] has been misinterpreted and mistreated because, coming to it from earlier frontiers where conditions were not unlike those of northern Europe, Anglo-Americans found it different, daunting, exhilarating, dangerous, and unpredictable, and entered it carrying habits that were often inappropriate and expectations that were surely excessive. (57)

Accustomed ways of seeing arid landscapes as barren and desolate can have devastating consequences. An aesthetics of dry places is therefore not merely an exercise in redefining perceptual states but also an exertion of social and environmental justice in abused places. For example, established in 1951, the Nevada Test Site, one hundred kilometres south of Las Vegas in the expanses of the American desert, was the proving ground for nuclear technology, leading to high rates of leukaemia among children in neighbouring states in the mid twentieth century (Clarfield & Wiecek, 1984). Similar nuclear testing sited in the arid lands of Western Australia had severe health effects on Aboriginal inhabitants, whose xeric countries were looked upon as uninhabited *terra nullius* (for an account, see Lindqvist, 2007).

Interlude XIV: First Kangaroo Paws

The Mangles Kangaroo Paw (*Anigozanthos manglesii*), also known as the Red and Green Kangaroo Paw, has been the floral emblem of Western Australia since 1960 (Corrick & Fuhrer, 2002, p. 83). Nyoongar people have known this iconic South-West plant by its traditional names *kurulbrang, krulbrang* and *kuttych*, and the tender rhizomes were consumed before the emergence of the flower (Hopper, 1993, p. 65). In 1834, botanist David Don published in *The British Flower Garden* the first formal description of a cultivated Mangles Kangaroo Paw specimen whose seeds were brought to England by Governor Stirling:

> This singularly beautiful species of *Anigozanthos* was raised in the garden at Whitmore Lodge, Berks, the seat of Robert Mangles, Esq. from seeds brought from Swan River by Sir James Stirling, the enterprising governor of that colony, by whom they had been presented to Mr Mangles (Don, 1835, p. 266).

Like the donkey orchid, the kangaroo paw draws its common name from an association with an animal. The new flower "capped in green" is concurrent with the onset of spring, the season in which "colour is gestated," but soon the complex blossoms will "resign to brown." Through colour symbology, the poem signals the passing of a season within itself. As a macrocosmic unit of change, a season consists of constant microcosmic instances of transformation that herald its overall passage:

> peddling their charms this way—
> up briskly from tawny earth
> candelabras of crayon red, capped in green,
> old tentacles darkening to crimson (ll. 1-4)

The final tercet conjoins the experience of touch and an attempt at smell to my understanding of the visual variations of the endemic plant (to read "First Kangaroo Paws" in full, see Hickman & Ryan, *in press*).

Figure 15 Kangaroo Paws, Anstey-Keane Damplands. *Image courtesy of the author.*

Figure 16 The Ravensthorpe Woman, September 2009. Created by local botanical artists, the Ravensthorpe Woman, as I call her, complements one of the largest wildflower shows in Australia by exemplifying the textural versatility of South-West plants. *Image courtesy of the author.*

Towards an Aesthetics of Xeriscapes: The Ravensthorpe Woman

> They say here the world's upside-down,
> And in summer it is true I find
> All the green lawn covered in the morning
> With the close pattern of what seem autumn leaves.
> The eucalypts, wiser than the trees of the old world,
> Ancient in sacrificing to the sun what is due,
> This way will find new strength to put out afresh
> Tawny young leaf-sprays when the first autumn rains come.
>
> Glen Phillips (1988, p. 8, ll. 1-8)

Several South-West Australian writers and scientists have responded to Hope and Stegner's call for a shift in the perception of arid landscapes. The works of botanist Alex George, ecologist Barbara York Main and ecologist and essayist George Seddon all evidence a grappling with green tropism as it has been invoked in the South-West. The historical representations of the flora in the journals of de Freycinet, Stirling, Darwin and Lesueur-area settlers show the tensions of acclimating perception. Dun-coloured places require revised vocabularies of seeing and expanded perceptual linkages between colour and ecology. By no means an exhaustive summary, Table 5 compiles some of the aesthetic qualities of the South-West, distributed across the five senses. An aesthetics of xeriscapes takes into account the peculiar qualities of dry landscapes in forging judgements of beauty or worthiness in matters of conservation (see the beginning of chapter 5 in reference to the Lesueur management plan).

New descriptive terminologies and neologisms have entered into the botanical language of the South-West to account for the colour variation of the flora. In his study of autumn foliage change and the variable green colouration of plants, botanist Alex George (2002, p. 2) proposes the term *diallagy* to describe colour reversibility in which "plants show an ability to change leaf colour and reverse it according to the weather." Diallagy is derived from the Greek *diallage* for interchange, and reflects Freud's notion of antithetical meanings of words, but through the ecological inherency of green within variable colour states. George argues that the unusual strategy of retracting greenness during dry conditions, then re-greening after the arrival of autumn rain, is indicative of the successful adaptations of South-West plants. The ecologies of the flora reflect their visual peculiarities. After a typically long dry summer, the autumn has been considered a time of dormancy or decline in the South-West—the onset of aesthetic monochrome—yet according to George (2002), it is a rich time for plants:

In south-western Australia, few people venture into the field during the autumn season. It is usually the end of a long, hot, dry summer, and there is still a widely held view that little is happening in the bush at that time; spring is the wildflower season, though some flowering continues into summer. Consequently, little has been written about the bush in autumn, or about the effect of drought. (1)

Table 5 Multi-Sensorial Qualities of South-West Landscapes. An embodied aesthetics of flora necessarily considers the non-visual attributes of a landscape and its life forms.

Tactile	dryness solar exposure prickliness
Visual	flat expansiveness harsh light dun colours
Olfactory	menthol sandalwood boronia
Auditory	sibilant wind gum nuts banksia birds
Gustatory	root starch quandong tang plum sweetness

After observations of autumn leaf colour change in response to rain levels at sites such as Lesueur, George (2002, p. 1) further defines diallagy as "the strategy of reversible change between the green and coloured states." Diallagy accounts for the autumn colouration of the flora, exhibiting a range of browns, oranges, purples, reds and yellows that would normally indicate dying plants, but instead signify a state of decreased metabolic activity. Leaf colour changes gradually as the summer progresses in response to diminished soil moisture, and is generally reversible after ten to fifteen millimetres of rainfall within a twenty-four hour period.

Thirty five years earlier than George, zoologist and ecologist Barbara York Main (1967) reported the phenomenon of diallagy in *Between Wodjil and Tor*, her study of a remnant parcel of native bushland in the Western Australian Wheatbelt. As with George, close recurring study of the vegetation reveals a pattern of visual transformation connected to ecological adaptation: "These bushes had now assumed their characteristic summer colour—no longer green, the copses were now splashed with russet and purple and burgundy…the change in colour of the 'foliage' of tamma copses is as dramatic as that of any deciduous tree" (Main, 1967, p. 9). The recognition of the drama taking place required close, prolonged engagement with the bush as a living community of organisms and land forms. Main puts into regional practice Stegner's claim that "our first and hardest adaptation was to learn all over again how to see." *Between Wodjil and Tor* suggests learning to see indigenous plants with enlivened eyes through an attitude of intrigue, rather than animosity or resentment. Main (1967, p. 145) goes on to say that "the sudden contrast in colour of the eucalypts accompanying the sloughing of their bark is as spectacular, as much anticipated, as the autumnal transformation of any truly deciduous forest of the northern hemisphere." Through new habits of perception, the Old World, as the default mode for seeing Australian flora, diminishes in influence.

Seddon also invokes a new sensibility for the Western Australian bush. *Sense of Place* (1972) and most of his writings in the subsequent thirty five years have addressed issues of perception. As constituting the region's character, plants figure prominently in Seddon's work in order to dispel cultural misconceptions:

> The colour range is highly distinctive: grey, grey-green, blue-green, and then translucent copper reds in the new flush of growth (because in nutrient-poor soils, the production of anthocyanin outstrips that of the more nutrient-demanding chlorophyll).
> …What we do not commonly find in this world of the sun are plants with large mid-green leaves, wilting as soon as their water-filled cells are thirsty…
> (Seddon, 2005, p. 20)

Seddon is selective with his choice of words and progression of ideas. The "highly distinctive" colour range of indigenous plants results from botanical adaptation: anthocyanin rather than chlorophyll. Eyes habituated to "universal chlorophyll" (Stegner, 1992b, p. 53) would perceive, by training, sinople shades of vegetation as verdure, as healthy, as productive, welcoming and reminiscent. In Seddon's terms, when we comprehend the physiological underpinnings of plants in dry climates, we begin to appreciate differences. The scientific basis of this idea recalls the "natural environmental" model of aesthetics (see chapter 5). Our eyes entrain to anthocyanin, and a transformation of seeing gains momentum. Table 6 lists a few ecological and morphological qualities of the South-West that, in part, also determine its aesthetic character.

Getting over the colour green in the South-West has required the united literary and scientific perspectives of regional botanists and writers. An appreciation for its globally significant diversity, rather than an excoriation of its barren appearance, its "drab green, desolate grey" vegetation, engages a retraining of eyes. Rather than the Green Man, the Ravensthorpe Woman (see figure 16) offers an archetype for an aesthetics of the post-colonial era. She offers a new colour paradigm, invoking reappraised associations of greenness and other-than greenness, and positing a regional aesthetics. Created from plants collected near the Fitzgerald River National Park, her imbricate dun-coloured image counterpoises that of the foliated Green Man. The perception suggested by the Ravensthorpe Woman is not of warring against, but symbiotically aligning to, the indigenous plants of the South-West through an interest in the distinct textures, sights, sounds, smells and tastes of the region.

Table 6 Select Ecological and Morphological Qualities of the South-West. The flora's evolutionary distinctiveness can figure into aesthetic assessments of South-West places.

Ecological	aridity endemism diallagy
Morphological	irregularities small leaves and flowers schlerophylls

References

Armstrong, P. (1985). *Charles Darwin in Western Australia: A young scientist's perception of an environment.* Nedlands, Australia: University of Western Australia Press.

Bailey, J. (n.d.). Power struggle: Unions' role in a conservation campaign Retrieved April 14, 2011, from http://www98.griffith.edu.au/dspace/bitstream/10072/1574/1/20646.pdf

Basford, K. (1978). *The Green Man.* Ipswich, England: D.S. Brewer.

Baudin, N. (1974). *The journal of post captain Nicolas Baudin Commander-in-Chief of the Corvettes GEOGRAPHE and NATURALISTE* (C. Cornell, Trans.). Adelaide, Australia: Libraries Board of South Australia.

Buddhist Council of Western Australia. (2009). Jade Buddha for Universal Peace exhibition Perth Retrieved April 8, 2011, from http://www.wa.buddhistcouncil.org.au/home/jade-buddha-for-universal-peace-exhibition-perth/

Burchell, J. (2002). *The evolution of green politics: Development and change within European green parties.* London, England: Earthscan Publications.

Clarfield, G. H., & Wiecek, W. W. (1984). *Nuclear America: Military and civilian nuclear power in the United States 1940-1980.* New York, NY: Harper & Row.

Corrick, M. G., & Fuhrer, B. A. (2002). *Wildflowers of Southern Western Australia.* Noble Park, Australia: The Five Mile Press.

De Vries, A. (1974). *Dictionary of symbols and imagery.* Amsterdam, Netherlands: North-Holland Publishing Company.

Department of Conservation and Land Management. (1995). *Management plan: Lesueur National Park and Coomallo Nature Reserve 1995-2005.* Perth, Australia: Department of Conservation and Land Management.

Don, D. (1835). Anigozanthos manglesii: Mr. Mangles's Anigozanthos Vol. 3. R. Sweet (Ed.) *The British flower garden* (pp. 265-266) Retrieved from http://books.google.com

Firestein, P. (2009). *Crisis of character: Building corporate reputation in an age of skepticism.* New York, NY: Sterling Publishing.

Freud, S. (1997). The antithetical meaning of primal words. In *Writings on art and literature* (pp. 94-100). Stanford, CA: Stanford University Press.

Gage, J. (1993). *Color and culture: Practice and meaning from Antiquity to Abstraction.* Berkeley, CA: University of California Press.

George, A. (2002). *The long dry: Bush colours of summer and autumn in South-western Australia.* Kardinya, Australia: Four Gables Press.

George, A. (2002). The south-western Australian flora in Autumn: 2001 presidential address. *Journal of the Royal Society of Western Australia,* 85, 1-15.

Giblett, R. (2011). *People and places of nature and culture.* Bristol, England: Intellect.

Gregory, A. C., & Gregory, F. T. (1884/1981). *Journals of Australian explorations.* Victoria Park, Australia: Hesperian Press.

Grey, G. (1841). *Journals of two expeditions of discovery in North-West and Western Australia during the years 1837, 38, and 39.* (Vol. 2). London, England: T. and W. Boone.

Harper, D. (2010). green. *Online Etymology Dictionary* Retrieved from http://www.etymonline.com/index.php?search=green&searchmode=none

Hassell, E. (1975). *My dusky friends.* Dalkeith, Australia: C.W. Hassell.

Herndl, C. G., & Brown, S. C. (1996). Introduction. In C. G. Herndl & S. C. Brown (Eds.), *Green culture: Environmental rhetoric in contemporary America* (pp. 3-20). Madison, WI: The University of Wisconsin Press.

Hickman, E., & Ryan, J. (in press). *Two with nature: Ellen Hickman and John Ryan.* Fremantle, Australia: Fremantle Press.

Hope, A. D. (1977). *Collected Poems,* 1930-1970 (2 ed.). London, England: Angus & Robertson.

Hopper, S. (1990). European Exploration. In A. A. Burbidge, S. Hopper & S. Van Leeuwen (Eds.), *Nature conservation, landscape and recreation values of the Lesueur area: A report to the Environmental Protection Authority from the Department of Conservation and Land Management* (pp. 9-14). Perth, Australia: Environmental Protection Authority.

Hopper, S. (1993). *Kangaroo paws and catspaws: A natural history and field guide.* Como, Australia: Department of Conservation and Land Management.

Hopper, S. (2004). Southwestern Australia, Cinderella of the world's temperate floristic regions. *Curtis's Botanical Magazine*, 21(2), 132-180.

Hutchings, J. (1997). Folklore and symbolism of green. Folklore, 108, 55-64.

Jung, C. G. (1959). *The archetypes and the collective unconscious* (R. F. C. Hull, Trans. 2 ed.). London, England: Routledge and Kegan Paul.

Kaptchuk, T. (2000). *The web that has no weaver: Understanding Chinese medicine.* Chicago, IL: Contemporary Books.

Keane, B. (2009). Pounamu – jade or greenstone - Pounamu – several names. *Te Ara: The Encyclopedia of New Zealand* Retrieved May 30, 2012, from http://www.TeAra.govt.nz/en/pounamu-jade-or-greenstone/1

King, H. (2004). *The disease of virgins: Green sickness, chlorosis and the problems of puberty.* London, England: Routledge.

Lindqvist, S. (2007). *Terra nullius. A journey through no one's land* (S. Death, Trans.). London, England: Granta.

Main, B. Y. (1967). *Between wodjil and tor.* Melbourne, Australia: Jacaranda.

McLeod, A. L. (1980). Maturity in Australian satire: The poetry of A. D. Hope. *Modern Language Studies*, 10(2), 42-48.

Moore, G. F. (1846). *Descriptive vocabulary of the language in common use amongst the Aborigines of Western Australia.* Nedlands, Australia: University of Western Australia Press.

Moore, G. F. (1884/1978). *Diary of ten years eventful life of an early settler in Western Australia,* and also *A descriptive vocabulary of the language of the Aborigines.* Nedlands, Australia: University of Western Australia Press.

Phillips, G. (1988). *Sacrificing the leaves: Poems.* Thailand: 10th World Congress of Poets.

Reich, C. A. (1970). *The greening of America.* Ringwood, Australia: Penguin Books.

Sarni, W. (2010). *Greening brownfields: Remediation through sustainable development* (1 ed.). New York, NY: McGraw-Hill.

Seddon, G. (1972). *Sense of place: A response to an environment.* Perth, Australia: University of Western Australia Press.

Seddon, G. (2005). *The old country: Australian landscapes, plants and people.* Cambridge, England: Cambridge University Press.

Sheldrake, R. (1991). *The rebirth of nature: The greening of science and god.* New York, NY: Bantam Books.

Stegner, W. (1992a). Living dry. In *Where the bluebird sings to the lemonade springs: Living and writing in the West* (pp. 57-75). New York, NY: The Modern Library.

Stegner, W. (1992b). Thoughts in a dry land. In *Where the bluebird sings to the lemonade springs: Living and writing in the West* (pp. 45-56). New York, NY: The Modern Library.

Stirling, J. (1827/2009). Narrative of operations - Expedition to Swan River, 1827. In J. Shoobert (Ed.), *Western Australian Exploration Volume One December 1826 - December 1835* (Vol. 1, pp. 20-41). Carlisle, Australia: Hesperian Press.

Tennyson, A. L. (1850/2009). In memoriam A.H.H. OBIT MDCCCXXXIII: 56 Retrieved April 8, 2011, from http://rpo.library.utoronto.ca/poem/2141.html

Thomas, D. (1952). The force that through the green fuse drives the flower Retrieved May 19, 2012, from http://www.poets.org/viewmedia.php/prmMID/15379

7

Ethnography

Abstract: Anthoethnography is the use of ethnographic practices for researching human relationships to flowers. This chapter will propose anthoethnography for researching wildflower tourism as a culture of flowers in the South-West. Through an analysis of its history, contemporary practice and aesthetic values, a direction for wildflower tourism leads to material conservation, spiritual engagement and multisensory embodiment. Rather than wildflower tourism, botanical appreciation suggests engagement with flora, including the whole plant throughout the seasons and through various human sense faculties. Anthoethnography responds to a consciousness amongst botanists and enthusiasts in the South-West of the need for the practices of wildflower tourism to engage with the realities of plant conservation in the twenty-first century.

> *We were travelling up north around the Murchison River area, and I came across all these tiny orchids about ten centimetres high. They were in a mass and for some reason the orchids just grabbed me and I thought 'Wow, these are amazing'.*
>
> Lyn Alcock, Wildflower Enthusiast (pers. comm., September 8, 2009)

Anthoethnography and the Culture of Wildflowers

The prefix *antho* is a linguistic keystone between flora, culture, history and science. Denoting a collection of literary pieces, *anthology* means "flower gathering." Yet, the convergences between texts and living flora are more than metaphors. A *flora* signifies both the descriptive treatise and the plants to which the writing refers, both the living organism and the medium in which a plant is depicted. The historical conjunctions between book and blossom suggest that words, plants and human affairs entwine ineluctably. The obsolete term *anthography* refers to the anatomical description of flowers, as in a study of eucalypts by Grimwade (1920). *Antho* also appears in contemporary plant science where *anthocyanins* are compounds that pigment flowers, and *anthoecology* is the study of the ecological interactions between flowers and their environments.

As agents of healing, purveyors of ornamentation, symbols of inspiration, inciters of attraction, and repositories of beauty, wildflowers have special roles in human societies worldwide (for example, see Goody, 1993). People bond emotionally to flowers. Thus, *antho* may be added insightfully to different human experiences to recognise the connections between flowers and culture. Concerning matters of the heart, *anthomania* denotes love, flowers and frenzy, such as in seventeenth century Holland where an over-zealous love of blossoms galvanised the social and economic furore over tulip bulbs known as *tulipmania* (Goldgar, 2007, p. 7). Analogous to biologist Edward O. Wilson's idea of biophilia, or the human bond to other species, *anthophilia* would refer to emotional bonding to blossoming plants (see Wilson, 1984). Engineered into hybrids and raised in greenhouses, cultivated flowers have particular affinities with people as common members of domesticated spheres. In contrast, wild-growing flowers invoke the non-cultivated natural world. By governing their own biological stabilities, wildflowers can appeal to us through their undomesticated beauty and resilience.

The "culture of flowers" points to the intricate aesthetic, spiritual, artistic, mercantile, economic, symbolic, material and therapeutic relationships between human societies and wild-growing or cultivated flowers. In its general senses, the phrase indicates an interface between the broad categories of nature and culture, people and landscapes, flowers and appreciators (Giblett, 2011; Rose & Robin, 2004; Williams, 1982). In nineteenth century Europe, the "*culte des fleurs*" invigorated a "new flower rhetoric" within French literature, highlighting the importance of flowers to cultural expression (Knight, 1986, p. 3). The social anthropologist Jack Goody (1993, p. 1) questions whether the culture of flowers portends a universal "interest in the natural world that in some form is found in all human societies," or whether it is an affinity for flowers specific to certain cultures and landscapes.

As chapter 1 argues, there are asymmetries between technical and humanities-based forms of enquiry into plants. Whereas empirical methods for studying plants are established as botanical science, humanities-based approaches for researching the cultural aspects surrounding wildflowers are less defined. Chapter 7 pursues this position and explores the South-West culture of wildflowers, deploying *anthoethnography* to accentuate regional connections between plants and people, through an analysis of wildflower tourism. As an emergent area of scholarship bringing the perspectives of cultural studies to the natural world, anthoethnogaphy applies humanities techniques to plants and the cultures of flora. Anthoethnography may be defined as the application of ethnography, including semi-structured interviews and participant observation, for producing accounts about the interactions, interdependencies and embodied engagements between people and flowering plants (see chapter 2). Such scholarship is necessary. Despite the increase in the scientific knowledge and international profile of South-West species, relatively little has been researched about commonplace engagements between wildflowers and people in the region. So, while the classification of wildflowers grows along specialist lines, the cultural dimensions of wildflowers, as well as flora out of flower, remain relatively uncharted. In order to describe a culture of flowering plants, this chapter will interrogate the history, contemporary practices and implied values of wildflower tourism. The field outcomes of the method will be presented as interview transcripts. Anthoethnography deployed at places of botanical prominence and touristic appeal brings to light the diverse intersections between settler, post-colonial and contemporary societies and wildflowers. Venues include Western Flora Caravan Park, in Eneabba WA near Lesueur National Park, the Ravensthorpe Wildflower Show near Fitzgerald River National Park, and Banksia Farm in Mount Barker, WA.

As described in chapter 2, the flora of Western Australia is vast, comprising between 9,000 and 10,000 documented vascular plant species distributed throughout 1,500 genera and 220 families, with the South-West corner comprising the bulk of the endemic flowering biodiversity. Nearly half of Australia's 25,000 flowering species occur in Western Australia (Hopper, 1998, p. 270). Throughout the South-West, botanical science continues to identify new species at an unprecedented rate, such as the Rock Mallee (*Eucalyptus petrensis*) within the outer reaches of Perth (Hopper, 1998). In a study of nature-based tourism in the Central Coast region, including Lesueur National Park, Priskin (2003, p. 518) notes that even though "the flowering plant diversity of Western Australia is a tourism drawcard," wildflower tourism represents "an unresearched form of nature-based tourism" in the region. Many tourism studies describe the field of nature-based tourism but overlook, or only slightly acknowledge, wildflower appreciation as a significant form of biodiversity appreciation. The book *Nature-Based Tourism, Environment, and Land Management* makes only parenthetical mention of wildflower tourism and largely excludes concerns over the management of botanical reserves and the impacts of tourism (Buckley, Pickering, & Weaver, 2003).

Chapter 7 will focus on South-West wildflower tourism, as well as the merits of *botanical appreciation*. Wildflower and botanical appreciation are seasonal forms of nature-based tourism that occur throughout the world wherever flowering spectacles or other compelling plant life occurs (for example, see Loubster, Mouton, & Nel, 2001). However, whereas wildflower tourism is bloom-focused and occurs mostly during spring, botanical tourism encompasses appreciation of plants out-of-flower, perhaps for their fruit, seed, aromas, tastes or sheer magnitude. Botanical appreciation could include plants of scale such as the massive tingle eucalypts popularised as the Valley of the Giants (Valley of the Giants, *n.d.*). In the South-West, wildflower tourism and botanical appreciation are significant, though barely researched, cultures of wild plants that bring local, national and international human appreciators into habitats, while also conveying species of flora into churches and community centres through wildflower shows (Priskin, 2003; Western Australian Tourism Commission, 2003).

Comparatively, how do tourists and experts perceive flowering plants? How is wildflower tourism marketed? What kinds of language forms are used to communicate experiences of wildflowers? A reading of the rhetoric of spring wildflower tourism seasons 2009 and 2010 provides some useful answers. Set within a broader discussion of the history and values of regional wildflower tourism, the reading of anthoethnographic interviews suggests the varieties of cultural intersections with plants, from distanced visual appreciation to proximal multi-sensorial engagement. The interviewees range through a spectrum of novices and experts who could provide insight into the culture of flora through their experiences as tourists or expertise as specialists (see chapter 2). Respondents put forward potential directions for wildflower tourism including proximal interactions with plants in habitats (Merle Bennett); tourism with a conservation ethos (Lyn Alcock); conservation through scientific understanding (Allan Tinker); corporeal engagement involving smelling, tasting, touching and listening closely (Kevin Collins); and Aboriginal knowledges and spiritualities towards the long-term wellbeing of plants and people (Noel Nannup). Participatory engagement with flora offers other modes of appreciation that build upon the experience of seeing.

What Does Wildflower Tourism Entail?

> We try to take wildflower tourists a step further from just something that has aesthetic value, just physical images in the mind, the 'yeah it looks pretty'. We add the dimension of how wildflowers function.
>
> Allan Tinker (pers. comm., August 29, 2009)

Aesthetic experience of plants usually translates to looking at flowers. In its regional manifestations, wildflower tourism tends to emphasise the appreciation of plants as affects of sight. Indeed, wildflower tourism may entail expectations of colour, form, scale and profusion focused on the static appearance of flowers. Anthoethnography draws theoretically from the critique of pictorialisation, which constructs the natural world as a static two-dimensional landscape (see chapters 1 and 2). Heidegger (1938/2009, p. 221) claims that "the fundamental event of modernity is the conquest of the world as picture." His later writings respond to the "vision-generated discourse" intrinsic to the emerging visual technologies of modernity (Levin, 1993, p. 186). Echoing Heidegger in some ways, Crary (1990, pp. 25-66) traces the linkage between technologies of sight and modernity's production of objective knowledge of the world. Through the lens of a camera, flowers may be apprehended objectively as a series of images. Owing to the contemporary profusion of mimetic technologies, wildflower tourism tends to value surface appearances, rather than corporeal methexis linked to deeper understandings of plants as nodes in broader ecological webs. Through the critique of vision-generated discourse, the culture of flora in the South-West may be broadened to comprise physical interactions with plants through the multiple senses. The guiding conceptual premise of anthoethnography is that vision offers a narrow range for experiencing flora that can be complemented and broadened by participatory aesthetics and Aboriginal traditions of plants.

Based in aesthetic perception, wildflower tourism is a diverse activity and a global phenomenon, occurring in such a variety of places as Namaqua National Park in South Africa (Loubster, et al., 2001), the town of Bohinj in Slovenia (Turizem Bohinj, 2010) and the state of Florida in the southern United States (Florida Wildflower Foundation, 2011). *Lonely Planet* acknowledges the importance of wildflower driving circuits to American tourism economies: "To Texans, wildflowers are a way of life. Wildflower tourism is so entrenched that the highway visitor centers can help you plan an entire trip around watching them bloom" (Lyon & Nystrom, 2002, p. 594). As visually marketed and practiced, wildflower tourism entails a quest for colours, forms, symmetries and morphological harmonies. Moreover, as an aesthetically motivated form of recreation and travel, its beliefs and practices embed perceptual values towards the botanical world. People travel thousands of kilometres to Western Australia to *see* wildflowers, not necessarily to smell, taste, touch, hear, know deeply or otherwise engage in somatic or spiritual ways.

A broadly inclusive and popular category of recreation, wildflower tourism invokes scenes of contemporary self-drive or escorted bus tours into the bush to encounter flowering plants in their habitats (Western Australian Tourism Commission, 2010). Yet, wildflower tourism also comprises flower shows, displays, and festivals in which plants are picked, transported and displayed in community and visitor centres, churches and botanical gardens. These venues often serve as nerve centres for peripatetic wildflower appreciators. The multi-directional flow of plants and people between urban, pastoral, agricultural and conservation areas is historically important to wildflower tourism. Yet, owing to the contemporary accessibility of cameras, flower tourism prioritises the visual appearance of flowering plants, rather than embodied experiences embedded in ecological webs (Tourism Australia, 2010; Western Australian Tourism Commission, 2003, 2010). The ascendancy of technologies of sight, combined with the protected status of many wildflowers, can prohibit physical interaction with wild plants.

In addition to walks and driving circuits, contemporary wildflower tourism in Western Australia includes a range of associated activities and desideratum like shows, displays, exhibitions, crafts, festivals and events related to the appreciation of plant life through the bringing of the flower into domestic settings. Indeed, "Wattle Days" are expressions of Australian nationalism. Since 1994, the shire of Dalwallinu has hosted an annual spring "Wattle Week Festival" to celebrate

Australia's floral emblem (see figure 17). The celebration of *Acacia* species as Wattle Days, however, is not unique to contemporary Western Australia but happened also during the colonial era (Robin, 2002). In 1838, participants in a Hobart regatta were encouraged to adorn themselves with wattle sprigs (Department of Environment and Conservation, Shire of Dalwallinu, & Australian Tree Seed Centre, 2011, Brief history of Wattle Day). Also, in 1891, a wattle flower banner was first displayed publicly in Adelaide in conjunction with Foundation Day (Department of Environment and Conservation, et al., 2011, Brief history of Wattle Day). The appreciation of flowering plants through walking appeared in the late nineteenth century Wattle Day outings, proposed annually for 1 September at popular bushwalking locations, such as Werribee Gorge near Melbourne (Department of Environment and Conservation, et al., 2011, Brief history of Wattle Day). In Western Australia in 2010, wildflower festivals occurred in Mullewa, Esperance, Wubin, Ongerup, Chittering, Mundaring, Busselton,

Figure 17 Wattle Week Festival in Dalwallinu Shire. This annual Wheatbelt festival celebrates *Acacia* spp. diversity. *Image courtesy of the author.*

Albany, Kojonup and Cranbrook, as well as Kings Park and the Botanic Garden in Perth (Wildflower Society of Western Australia, 2010).

As appreciation that takes visitors into habitats, wildflower tourism crosses into nature-based tourism, a general category comprising action-oriented adventure tourism and conservation-minded ecotourism. However, the assertion that wildflower tourism is automatically a form of ecotourism is generally off-kilter. Nature-based tourism is "primarily concerned with the direct enjoyment of some relatively undisturbed phenomenon of nature" (Priskin, 2003, p. 501). In a study of tourism along the Central Coast of Western Australia, Priskin (2003) notes three variations of nature-based tourism: activities, such as bird-watching and wildflower-hunting, that depend on nature; activities enhanced by nature, such as camping; and activities in which the setting can vary from natural to artificial, as with swimming. Adventure tourists, on the one hand, are characterised by their interest in four-wheel driving and other adrenaline-inducing forms of sport. On the other hand, ecotourists aim to experience nature as a whole and to learn about its conservation (Blamey, 2001).

Wildflower tourism crosses into adventure tourism and ecotourism, but its parameters and ethics are not clear. What constitutes wildflower tourism? Who participates? And what kinds of values and expectations do visitors bring to the experience? Demographically, wildflower tourists tend to be mostly "older empty nesters" (Western Australian Tourism Commission, 2003, p. 8). In Australian snow country as indicated in a book on the field of nature-based tourism, "wildflower appreciation and *other forms of ecotourism* tend to occur in the summer [italics added]" (Buckley, et al., 2003, p. 9). Wildflowering can be combined with activities usually considered adventure tourism, such as backcountry four-wheel drive exploration, as those activities are not explicitly connected to conservation values. According to the Western Australian Tourism Commission (2003), "wildflower enthusiasts" are focused on seeing different species of wildflowers, whereas "nature lovers" are more independent, interactive and inclined towards diverse natural and cultural attractions. Considering this, wildflower tourism is not intrinsically linked to practices of botanical conservation as ecotourism or nature-based tourism, although it can be. For example, the picking of wildflowers has had a troubled relationship to the longevity of South-West plants. In 1926, a request for the protection of flora and fauna within national parks was rejected by the West Australian Department of Lands and Survey in order to defend the interests of wildflower gatherers: "The primary inducement for people to go into reserves...is to gather the wildflowers with the object of adorning their homes and taking part in the wildflower shows" (cited in Hall & Page, 2006, p. 56).

In contemporary settings, not all biodiversity is housed in conservation areas where vehicular or pedestrian access is regulated. Flowering plants occur alongside road verges or railway corridors and on private lands outside the auspices of legal protection. As some ecotourism researchers argue, the exposure of wildflowers as swathes of flora in roadside areas can amplify their appreciation: "Strips of natural vegetation along roadsides in Western Australia have become a significant wildflower tourism resource" (Newsome, Moore, & Dowling, 2002, p. 47). However, plants growing along such corridors are exposed to greater stress and exploitation. Hence, whether overtly or implicitly, cooperatively or antagonistically, intentionally or unwittingly, wildflower tourism interfaces with conservation because many flowering species are endangered, threatened or their habitats compromised as remnant populations along roads or in isolated pockets of degraded land. Since wildflower tourism depends on the protection of habitats, proprietors of tourism venues will need to consider conservation in order to sustain viable living botanical heritages.

Interlude XV: Dispersion of Seeds

The South-West region is a mosaic of changing places, with extensive tracts of indigenous bushland converted to wheat, canola and stock pasture during the nineteenth and twentieth centuries (see chapters 8 and 12). Relatively recent groundwork to re-establish plants in national parks represents an effort to restore balance to a landscape in transition. "Dispersion of Seeds" is a meditation on the ephemerality and changeability of place as measured by its flora. I wrote the poem during a wildflower tour by bus led by Ayleen Sands, the knowledgeable proprietor of the Stirling Range Retreat in Borden, WA, and a participant in my spring 2009 wildflower trail interviews (see chapter 8). Chester Pass Road bisects the national park and was the main artery for the wildflower explorations of the morning. Where the ascending road from Albany begins to level out, an expanse riddled with slender kingia stalks comes into view. *Bullanock* is the Nyoongar name for *Kingia australis*, for many years considered by scientists a female balga, but, despite its visual congruence, revealed recently as more closely akin to kangaroo paws (S. Hopper, pers. comm., September 9, 2009). I attend to the intersections between indigenous and naturalised plants—between island pockets of original flora and the surrounding agricultural matrix which engulfs it—through the metaphor of the displaced bag of wheat seed. A theme of intrusions is given impetus by a riven bag intervening in a touristic experience of pristine bush. But the poem is also about habituations—"habitats become habits" (l. 25)—and concurrences. Spring, as we know it in Western thought, can be amplified by the Nyoongar understanding of the season of *djilba* and its alignments to a nest of ecological understandings. Hopefully, the poem conveys that I am not meaning to romanticise the bush as it was, but rather to observe the often unexpected and sometimes shocking confrontations that characterise its everyday experience.

Dispersion of Seeds

a gap in the range
along Chester Pass Road
where *bullanock* grow
on charcoaled soils;

we have come to see
wild things, anxious
as taut buds; instead
lying on the road verge

the sun-baked corpse
of an old kill, wincing
its last breath, grains
sprawled about, its guts;

umber burlap bleached
tawny brown and riven,
spilling maggots and stones
in difficult morning light;

Hoodini escapee from
roadtrains, persona non
grata at an ill-fated wedding
where wheat sullied the rice;

interloper on an island
of mallee and wattles,
sweeping the understory
in a bile yellow peristalsis.

habitats become habits —
all the shades of saffron and
sinopia signify the storm
of spring as we know it, or

djilba as it has been called;
these worlds, sharp with edges
that overlap and tatter,
curl up like kingia fronds

shrivel the next, dispersing.

Breath of the Bush: The History of South-West Wildflower Tourism

Since the mid-1800s, Western Australia has been recognised by botanists, horticulturalists, artists and plant enthusiasts globally for its flowering diversity (see chapter 4). Former director of the Western Australian Herbarium, botanist Neville Marchant (2005) remarks on the congruence between species numbers and flowering forms:

> Apart from high species numbers, this diversity is also expressed in the myriad of plant forms that abound in the south western flora; a huge array of plant form, branching patterns, stem shapes, leaf types and flower shapes go hand in hand with species diversity. (19)

Throughout the region, the profusion of species, combined with the particular adaptive ecologies of the flora, has given rise to a richness of colours and forms. Botanically-minded visitors often comment on the tenacity and beauty of the flora, enduring and adapting to extreme conditions of dryness and heat: "Travelling across the most barren area, suddenly you come across a Stuart [*sic*] Pea growing out of this red soil…you just can't believe that's come out of this red dust" (Western Australian Tourism Commission, 2003, p. 17) (see figure 18).

Figure 18 Sturt's Desert Pea in Kings Park and Botanic Garden Near Perth. A member of the legume family, Sturt's Desert Pea is one of the most recognisable Australian wildflowers. It grows in the arid inner reaches of Western Australia. *Image courtesy of the author.*

The appreciation of South-West wildflowers is evident in writings from the Swan River Colony era. Spring walks to Mount Eliza, the present-day site of Kings Park, were important to the palette of social activities for Perth residents. In 1856, Sophia Phillips, the daughter of John Septimus Roe, reported the "bush lovely with flowers" on a trip to Toodyay (cited in Summers, 2011). In his diary kept between 1877 and 1884, Alfred Hillman wrote of wildflower jaunts to Mount Eliza (Hillman, 1990, p. 419). However, Hillman (1990, p. 424) comments that many wildflowers lack a smell: "They can scarcely be said to waste their sweetness on the desert air as they are for the most part scentless." Moreover, recalling her visit to Western Australia in the late nineteenth century, including Albany and the Goldfields, May Vivienne published one of the first travelogues of the state. She travelled by train and horse, witnessing wildflowers. Describing Serpentine Falls, south of Perth, as the "garden of the colony," the author depicts flowers as elements of a picturesque painting, evidencing an aesthetic of spectatorship: "Sweet flowers and ferns form an idyllic picture" (Vivienne, 1901, p. 98). Her framework for experiencing flowers is comparable to some contemporary wildflower tourism models that emphasise the visuality of expansive carpets of multicoloured everlastings viewed from motorised vehicles: "As the train sped past the idea struck me that these flowers—lovely immortelles, white, pink, and yellow, growing in countless millions—could be turned to good account" (Vivienne, 1901, p. 28). She intimates the conversion of the idle—though idyllic—living wildflowers into lifeless products, including perfumes, "crosses, wreaths, anchors, screens, fans, and other decorations" (Vivienne, 1901, p. 28).

In the late nineteenth and early twentieth century, technological developments afforded greater avenues for visualising the bush. West Australians embraced rail and car travel because of the distances between locales. Emerging forms of transport augmented opportunities for public interaction with wildflowers. This period marked the emergence of flower tours by train, departing metropolitan Perth for outlying areas such as Gingin and Wongan Hills. Wildflower trains became desirable forms of tourism, providing regular excursions to the bush and promoting Western Australian countryside as an "Arcadian idyll" (Summers, 2011). A 1905 issue of *Western Mail* features an article titled "Along the Midland Railway, Perth to Geraldton, A Lady's Journey." Outside of Gingin, the author reported everlastings, mistletoes, banksias and the "monotonous grandeur" of gums. She observed "bushes and masses of colour, scarlets, reds, pinks, and yellows of every shade...each tree is a mass of shining leaves, moving in the breeze. They flash back the sunlight, like a myriad mirrors" (Bickerton, 1905, February 18, p. 39). Around this time, botanical illustrator Emily Pelloe (1921) also commented on the flower trains:

> Special 'flower trains' are run. These and the usual week-end and holiday trains are always packed with city folk eager to explore the bush and gather the lovely flowers. Glen Forrest (late Smith's Mill), Darlington, Gooseberry Hill, Kalamunda, and Serpentine are all favourite haunts of the flower-seekers. (29)

However, the trains also brought both country-dwellers and wildflowers from the countryside into Perth for wildflower shows. During the late 1800s, the Perth Wildflower Show served as a plenum for the collective botanical diversity of the state. The railways transported plants to Perth committee members at no cost. Train transport galvanised the collection of species from remote areas. An 1899 issue of *The West Australian* explains that the committee of the Perth wildflower show hoped that flowers "will be sent from all parts of the colony to them to make the show as representative as possible" (The West Australian, 1899, August 30, p. 4). *The West Australian* from 1893 also mentions special excursion rates offered by the Railway Department to "enable country residents to see the floral exhibitions which are to be open, and to spend a few days of the most charming season of the year in the capital and at the seaside" (The West Australian, 1893, September 2, p. 4). Hence, the trains became related inseparably to early modes of flower touring. In the late colonial history of Western Australia, the viewership afforded by rail transport moulded the perception of plants as aesthetic elements of the landscape through which new forms of locomotion shuttled burgeoning settler populations.

During this era, flower shows and other forms of tourism conflicted with values of biodiversity conservation, especially when wildflower picking was an acceptable form of appreciation. By the late 1890s, discussions about the future of Western Australian flora commenced amidst pressures exerted by wildflower collectors. By the 1920s, Pelloe (1921) extolled the virtues of South-West wildflowers as the "breath of the bush" but decried their despoliation at the hand of pickers:

> Flowers should ever be regarded as something too beautiful and precious to be wasted…in the city they bring a breath of the bush and recall happy memories to many a tired heart, in the home they add a dainty charm to the rooms, and as a gift to the sick are always welcomed. (16)

The 1912 Native Flora Protection Act was implemented to protect flora from destruction by excursionist pickers, as well as commercial exploitation from floral decorators (Summers, 2011). Moreover, the 1935 Native Protection Act increased the schedule of species protected from collection. Local legislation also began to take place along popular touring routes. For example, Mundaring Weir Road between Mundaring and Kalamunda was constructed as a tourist loop through forested land. Community concerns about the recreational picking of wildflowers prompted early botanical protection legislation in the Hills outside Perth. Further along, in 1953, botanist Charles Gardner reported a rare endemic Mountain Bell, only known to exist on the summit of Coyanarup Peak in the Stirling Range, at the Kalgoorlie Wildflower Show (Summers, 2011). Thus, the aesthetic appreciation of visible forms and the procurement of wildflowers began to put demands on biodiversity conservation (see figure 19).

Mediation of Presence: Contemporary Wildflower Tourism in the South-West

As they attract bees and other pollinators, wildflowers similarly have garnered the attention of an international contingent of contemporary tourists. Mid twentieth century colour posters, promoting the state's wildflower season, illustrate exuberant, youthful visitors on foot, immersed in the idyllic landscape (Western Australian Tourism Commission, 2010). During the 1960s, due largely to the personal interest of Premier David Brand, the tourism industry promoted Western Australia with car licence plates bearing the slogan "The Wildflower State." Presently, the state's wildflower season is locally, nationally and internationally renowned. The Western Australian Tourism Commission (2003) study identifies the state as the best destination for wildflowers in Australia and among the premier destinations for wildflower tourism worldwide.

Figure 19 Gathering Wildflowers, Bardoc. Photographer: John Joseph Dwyer (1869-1928). From image caption: "Unidentified young woman picks a posy of wildflowers, Bardoc countryside." *Image courtesy of the State Library of Western Australia.*

Visitors with diverse cultural backgrounds and varying levels of scientific expertise participate in regional wildflower tourism each year. They seek the colourful vistas and unusual flowering forms of the region in order to have an aesthetic experience of wild, uncultivated flora. Between 2001 and 2003, nearly one in five Perth residents visited wildflower reserves (Western Australian Tourism Commission, 2003). Moreover, of the 6000 annual visitors to Banksia Farm, the premier destination for photographing all known species of banksia, nearly two-thirds are from eastern Australia and 7% from overseas (pers. comm., K. Collins, June 4, 2012). In 2009, the annual wildflower show in rural Ravensthorpe reported a broad spectrum of visitors from Asia, North America and Europe (pers. comm., M. Bennett, September 13, 2009). Kings Park and Botanic Garden in Perth reported over 500,000 visitors to its Wildflower Festival held throughout September (Botanic Gardens and Parks Authority, 2009).

The South-West wildflower season satiates a visual longing for intriguing shapes and variegated colouration. Primarily from June through September each year, flowers "explode across the state," and wildflower appreciators, hungry for stimuli, are beseeched to "forget drab days and bleak landscapes" and "add colour to...winter blues" (Tourism Australia, 2010, para. 1). The emerging colour and forms of the spring blossoming are represented in the rhetoric of portraiture. The painterly act animates and infuses a drear, slumbering landscape with artistic elements: "The state's south-west springs to life with grand, Picasso-style strokes" (Tourism Australia, 2010, para. 5). The spectacle of flowering casts the plant in the role of the performer, and the tourist becomes the audience, as "wildflowers continue to dance south along the coast" (Tourism Australia, 2010, para. 6).

In images, the colourful succession of flowering in South-West Australia circulates beyond the state's geographical boundaries. In one of the few published reports on Western Australian wildflower tourism, by the Western Australian Tourism Commission (2003, p. 5), the sole suggestion for increasing the "motivational appeal" of the annual Wildflower Holiday Guide is to "focus more on larger photos of expansive, scenic and experiential wildflower imagery rather than micro photos of individual flowers." Imagery, as a marketing tool, is based on visual enticement and the prospect of having an experience of sight. It can be said that the experience of tourists is defined by seeing flowers and taking images. To return to a relevant phrase by Warwick Mules (2006, p. 2), wildflower imagery is "the mediation of presence through images." The presence mediated by the images is of living plants in their multi-sensorial habitats and in their complexity of relations to other life forms. Instead, the spectator is lured into the appearance of performative flowering. The spectacle of fantastic form and compelling colour is optical and evanescent; the landscape—that which is seen as scene—is construed as a decorative setting or theatrical backdrop.

Visual affordance from the detachment of trains and cars, as well as from behind image-making instruments, can create an objective space between people and plants, reinforcing nature/culture dualisms. A survey of postcards at Kings Park and Botanic Garden in Perth identified only two images that depicted something other than flowering parts. The most suggestive exception is of two curvaceous tawny wattle seed pods split open and exposing an array of black seeds. The caption on the postcard notes the "diversity of form and unique colour" of South-West plants (Aspects of Kings Park, 2007). Through an ahistoric kind of visual aesthetics, the image conceals the rich history of wattles as a staple food for Nyoongar people who have baked the ground powder of the seeds into cakes (Daw, Walley, & Keighery, 1997, pp. 30-31) (see "Balance Lost and Found," Interlude III). On a different postcard, the Mangles Kangaroo Paw, the emblematic state flower of Western Australia, is washed in the pure azure blue of the sky (Aspects of Kings Park, 2007). The Western Australian photographer whose image appears on the wattle postcard, Jen Grey Wilson (cited in Australian Broadcasting Corporation, 2011), attributes visual appearance (colour, flamboyant display, exquisite visual detail, grace and harmony) to the evolutionary survival strategies of the flora:

> They are the oldest plants on earth, and when you get close...you find the most exquisite detail. They had to attract the pollinators because there were so few, they had great competition and they had to get more attractive colour-wise and more flamboyant in displaying themselves. (para. 2)

The profusion of flowering images on postcards points to the aestheticisation of flora in which multi-sensorial complexity (i.e., experience, sensation, context) is simplified to form and colour. Cultural theorist Mike Featherstone (1992, pp. 268-275) posits three definitions or "senses" of aestheticisation: (1) the effacement of the boundaries between works of art and everyday life, originating in the Dada and surrealist movements of the 1920s; (2) the transformation of life experience (or a living thing) into a work of art, entailing "a dual focus on a life of aesthetic consumption and the need to form life into an aesthetically pleasing whole" (p. 269); and (3) the saturation of everyday life with images of desire, central to the postmodern consumer engine of "sign-value," a term drawn from Baudrillard (p. 270). Reflecting Featherstone's first definition, chapter 5 called into question the conflation of plants and objects of art. Aestheticisation, in the context of tourism, is the process that drives focus on the visual beauty of wildflowers to the exclusion—or marginalisation of—immediate sense experiences and ecological contexts. The production of material culture—here, the wildflower postcard as a memento of ecotourist values—underpins this process of aesthetics, as a visual science.

As touristic commodities based on aestheticised plants, as per Featherstone's third definition, photographic images of botanical forms and colours saturate the visitor market and the public perception of flora. In these instances, plants *are* their aesthetically pleasing parts: their flowers. The Tourism Commission study (2003, p. 15) states that many travellers "often cited previous visually stunning experiences as their motivators for thinking about taking another wildflower holiday" and one participant described, with especially strong emotion, seeing a "carpet of flowers" engulfing visitors. The carpet of expansive everlastings is a high expectation of most respondents, who used rhetoric like "variety, colour, and vast expanses" as well as "carpet of colour" and "perfectly formed" to describe the wildflowers (Western Australian Tourism Commission, 2003, p. 16). Respondents suggest that the appeal is pictorial and Arcadian, including the vastness of "untouched" flowers as far as the eye can see, colour varieties and contrasts of colours against "barren" backdrops (Western Australian Tourism Commission, 2003). One recommends more "bright, colourful, scenic shots with scope and magnitude" to improve the *Wildflower Holiday Guide* (Western Australian Tourism Commission, 2003, p. 39).

Contemporary images of wildflower tourism include four-wheel drive vehicles in the background with visitors ambling among or kneeling down near flowers (Western Australian Tourism Commission, 2010). The visuality of the experience may be determined by differences of preference and comfort level amongst tourists. As Section II of this chapter alluded, enthusiasts are mainly interested in seeing different kinds of wildflowers and are prone to patronise shows and guided tours. Nature lovers are more inclined towards self-directed activities like bushwalking (Western Australian Tourism Commission, 2003) (see chapter 12). Wildflower enthusiasts are less motivated to walk, whereas nature lovers engage bipedality as a mode of intimate connection to place through the physicality of bushwalking and camping. So, while the activities of wildflower tourism are variable and unspecified, so too are the identities and demographics of contemporary wildflower tourists diverse.

Interlude XVI: Orchid Anima

Desire, delicacy and shape-shifting characterise orchid flowers. I found an intriguing expression of these qualities in the term *anima*, derived from the Latin *animare* meaning to enliven or inspirit. Carl Jung described anima as the feminine aspect of a man's psyche, and animus as the male part of a woman's psyche. In Jungian psychology, both archetypes constitute the wholeness of a personality (Sanford, 1980, p. 6). "Orchid Anima" was written in response to the profusion of donkey orchids (*Diuris* spp.) in the Lesueur-Eneabba region during the spring of 2009. The poem was inspired by a visit to Western Flora Caravan Park and an interview with local expert Allan Tinker. *Diurnis* stems from Greek for "double tail." The common name donkey orchid reflects the two prominent lateral petals' resemblance to a donkey's ears. John Lindley (1840, p. li) related to donkey orchids zoomorphically as birds: "The two lateral petals are broad and spreading like a pair of wings, the back sepal has the form of a bird's head and neck, and the whole resembles something from fairy land upon the wing." Crouching beside an orchid and waiting, I began to feel the harmonic interplay of opposites towards a sense of dynamic wholeness and transubstantiation: orchid anima | male animus; animal | plant; ether | matter; hammer | dulcet. In the second stanza, the orchid's beingness is neither harsh nor delicate, but necessarily defended by plectra:

> sanguine-yellow tremors in air,
> stammerings of petal-syllables
>   cheeks animated by the vivid flush
> of pigments bladdered in downpour,
> indelicate elementals, entirely
>   guarded by scorpion plectra (ll. 7-12)

Although the orchid can appear diaphanous, it is tenacious while insubstantial (to read "Orchid Anima" in full, see Hickman & Ryan, *in press*).

Anthoethnographic Themes at Wildflower Venues

Thus far, this chapter has treated the definition, history and contemporary visual rhetoric of wildflower tourism. The discussion now turns towards the outcomes of anthoethnography. Participant observation and semi-structured interviews were used to identify visual and embodied evaluations of flowering plants during the spring wildflower seasons of 2009 and 2010. Towards this purpose, I participated in wildflower tours and attended wildflower shows at churches and community centres. Interviewees included professional and amateur botanists, horticulturalists and wildflower enthusiasts or tourists (see chapter 2). This section highlights prevalent themes and specific settings: the everlasting and orchid effects; the role of science in the aesthetics of flora and wildflower tourism in the Lesueur area; embodied experience of the Proteaceae family of plants at Banksia Farm; and Fitzgerald River area wildflowers on display at the Ravensthorpe Wildflower Show.

According to interviewees, there are two predominant modes of viewing wildflowers and flowering landscapes: the orchid effect and the everlasting effect. Any mass of flowering conveys an everlasting effect in the expansiveness of the view, the accessibility and consistency of the colour, and the oceanic experience of the rippling distance consumed by red, yellow and pink hues. Orchid flowers of the Fitzgerald River National Park, however, are subtle, delicate and sometimes unnoticeable arrays of intricate colourations and aerodynamic shapes. They beckon the flower-seeker to get down on hands and knees and enter into the vegetal corpus, edging through *Daviesias* and *Melaleucas* with the proboscis of a lens. Similarly, the bold, drooping rose-like blossoms of the Qualup Bell are elegant and striking but require proximity, while carpets of everlastings, especially near Morawa and Mullewa in the north central area of the South-West, enfold the human viewer in an antediluvian sea of colour. Broadly speaking, carpets of everlasting flowers prevail in the northern sections of the South-West only and are not prevalent south of York, WA (see figure 7, chapter 2).

The everlasting and orchid effects are divergent modes of visually interpreting plants (see figures 20 and 21). The orchid effect in the southern part of the region requires a subtlety of perception and the willingness to interact viscerally with the bush by bending down and using magnifiers to perceive the architectures of morphologically minute flowers. Local botanist and coordinator of the Ravensthorpe Wildflower Show near Esperance, Merle Bennett comments that the everlasting effect is unrepresentative of the southern members of the South-West flora, particularly in Ravensthorpe Shire and the Fitzgerald River National Park:

> You sometimes see mass flowering of *Leptospermums* or *Kunzeas* at the side of the road. But on the whole, you don't get great masses of one thing flowering. You do in scattered areas but not to the same extent as the everlastings up north. (pers. comm., M. Bennett, September 13, 2009)

Figure 20 The Orchid Effect. Orchids require proximal attention that is conducive to multi-sensorial interaction and gestural outreach. *Image courtesy of the author.*

Figure 21 The Everlasting Effect. The expansiveness of everlastings in the Eneabba area is amenable to the picturesque mode of landscape appreciation. *Image courtesy of the author.*

Although the everlasting effect is the most panoramic, the orchid effect holds more biodiversity and requires proximity between vegetative and human bodies during appreciative encounters. Even at the apex of flowering, the landscape of the Fitzgerald River National Park outside of Ravensthorpe, perceived from a car, can appear uninviting. The everlasting effect comprises panoramic visualism and encourages disembodied viewership. It evokes theories of the imperial gaze of colonial explorers put forward by Carter (2010, pp. xx-xxii) and other post-colonial scholars. Carter (2010, p. 82) argues insufficiently, however, that the imperial gaze of the explorer is "phenomenological in nature. It is grounded, that is, in his recognition that he, the observer, does not gaze on the world as through a window, but rather inhabits it." Unlike Carter's concept of phenomenological encounter, the aesthetic gaze of the everlasting effect implies no inhabitation of a habitat. It supports a detached mode of spectatorship rather than bodily immersion. In contrast to the everlasting effect, the orchid effect makes appreciation possible through curiosity, attention and close physical engagement, diminishing the ordering powers of visual objectification by invoking the participatory senses. A comprehensive form of appreciation would shift between everlasting and orchid modes of perception: from seeing and gazing to feeling and tasting.

Proprietor of Banksia Farm in Mount Barker, Kevin Collins summarises the orchid and everlasting effects and consequences of these two modes of appreciating flowering landscapes for regional tourism:

> The tourism commission promote masses of flowers that occur north of Perth in the drier regions in good seasons when you receive rain. That's what most people are looking for, masses of everlastings. If it's a dry season without the rains, there's very little in flower up there and people head south into the more reliable rainfall areas. The flora is different and there are many more species but not generally massed and scattered through the bush. When flowering is poor up north, we get more visitors here. (pers. comm., K. Collins, June 4, 2012)

As a category of visual appreciation, the everlasting effect, though applicable to some regions of the South-West, is not a widely applicable mode of perceiving the flora of the region. Insofar as it challenges the conventions of panoramic visualism and galvanises physical engagement, the orchid effect affords opportunities for embodied appreciation, whereas the everlasting effect upholds the model of motorised viewership in which the plantscape is constituted as a scene.

Anthoethnographic enquiry further elucidates some of the dynamics between objective science and appreciation of plants through visual and multi-sensorial modes (see chapter 5). As mentioned in chapters 2 and 3, the Lesueur-Eneabba area comprises the species-rich low heaths of the sandplains, or the *kwongan* (Hopkins, Keighery, & Marchant, 1983). Although the area's species diversity results in year-round blossoming, flowering is most intensive during the spring months of August, September and October (Priskin, 2003, p. 504). Proprietor of Western Flora Caravan Park in Eneabba near Lesueur National Park, Allan Tinker states that the kwongan now comprises pockets of biodiversity in a land severely altered by clearing:

> We're down to 3% of the richest heathlands in the world left intact. That's very, very noticeable. Places like Lesueur are now termed *vegetative islands*. That's what they've become. They're islands in a landscape that is now being used for other purposes. (A. Tinker, pers. comm., August 29, 2009)

As botanists have noted, the Nyoongar term *kwongan* can be defined as "sandy country with open scrubby vegetation" (Beard & Pate, 1984, p. xvii). However, since *scrub* is used pejoratively to denote worthless bushland, Allan's interpretation of kwongan as bush that is the height of the eye is more nuanced. The Aboriginal etymology, mentioned by him during a group tour, connects bodily metaphor, scale and measurement to the botanical world.

The conjunctions between visuality, science and scenic values extend to Allan's conceptualisation of the aesthetics of flora. His discourse typifies the "natural environmental" model espoused by Allen Carlson as a scientific basis for an aesthetics of the natural world (Carlson, 2000) (see chapter 5). Accordingly, he explains in lavish detail the symbiotic relationship between the smoke bush

(*Conospermum* spp.) and the bee that pollinates it. Allan then illuminates the anatomical differences in seed structure between *Hakea* and *Grevillea*, even though the majority of tourists have come to view pleasing wildflowers rather than learn about scientific botany *per se*. For him, science provides to visual appreciation "the reality of what visitors are missing out there." The conservation outcomes of using scientific exegesis to broaden aesthetic judgments can be long-lasting, according to Allan:

> I would say 80% of the people have just come to see pretty wildflowers. It's our intent to send them away just a little more open minded about what they're seeing. We want to give them a better experience of wandering around in nature. (pers. comm., A. Tinker, August 29, 2009)

An emphasis on appearances—the prettiness or suitability of plants for photographs—constitutes shallow engagement with flora, in Allan's view. Through an account of physiology and ecology, ocularcentric apprehension is given meaning, substance and complexity. Tinker comments that the effects of using scientific models to broaden surface-oriented visual judgements are more enduring:

> My opinion is that if people know more about the reality of the system around them, they'll have more respect. And it pays off. We have a lot of people who come back who start off just coming here to look at the pretty wildflowers. They come back a couple years later and say 'I saw one of those over the road. I used to dump my rubbish over there. Now, we don't do that anymore because we know there are things there'. So this way of educating does have an impression on some. (pers. comm., A. Tinker, August 29, 2009)

Knowledge and appreciation of pollination, ecology of fire and the relation between geomorphology and plant life deepen how we regard, and ultimately conserve or destroy, the botanical world, according to Tinker's philosophy of teaching the public about plants. He goes on to claim that appreciation, as either visual aesthetics or participatory engagement, interferes with the attainment of sound scientific comprehension of the role of flowering plants in ecological systems:

> Human appreciation of flora is either aesthetic or edible. They're our two major forms of appreciating plants. It's about aesthetic beauty or whether the plant can be consumed. We forget that the only reason a plant has a flower is for its own reproduction. (pers. comm., A. Tinker, August 29, 2009)

Allan here reduces the appreciation of plants to sight and taste; interestingly, from his perspective, aesthetics is synonymous with visuality and edibility. Instead of more extended sense relationships, he uses scientific epistemology to instil in visitors a conservation ethos towards wildflowers. Although Allan formerly included information about the palatability of certain species, such as the edible tubers of orchids, he describes himself as "cautious...I very seldom talk about the edible side of the plant communities" and largely limits his discussions to their ecological adaptations (pers. comm., August 29, 2009). Part of Allan's inhibition in using information about the edibility of plants relates to concerns about visitor safety and the possibility that guests might try the wrong plant or have an adverse reaction to sampling a species.

Rather than setting up a binary opposition between vision and embodiment, participatory relationships based in bodily interaction can work dynamically with objective knowledges towards comprehensive forms of appreciation. Whereas Allan Tinker emphasises scientific rigour as the primary springboard from which to appreciate plants and the main impetus for deciding to conserve them, Kevin Collins fosters multi-sensorial interaction as a mode of educating the public about South-West species. Banksia are icons of the South-West, its life forms known for a variety of ecological adaptations to intense dryness and nutrient-poor soils (Collins, Collins, & George, 2008). Kevin Collins, founder and proprietor of Banksia Farm in Mount Barker, WA, epitomises the use of body-engaged (as compared to vision-engaged, body-disengaged) and multi-sensorial interaction. A walk with Kevin among his plantings and in his gallery occurs at the conjunction of flora, aesthetics, embodiment and anthoethnography. Rather than austere and inanimate, the Proteaceae—the hardy ancient family of plants to which banksia belong—are replete with perfumed aromas, enrapturing textures, nectareous tastes and the sound of wind whipping through the needle-like leaves of some species. Kevin crushes a handful of plant parts and passes his fist under my nose to inhale what might be called "Proteaceae Perfume," a blend of the aromas of three notoriously prickly plants. He speaks of the fragrance of the crushed leaves jocularly, considering the poetic irony that a family of plants known for its toughness and harshness—plants that symbolise the inhospitable appearance of the region—could produce sweet and elaborate aromas. Kevin's appreciation is not constrained to the visual qualities of the indigenous flora—their spatial assemblage in the garden according to colour and form—but is instead inclusive of their olfactory, gustatory, auditory and palpable characteristics (chapters 8 and 13 further discuss the approach of Collins).

Whereas strolling with Kevin yields an embodied experience of banksias *in situ*, a wildflower show is a static display of flowers removed from their habitats. A show is a collaborative, community-wide celebration of local botanical diversity, strongly focused on the visual qualities of flora. In contrast, a wildflower walk tends to engage multi-sensoriality (see chapter 12). The Ravensthorpe Wildflower Show showcases the flora of the Fitzgerald River National Park, and the species endemic to Hopetoun on the eastern boundary and Bremer Bay on the western side (see figure 22). Since 1982, Ravensthorpe has been home to one of the largest wildflower displays in the world, according to the number of species shown. Merle Bennett (pers. comm., September 13, 2009) comments that the committee members "normally advertise our show as having about 700 species. We have actually had up to 800 in one year, but years vary." The number of species represented in the show reflects the biodiversity of the Fitzgerald River area.

A wildflower show requires coordination of resources and cooperation between enthusiastic volunteers and professional local and state-wide botanists. At Ravensthorpe, flower specimens are brought in from the surrounding bushland by a team of volunteers, then meticulously placed in bottles, transformed into imaginative creations such as the Ravensthorpe Woman (see chapter 6), and arranged according to taxonomic family. A pleasing presentation is equally important to the taxonomic layout on the tables. Bennett, the central coordinator of the show and an elder figure in local botany circles, refers to the importance of using "visual images" and "species which are colourful and eye-catching" (pers. comm., September 13, 2009). Within the wildflower show committee, members specialise in species or families, demonstrating that the breadth of the flora necessitates an orchestration behind the show: "Helen does the hakeas and the

Figure 22 Royal Hakea on Display at the Ravensthorpe Wildflower Show. This iconic plant is one of up to 800 local plant species on display each year in Ravensthorpe. *Image courtesy of the author.*

banksias. I tend to do the grevilleas and orchids and Enid does other ones. We all work together to try to put names on." As mentioned in the discussion of Lindley in chapter 4, naming is an important first phase in classifying collected specimens and putting them out for public display.

Beyond the Flower: From Wildflower Tourism to Botanical Appreciation

Despite contemporary popularity as suggested by the Ravensthorpe Wildflower Show, South-West plants have not always been extolled for their beauty or represented as integral to the landscape and human settler activity. Seddon (2005) remarks:

> Appreciation of the unusual quality and beauty of the Australian flora has a long history, but not always a continuous one, and not one that has ever been fully translated into effective conservation practice as the current status of the *Banksia* genus in Western Australia indicates all too well. (147)

Some nineteenth century depictions portray Swan River Colony plants as artefactualised objects for the appreciation of discerning human subjects, rather than as autonomous forms of life worthy of conservation (see chapter 4). The wildflower tourism of the future, however, will encompass concepts of conservation, Aboriginal knowledges and connectivity, and the appreciation of plants beyond objects of art and through multi-sensoriality.

In augmentation of wildflower tourism, botanical appreciation suggests engagement including whole plants—bark, leaves, fruits and seeds—throughout the year and not only during the height of flowering. Whereas wildflower tourism as it has been practiced and conceptualised poses an aesthetics of static objects and scenes, botanical appreciation offers the possibility of appreciation based in diachrony, seasonal awareness, bodily sensation and ecological processes. Such a mode of appreciation especially reflects Thoreau's practice of cultural botany in which sustained physical experience of the land engenders a deeper relationship between human and botanical bodies (see chapter 1). Pelloe (1921, p. 15) comments that "all over Western Australia the display of wildflowers is especially extensive in the springtime, but the bush always holds something to interest the collector." In contemporary conservation language, a collector is an appreciator; hence the bush offers something to appreciate at all times of year. However, unlike scientific collectors of the colonial era for whom classificatory knowledge was key, contemporary appreciators are free to draw from a spectrum of appreciative means, including bodily experiences through the practice of eating wild plants (Clarke, 2008, pp. 81-90).

Due to their ecological adaptations, especially the general smallness of their foliage and flowers, South-West plants call for proximal appreciation to complement visual spectatorship. Pelloe (1921, p. 29) encourages the flower-seeker to "turn your back on the view. Otherwise, you will spend a lot of time gazing out over that wide expanse of country with the ocean gleaming in the distance, and possibly miss many of the floral treasures at your feet." Further, Priskin (2003, p. 518) observes that "wildflower appreciation requires walking in close proximity to plants, as numerous species require one to be less than one metre away for clear observation" (see chapter 12). Moreover, Hobbs and Hopkins (1990, pp. 93-114) argue that wildflower tourism has the potential to promote biodiversity conservation, threatened by habitat loss, the clearing of flora and over-harvesting. However, the embodied elements of wildflower appreciation, to include flora as a whole rather than flowers in isolation, need development in order to achieve the long-term benefits suggested by Hobbs and Hopkins.

In conjunction with technical understandings, Aboriginal spiritualities and material relationships to plants prompt directions for regional botanical appreciation. Nyoongar and Injabarndi elder Noel Nannup (pers. comm., July 21, 2010) explains that the aesthetics of flora can be deepened beyond visual appreciation towards the engagement of other senses and the understanding of the place of flowers in Nyoongar cosmologies. Aboriginal conceptions of flora are broader than the wildflower itself and encompass long-standing traditions between people and plants (see chapter 3). Knowledges of land and communities position flowers in ecological and cultural networks, from which a flower might be extracted as an object of contemporary wildflower focus. Habitats, flowers, animals and spirit intertwine mutualistically in traditional Nyoongar totemism (see chapters 3 and 13). For Noel, the aesthetic experience of flora can be deepened through educating tourists about Aboriginal understandings of flora as foods, medicines, fibres and totems, rather than as aestheticised items alone:

> If you look at a plant aesthetically, it's very pleasing. But you can add a new dimension to the discussion by talking about how you could make string out of the plant, how it was poisonous or how you used it as food or medicine. Suddenly there's a whole different understanding. (pers. comm., N. Nannup, July 21, 2010)

Noel articulates appreciation, but in different terms to Allan Tinker. Whereas botanical science fosters a conservation ethic for Allan, Aboriginal spiritual relationships impart a concern for the shared long-term wellbeing of plants and people through the concept of totems:

> You're also bringing in Aboriginal spirituality which is connected to the appreciation of plants because these plants we're talking about are all totems. In the old way they were someone's plant. (pers. comm., N. Nannup, July 21, 2010)

Through a framework of spirituality and physical interaction related to cultural heritage, Noel encourages the deepening of the nodes between the appreciation of wildflowers and cultural wellbeing. His statements promote the interconnectedness between plants, people and place, rather than the isolation of images as the aesthetic outputs of the tourism industry:

> As Aboriginal people looked after their totems, they had to know everything that is possible to know about that plant. That's what a totem is. You know the totem plant intimately. You know the relationship it has with other plants, animals, birds, even what pollinates it. (pers. comm., N. Nannup, July 21, 2010)

Somatic interactions with plants throughout the seasons have been part of Aboriginal cultural legacies in the South-West (see chapters 2 and 3). Intimacy results from participation in habitats. Hence, spirituality, embodiment and ecological knowledge are dialogical rather than in conflict. Networked botanical knowing can inspirit the detached approach of tourism and its emphasis on flowering. Noel's vision of replanting the traditional Dreaming Trails expresses botanical reconnection and the reformation of land-based cultural networks. It is also a statement of collective solidarity and the healing of communities: "As we follow the trails along, our youngsters are going to want to do something for their land so we can get them to plant back on the Dreaming Trails" (pers. comm., N. Nannup, July 21, 2010). If such trails would be accessible to the tourism populace, spirituality would be brought to bear on the appreciation of flowering beauty. Whereas Allan Tinker encourages the metanarrative of science and Kevin Collins emphasises bodily participation, Noel Nannup holds that deeper appreciation can occur through Aboriginal frameworks in which the flower is networked broadly within habitats, spiritualities and sustenance.

The Promise of Anthoethnography

How does anthoethnography contribute to the development of understandings of aesthetic experiences of wild plants and wildflower tourism? As exemplified by the quintessential aesthetic industry of wildflower tourism, the culture of flora represents diverse engagements between people, plants and place. Such complexities offer further avenues for research into the cultural botany of the region and elsewhere. The critical approach of anthoethnography has been one such pathway to circumscribing the values, practices and rhetoric of wildflower tourism. Interviews reveal perceptual phenomena, such as the orchid and everlasting effects as two modes of visual aesthetics. For appreciators such as Allan Tinker, botanical science substantiates appearances by showing the functional role of plants. However, the taxonomic eye is not the only judge of the value of flowers. The approach of Kevin Collins suggests an embodied aesthetics in an educational setting. Furthermore, as underscored by Noel Nannup, Aboriginal perspectives proffer complex cultural modes of engagement and promising directions for tourism based in haptic experience and spirituality. Conceptualisations of plants as food, medicine and fibre, conjoined to totemic values, will be integral to the development of sustainable tourism.

An anthoethnographic approach produces accounts of the spectrum of human perceptions of wildflowers in order to outline directions for tourism of the future. Through a participatory aesthetics in contemporary Australian landscapes, appreciative interactions will occur not only through visual values, but also through the smell, taste, sound, or feel of plants and how one moves through communities of flora (see chapter 13). Scientific knowledge can amplify ocular and corporeal modes. However, as an anthoethnographic lens elucidates, wildflower tourism in the region is weighted towards visuality. Indeed, the history and contemporary practices of wildflower tourism encode ocular values that pose separation between post-colonial cultures and indigenous flora. A hopeful direction is towards participatory relationships beyond the aestheticisation of the surface qualities of flora and beyond the "conquest of the world as picture" (Heidegger, 1938/2009, p. 221). In an era of rapid species loss, wildflower tourism will need to embrace concepts of conservation, Aboriginal knowledges and the recognition of spiritual heritages, and the appreciation of plants beyond their visual impact towards appreciation of whole plants over time. The expression of human sensory capacities joined to an ethos of botanical conservation, drawing from scientific thought, can better ensure the longevity of flowers through the evolution of the culture of flora in the region.

References

Aspects of Kings Park. (2007). *Kings Park and Botanic Garden, Western Australia postcard.* Perth, Australia: Kings Park.

Australian Broadcasting Corporation. (2011). Transcript: Capturing the beauty of wildflowers in photos, broadcast 09/09/2005 Retrieved April 10, 2011, from http://www.abc.net.au/stateline/wa/content/2005/s1458437.htm

Beard, J., & Pate, J. (1984). Foreword. In J. S. Beard & J. S. Pate (Eds.), *Kwongan, plant life of the sandplain: Biology of a south-west Australian shrubland ecosystem* (pp. xvii-xxi). Nedlands, Australia: University of Western Australia Press.

Bickerton, E. (1905, February 18). Along the Midland railway, Perth to Geraldton, a lady's journey [Australia Trove]. *Western Mail*, 39-40. Retrieved from http://trove.nla.gov.au/

Blamey, R. (2001). Principles of ecotourism. In D. Weaver (Ed.), *The encyclopedia of ecotourism* (pp. 5-22). New York, NY: CABI Publishing.

Botanic Gardens and Parks Authority. (2009). Kings Park Festival over all too soon! Retrieved 15 May, 2009, from http://www.bgpa.wa.gov.au/o/content/section/14/74/

Buckley, R., Pickering, C., & Weaver, D. (2003). *Nature-based tourism, environment, and land management*, Volume 2001. Wallingford, England: CABI Publishing.

Carlson, A. (2000). *Aesthetics and the environment: The appreciation of nature, art, and architecture*. New York, NY: Routledge.

Carter, P. (2010). *The road to Botany Bay: An exploration of landscape and history*. Minneapolis, MN: University of Minnesota Press.

Clarke, P. (2008). *Aboriginal plant collectors: Botanists and Australian Aboriginal people in the nineteenth century*. Dural Delivery Centre, Australia: Rosenberg Publishing.

Collins, K., Collins, K., & George, A. (2008). *Banksias*. Melbourne, Australia: Blooming Books.

Crary, J. (1990). *Techniques of the observer: On vision and modernity in the nineteenth century*. Cambridge, MA: MIT Press.

Daw, B., Walley, T., & Keighery, G. (1997). *Bush tucker plants of the South-West*. Kensington, Australia: Department of Environment and Conservation.

Department of Environment and Conservation, Shire of Dalwallinu, & Australian Tree Seed Centre. (2011). Wattle Day (1 September) Retrieved April 9, 2011, from http://www.worldwidewattle.com/infogallery/symbolic/wattleday.php

Florida Wildflower Foundation. (2011). Florida's native wildflowers Retrieved April 9, 2011, from http://flawildflowers.org/

Giblett, R. (2011). *People and places of nature and culture*. Bristol, England: Intellect.

Goldgar, A. (2007). *Tulipmania: Money, honor, and knowledge in the Dutch Golden Age*. Chicago, IL: University of Chicago Press.

Goody, J. (1993). *The culture of flowers*. Cambridge, England: Cambridge University Press.

Grimwade, R. (1920). *An anthography of the eucalypts*. Sydney, Australia: Angus & Robertson.

Hall, C., & Page, S. (2006). *The geography of tourism and recreation: Environment, place and space* (3 ed.). New York, NY: Routledge.

Heidegger, M. (1938/2009). The age of the world picture (J. Veith, Trans.). In G. Figal (Ed.), *The Heidegger reader* (pp. 207-223). Bloomington, IN: Indiana University Press.

Hickman, E., & Ryan, J. (*in press*). *Two with nature: Ellen Hickman and John Ryan.* Fremantle, Australia: Fremantle Press.

Hillman, A. (1990). *The Hillman diaries 1877-1884: The personal diaries of Alfred James Hillman for 21 December 1877 to 24 April 1884.* Applecross, Australia: F V Bentley.

Hobbs, R., & Hopkins, A. (1990). From frontiers to fragments: European impact on Australia's vegetation. *Proceedings of the Ecological Society of Australia,* 16, 93-114.

Hopkins, A., Keighery, G., & Marchant, N. (1983). Species-rich uplands of south-western Australia. *Proceedings of the Ecological Society of Australia,* 16, 93-114.

Hopper, S. (1998). An Australian perspective on plant conservation biology practice. In P. Fiedler & P. Kareiva (Eds.), *Conservation biology for the coming decade* (2 ed., pp. 255-278). New York, NY: Chapman Hall.

Knight, P. (1986). *Flower poetics in nineteenth-century France.* Oxford, UK: Oxford University Press.

Levin, D. (1993). Decline and fall: Ocularcentrism in Heidegger's reading of the history of metaphysics. In D. Levin (Ed.), *Modernity and the hegemony of vision* (pp. 186-217). Los Angeles: University of California Press.

Lindley, J. (1840). *A sketch of the vegetation of the Swan River colony.* London, England: James Ridgway.

Loubster, G., Mouton, P., & Nel, J. (2001). The ecotourism potential of herpetofauna in the Namaqua National Park, South Africa. *South African Journal of Wildlife Research,* 31(1-2), 13-23.

Lyon, J., & Nystrom, A. D. (2002). *USA* (2 ed.). Melbourne, Australia: Lonely Planet Publications.

Marchant, N. G. (2005). Documenting the diverse Western Australian flora. In *Black solander, Gregory Pryor.* Perth, Australia: Perth Institute of Contemporary Arts.

Mules, W. (2006). Contact aesthetics and digital arts: At the threshold of the Earth. *Fibreculture, 9* Retrieved from http://nine.fibreculturejournal.org/fcj-058-contact-aesthetics-and-digital-arts-at-the-threshold-of-the-earth/

Newsome, D., Moore, S., & Dowling, R. (2002). *Natural area tourism: Ecology, impacts and management.* Tonawanda, NY: Channel View Publications.

Pelloe, E. (1921). *Wildflowers of Western Australia.* Melbourne, Australia: C.J. De Garis.

Priskin, J. (2003). Characteristics and perceptions of coastal and wildflower nature-based tourists in the Central Coast region of Western Australia. *Journal of Sustainable Tourism, 11*(6), 499-528.

Robin, L. (2002). Nationalising nature: Wattle days in Australia. *Journal of Australian Studies*, 73, 5-23 Retrieved from http://www.api-network.com/jas/pdf/libby_robin.pdf

Rose, D., & Robin, L. (2004). The ecological humanities in action: An invitation. *Australian Humanities Review, 31-32* Retrieved from http://www.australianhumanitiesreview.org/archive/Issue-April-2004/rose.html

Sanford, J. (1980). *The invisible partners: How the male and female in each of us affects our relationships.* New York, NY: Paulist Press.

Seddon, G. (2005). *The old country: Australian landscapes, plants and people.* Cambridge, England: Cambridge University Press.

Summers, L. (2011). Wildflower season: The development of native flora protection legislation in Western Australia, 1911-1975. In A. Gaynor & J. Davis (Eds.), *Studies in Western Australian History: Environmental Exchanges, 27*, 31-44.

The West Australian. (1893, September 2). News and notes [Australia Trove]. *The West Australian* Retrieved from http://trove.nla.gov.au/ndp/del/home

The West Australian. (1899, August 30). News and notes [Australia Trove]. *The West Australian* Retrieved from http://trove.nla.gov.au/ndp/del/home

Tourism Australia. (2010). Western Australia's wildflowers Retrieved April 9, 2011, from http://www.australia.com/articles/wa_wildflowers.aspx

Turizem Bohinj. (2010). 5th international wildflower festival Retrieved April 9, 2011, from http://www.bohinj.si/alpskocvetje/eng/index.php

Valley of the Giants. (n.d.). Denmark Walpole wilderness forests and ocean coastline scenic tours Retrieved April 9, 2011, from http://valleyofthegiants.com.au/information.php

Vivienne, M. (1901). *Travels in Western Australia: Being a description of the various cities and towns, goldfields, and agricultural districts of that state.* London, England: William Heinemann.

Western Australian Tourism Commission. (2003). *Research into the wildflower tourism industry: An exploration of consumer and industry perceptions.* Perth, Australia: Western Australian Tourism Commission.

Western Australian Tourism Commission. (2010). *Wildflower holiday guide Western Australia: Your guide to the largest collection of wildflowers on Earth.* Perth, Australia: Tourism Western Australia.

Wildflower Society of Western Australia. (2010). Country wildflower shows. *Wildflower Society of Western Australia Newsletter, 48*(2), 41.

Williams, R. (1982). *Socialism and ecology.* London, England: Socialist Environment and Resources Association.

Wilson, E. O. (1984). *Biophilia: The human bond with other species.* Cambridge, MA: Harvard University Press.

8

Memory

Abstract: This chapter will uncover the nexus of ecology, sensory experience and emotion through the concept of botanical memory. Building upon theoretical precedents in environmental memory, collective memory, sensory memory, bodily memory, and emotional geography, botanical memory will be described as an important cultural convergence between plants and people. An ethnographic approach to the description of botanical memory draws from transcripts of interviews conducted during the spring wildflower seasons of 2009-10 in the South-West. The discussion will conclude by stressing that botanical memory is a heterogeneous blend of emotionality, sensuousness and embodiment. Research into this form of human memory provides a promising wellspring for revealing sense-rich emotional connections to flora.

> *Red gum everywhere! Fringed leaves dappling, the glowing new sun coming through, the large, feathery, honey-sweet blossoms flowering in clumps, the hard, rough-marked, red-bronze trunks rising like pillars of burnt copper, or lying sadly felled, giving up the ghost. Everywhere scattered the red gum, making leaves and herbage underneath seem bestrewed with blood.*
> D.H. Lawrence and Mollie Skinner, *The Boy in the Bush* (1924/2002, pp. 92-93)

Remembering Plants

This chapter theorises botanical memory through the exploration of environmental memory, multi-sensoriality and human emotion in the biogeographically diverse terrain of the South-West. As chapter 7 discussed, since the establishment of the Swan River Colony in the 1800s, the South-West of Western Australia has been a popular wildflower tourism destination during the spring months of September and October in particular when iconic species, such as kangaroo paws, wreath flowers and everlastings, blossom (Summers, 2011). However, despite relatively recent recognition of the global importance of its biodiversity, the South-West has been dramatically altered through modern agricultural expansion, urban development, and anthropogenic plant diseases (Beresford et al., 2001).

Connecting individual remembrance to collective remembering about thriving or declining plant environments, botanical memory, as will be discussed here, entails bodily and cultural memory. The fragrances, sounds, tastes and tactile sensations of plants summon this form of environmental memory that may be both shared amongst individuals and invoked through physical interaction with flora. Characterised by collective and bodily traces, botanical memory broadens the study of individual sense-based recollection of flora to the human communities living in proximity to indigenous plants. Memories can be of single plants or wildflowers (e.g., wreath flowers); communities of plants (e.g., an everlasting field); or a landscape, region or broader botanical scale (e.g., the Wheatbelt or the South-West region itself). This chapter will refer to all three scales. Recollections of flora are infused with sensory overtones, and are created by bodies and emotions. Such memories are derived not solely from inner imagery but from the sensuousness of plants and places. For example, in a study of the nexus between olfaction and memory, Waskul, Vannini and Wilson (2009) observe the capacity of odours to catalyse memories through nostalgic feelings. One respondent notes that wild roses remind her of childhood visits to her grandmother's house in Alberta, Canada, while another interviewee relates the smell of lavender to the comfort of "family, home, and safety" experienced as a child (Waskul, et al., 2009, p. 11). These statements suggest that botanical memory can be defined as remembrance of plants in the context of sensory, cultural, environmental and familial memories.

As propounded here, botanical memory derives a theoretical framework from related precedents in environmental memory (Chawla, 1994); sensory memory (Seremetakis, 1994); sensory ethnography (Howes, 2003; Pink, 2009; Stoller, 1989, 1997); bodily memory (Casey, 2000; Connerton, 1989); community and collective memory (Boyer & Wertsch, 2009; Hua, 2009); and emotional geography (Bondi, Davidson, & Smith, 2005; Jones, 2005). This genre of environmental memory involves sensory and emotional interactions with plants. Yet, whereas allusions to memories of plants appear in ethnobotanical literature, they are nearly unheard of in emotional geography, community memory research or ecocultural studies where, for example, research into the plants-people nexus is governed by the present tense of the ethnographic encounter (for example, Hitchings, 2003; Hitchings & Jones, 2004; Hoffman & Gallaher, 2007; Martin, 2004).

Despite this underemphasis, memories of plants strongly inform human emotional and collective cultural bonds with environments. For early travellers in Western Australia like May Vivienne (1901) or contemporary wildflower tourists to the South-West, the memory of a place may be indistinguishable from the memory of its plants. Amateur botanist Ayleen Sands expresses the intertwining of memory and flora as she recalls her first encounter with local orchids early in her proprietorship of Stirling Range Retreat, near the Stirling Range National Park:

> So we came in February 1996 and in April one of our first orchids comes out. It's a little orchid which in the years it flowers doesn't have a basal leaf (Leafless Orchid, *Praecoxanthus aphyllus*). Someone came to the office and said to me 'Oh, your orchids are out, did you know?' and took me down to show me and I was absolutely rapt. (pers. comm., A. Sands, September 8, 2009)

Sketched as emotion-rich memories, Ayleen's rapture exemplifies the felt engagements between people and flora. Initially, her arrival in the mountainous and biodiverse Stirling Range area from urban and suburban Perth, four hundred kilometres away, entailed anxiety about displacement: "I have to confess that I was really unhappy about coming this far away from Perth initially." Ayleen's statements recall the colonial botanist Georgiana Molloy's experience of living in the isolated settlement of Augusta, Western Australia, in the nineteenth century. For both women, wildflowers attain emblematic significance as consolations for displacement, distance and loss (Lines, 1994). Like Molloy, through the flowering of orchids, Ayleen has developed an emotional bond and a sense of place, gestated in part by the local flora of her immediate surrounds. With relief, she describes herself now as "very comfortable with the bush. I just absolutely love it. I love the way everything is interrelated."

This conceptualisation of botanical memory will draw primarily from interviews with amateur botanists and wildflower tourists (see chapter 2). The interviews suggest the recursive relationship between mourning and celebration, as well as the role of sensory memory of plants. Some conversations show significant emotional and embodied affinities, including mourning the loss of species and the homogenisation of biodiverse habitats through suburban overdevelopment or invasive exotic plants. Whereas wildflower tourists tend to emphasise the beauty of flowers, local residents—who can be scientific, amateur, eclectic and Aboriginal botanists—tend to convey a sense of despair over imperiled plant species and diminishing habitats. This is particularly the case for land owners and others who are materially attached to a specific property. The emotion of grieving surfaces more for interviewees, such as David James, Noel Nannup and Don Williams, who have lived for an extended time at a locale, property or multiple places in the South-West. These individuals exhibit a well-developed sense of place and have endured more long-term changes and floristic extinctions than seasonal tourists have. Indeed, interviewees do tend to construct and communicate memories differently depending on the longevity of their emotional bonds to the changing landscapes of which the plants they know are part.

As a caveat, however, the firm distinction between mourning and celebration—and between residents and tourists—is fuzzy at best and emotional complexities are to be expected. For instance, long-term visitors—such as Jack and Lena discussed in the context of the Stirling Range National Park and tourism proprietor Ayleen

Sands—can express emotions of loss over changes in the landscape tinged with feelings of appreciation for the rare orchids that remain in remnant parcels of bush. Thus, for the sake of structuring the analysis of interviews, the categories of mourning, celebration and embodiment will emphasise the themes stressed by the speakers themselves, although these categories are not mutually exclusive. Interviewees who are not included were perhaps less inclined to broach emotional or experiential matters when the interview format appeared formal, where trust was not developed fully, or where the paradigm of scientific objectivity—the pre-eminent discourse of botany—held back the expression of emotional bonds to flora. Such concerns should be scrutinised in further studies of ethnography as a method for researching botanical memory.

Interlude XVII: Three Moments – After Nerved Hakea

A calligram is writing in which the visual design of the words relates directly to the content or meaning of the text. A taxonomic description is calligrammatic. Botanical keys structure plant knowledge flows towards the Latinate epithet. "Three Moments - Nerved Hakea" attempts to revision the scientific plant calligram and implement the bodily plant calligram. As my conversations with botanists and plant enthusiasts suggest, the dominant modern language of plants is based on families, genus, species, biochemical constituents and geographical distributions. A guiding purpose of this field approach is to find the sensuous language of plants through corporeal experience, memory and the re-conceptualisation of plants as culturally embedded. Known as *Djanda* in Nyoongar, Nerved Hakea (*Hakea neurophylla*) is a localised indigenous species that only grows in Lesueur National Park and the Eneabba area. According to *FloraBase*, the online repository of Western Australian botanical data, the species name *neurophylla* derives from Greek *neuro* for nerve and sinew, and *phyllon* for leaf, denoting the prominent nerves in its leaves (Spooner, 2003). In the sunlight, the veins of Nerved Hakea glow like its cousin Royal Hakea (see "A Colony of Royal Hakea," Interlude XXII). The electricity of the nervous system was the first sensation I associated with the plant, inducing memories while camping near a grove of plants near Lesueur National Park.

Three Moments – After Nerved Hakea

I.
in a meadow amongst the gum-nuts
slick underfoot like marbles on tiles,
a campsite—cold water shower upslope
loo improvised from a trip to the tip—
faucet with drain plugged up—a moat
of soapy scum water—a minor calamity of

saponin—rubbish bins tucked under wispy
eucalypts—outlines of tents disappearing
into tall grass—looking for level ground
and the promise of repose in the late
winter pith—condensing as the sun dips
below the heathland hills of Lesueur

II.
awakened by muffled explosions,
in chiaroscuro between tender reverie
and stark reality—a flurry of sparks, like logs
thrown in the fire—a pyrotechnics—
startled like Dowager Gong Sheng in the 11th C—
lying supine, vulnerable as a newborn
lamb— encased by the hissing arc—
half-thinking the hubbub is a noctural
marsupial snuffing nearby—rooting
for ants or worms—sticking its snout rudely
into my centre belly chasm

III.
dew enters a worn out
power cord patched
with blacktape

The Physiology of Human Memory

What is the physiology of botanical memory in the context of contemporary research into memory studies, some of which emphasizes sensory or bodily memory? Memory has been well explored in literature, philosophy and art, but some accounts of memory are acutely visual and cognitive. Gaston Bachelard's (1971, p. 101) notion of "reverie," for example, involves "the beauty of the first images," rendering the immediate world "completely colourless." Similarly, for John Dewey (1967, p. 154), the mind produces memory as "knowledge of particular things or events once present, but no longer so." Created by the intellect, memory is an "active construction by the mind of certain data" (Dewey, 1967, p. 157). However, memory is not simply a visual or cognitive process, but can also be understood phenomenologically as the somatic extension of feeling into the world (Casey, 2000; Connerton, 1989). We recollect plants through corporeal sensations as smells, tastes, textures and sounds, as well as through the rhythm of walking in botanical sanctuaries (Hitchings & Jones, 2004). Thus, bodies are physiological and emotional—as well as natural and cultural—sites of memory expression.

Research of the last twenty years in the field of memory studies argues similarly that memory is not purely a cognitive construct of psychology or history, but a multi-sensorial faculty of the body, environment and culture (Boyer & Wertsch, 2009; Casey, 2000; Chawla, 1994; Connerton, 1989; Hamilakis, 2010; Hua, 2009; Jones, 2005; Seremetakis, 1994; Stratford, 1997; Waskul, et al., 2009). The embodied aspects of memory are particularly important to the term "memory work," introduced in the 1980s by the Frauenformen, a group of West German feminists interested in how various forms of femininity are imposed on female bodies (Stratford, 1997, p. 207). Memory work links bodily experience to memory research as a method for examining "how we are socialised in and through our bodies—complex sites that are 'natural', 'cultural', and intimately connected to how we experience and construct our worlds" (Stratford, 1997, p. 214). As an interrogation of knowledge production derived through privileged scientific and social discourses, memory work entails the exploration of embodied subjectivities: sensory experiences, bodily knowledges, first-person accounts of the world, placed-based stories and narratives of interconnectedness between natural and cultural milieus. Thought to "elicit 'only' the anecdotal and folkloric," place memory in the context of embodied subjectivity instead provides "richly heterogeneous stories" that attain their own validity and importance (Stratford, 1997, p. 214).

Botanical memory can also comprise bodily memories. Casey (2000, 147) defines bodily memory as "memory that is intrinsic to the body, to its own ways of remembering: how we remember in and by and through the body." While bodily memory is based in the senses, memories of the body can often be visual because sight "subtends most acts of recollection" and "has the effect of blurring the distinction between body memory and memory of the body" (Casey, 2000, p. 147). In *How Societies Remember*, Paul Connerton (1989, pp. 72-73) distinguishes between "incorporating" and "inscribing" practices of memory. Incorporating practices involve transmissions between bodies—for example, smiles, handshakes or words spoken in conversation—that occur when bodies are co-present. These practices produce "a mnemonics of the body" (Connerton, 1989, p. 74). Thus, for example, the memorisation of the correct posture for a ceremony and the subsequent transmission of a cultural practice involving that posture occur through words or gestures between people present to one another. In contrast, inscribing practices require memory storage technologies, such as books, photographs, and computers, that "do something that traps and holds information, long after the human organism has stopped informing" (Connerton, 1989, p. 73). Yannis Hamilakis (2010, pp. 191-192) problematises Connerton's categories, arguing instead for incorporating practices of memory that necessitate material culture. She cites eating and drinking as examples of mnemonic behaviors involving habits and gestures transmissible between individuals as well as the inscriptive materials of memory (i.e., the table, utensils, food, etc.). Hamilakis (2010, p. 192) argues that it is "time to collapse his binarism, which will also mean the collapse of his implied distinction between habitual, 'internal' memory which supposedly leaves no traces [as material artifacts, such as books, audio recordings or memorabilia], and 'external' material memory with its associated recording devices."

The dynamics between visual and bodily memory are further taken up by some cultural scholars. The anthropologist Nadia Seremetakis (1994) in *The Senses Stilled* argues that memory is not constrained to ocularcentric cognition, but is rather mediated by ongoing cultural and bodily (re)enactments:

> Memory cannot be confined to a purely mentalist or subjective sphere. It is a culturally mediated material practice that is activated by embodied acts and semantically dense objects. This material approach to memory places the senses in time and speaks to memory as both meta-sensory capacity and as a sense organ in-it-self. (9)

A material practice of memory through "semantically dense objects" reflects the notion of the indivisibility of "objects, settings, and moods" as integral to the formation of environmental memory (Chawla, 1994, p. 1). Seremetakis's notion of "sensory memory" augments environmental memory by emphasising its embodied nestings. Seremetakis (1994, p. 9) explains that the senses and memory are coterminous and "co-mingled." The senses coalesce to produce botanical memory, which is not purely a recitation of aestheticised plant images, constituting what Connerton (1989, p. 73) refers to as an "inscribing practice" of memory. Along with pictorialisations, the smells, tastes, tactile sensations and sounds of plants impart corporeal qualities to the act of remembrance (for example, Lawrence & Skinner, 1924, pp. 92-93). Paul Stoller (1997, p. 85) also suggests that "the human body [and non-human bodies, I would add] is not principally a text; rather, it is consumed by a world filled with smells, textures, sights, sounds and tastes, all of which trigger cultural memories." Sensory research into cultural practices, for Stoller (1989, p. 8), entails a post-Kantian critique of the alignment between vision and reason characteristic of much ethnographic writing. Research into human sensory experience in its fullness attends to "the sensual aspects of the field," making us "more critically aware of our sensual biases and [forcing] us to write ethnographies that combine the strengths of science with the rewards of the humanities" (Stoller, 1989, p. 9).

Memory furthermore engages collective abilities to interpret embodied events and experiences. Hence, memory is not only the experience of a body in a place but of bodies in places; the content of memory varies with collective cultural meanings and values. Whereas individual memory is accepted in psychology and physiology, collective memory has been introduced relatively recently into human geography, cultural studies, sociology and anthropology to describe memories held in common by a community or social group (Boyer & Wertsch, 2009; Hua, 2009). Collective memory "requires a public re-interpretation of personal memories that are placed at the service of the collectivity" (Hua, 2009, p. 137). Boyer and Wertsch (2009) observe that collective memory's primary shortcoming is its assumption of monolithic collectivity in the form of homogenised countries, communities or social groups. Collective memory raises the question of "whose voice, experiences, histories and personal memories are being forgotten" by the community, institution or nation (Hua, 2009, p. 137).

Memory can be bodily and collective as well as emotional and topographical. Collective memory, in particular, offers interesting intersections with emotional geography, a field which attempts to understand the spatial aspects of memory and its associations with environments (Bondi, et al., 2005). The emotional content of memory suggests that "memories always will have a spatial frame (even if it is unremembered or latent) and they will be always emotionally coloured in hues ranging from pale to vivid" (Jones, 2005, p. 210). Emotions construct how people make sense of the natural world through memory. Jones (2005) argues that the call towards emotion in landscape research is

> part of the movement away from the claim that knowledge is, and should be, an abstract, disembodied, purely rational and objective construct. It recognises the role of emotions in the *construction* of the world, and in *interpretations* of the world [italics in original]. (207)

Emotional geography responds to the objective imperatives of empirical research by expressing "something that is ineffable in such objectifying languages, namely a sense of emotional involvement with people and places, rather than emotional detachment from them" (Bondi, et al., 2005, p. 2). Urry (2005) further argues that the emotions of place have been given visual form through modern ocular technologies. In the mid nineteenth century the popularity of the Claude glass, advances in sketching techniques and developments in pre-photographic technologies spurred a "particular visual structure to the emotional experience of place" (Urry, 2005, p. 78). Indeed, when conceptualised as landscape, place and its associated emotional hues assume a visual structure. Stephen Daniels and Denis Cosgrove (1988, p. 1) define the term "landscape" as "a cultural image, a pictorial way of representing, structuring or symbolising surroundings." An iconographic orientation toward landscape values images of places and historically conceived such "pictures as encoded texts to be deciphered by those cognisant of the culture as a whole in which they were produced" (Daniels & Cosgrove, 1998, p. 2) (see chapter 10).

Yet, not solely a space or landscape demarcated by sight, place is a sensory and corporeal topography. Sensory scholarship of the last twenty years in the anthropology of the senses recognises this complexity and responds to the historical hegemony of vision in structuring sense of place and human perception (Classen, 1997; Hamilakis, 2010; Howes, 2003; Stoller, 1989, 1997; Synnott, 1991). This scholarship acknowledges the importance of the senses to human experience and stresses that cultural research should acknowledge the body as a sensorium in which sense interplay moulds cultural values and practices. Sensory experience, as such, is "the basis for bodily experience. We experience our bodies—and the world—*through* our senses [italics in original]" (Classen, 1997, p. 402). In *Sensual Relations*, David Howes (2003, p. xi) argues that sensation is a cultural locus and that, in preferencing vision, Western anthropologists of the twentieth century also tended to study each of the five senses in isolation "as though sight, hearing, smell, taste, and touch each constituted a completely independent domain of experience." Howes (2003, p. xi) suggests that dynamics between the senses create culturally important "combinations and hierarchies" of sensory experience and expression. In contrast, ocularcentric research obscures "sensory meaning—the associations between touch and taste, or hearing and smell—and all the ways in which sensory relations express social relations" (Howes, 2003, p. 17).

Botanical Memory at the Intersection of Embodiment, Environment, Emotions and Ethnography

> When from a long distant past nothing subsists, after the people are dead, after the things are broken and scattered, taste and smell alone, more fragile but enduring, more substantial, more persistent, more faithful, remain poised a long time, like souls, remembering, waiting, hoping, amid the ruins of all the rest; and bear unflinchingly, in the tiny and almost impalpable drop of their essence, the vast structure of recollection.
>
> Marcel Proust (1913/2008, p. 30)

Proust argues that smell and taste affect memory with particular poignancy and primeval endurance. As such, memory not only comprises a visual theatre, but a sensory interweaving of tastes, smells, tactile sensations, sounds and visible features. Jonah Lehrer (2007, p. 80) argues that "one of Proust's deep insights was that our senses of smell and taste bear a unique burden of memory." The potent connection between smell and memory, exemplified by Proust, has a physiological basis and has often been labelled the "Proustian hypothesis of odor memory" (Engen, 1982, p. 98). Receiving brain stimuli in the form of odours, the olfactory system of mammals regulates and integrates emotional responses, social behaviors and bodily memories (for example, Engen, 1982, pp. 25-29, pp. 97-112; Engen, 1991; Herz & Cupchik, 1995; Lledo, et al., 2005). In this passage from *Swann's Way*, Proust articulates the phenomenon of odour recognition or "knowing that an odor being experienced is one that was experienced on a earlier occasion" and, importantly, he intimates the possibility of odour recall, debated amongst physiologists and defined as the bringing back of "odor sensations from memory storage without any external aids" (Engen, 1982, pp. 14-15).

Extending Proust's metavisual mnemonics and the strong association between smell and memory he describes in particular, botanical memory can be said to comprise sensory, bodily and emotional modes of remembering. Botanical memory goes beyond the individual psyche and the imagistic proclivity of the aesthetic imagination. Anh Hua (2009, p. 137) characterises memory as "a construction or reconstruction of what actually happened in the past," but memory in practice is a composite of bodily experiences and sensations, mental images, cultural values and environmental stimuli in addition to a chronology of occurrences. Indeed, memory is beyond individual capacities and faculties. It invokes all mnemonic modes to produce "tangled memory" in which collective desires and needs are expressed (Hua, 2009, p. 139). Hence, memory is a nexus of environmental, social and personal factors, rather than purely a mental function of recall.

Chawla (1994) defines environmental memory as the convergence of the natural world, poetic deliberation and childhood memory, including:

> All the fittings of the physical world that surround us: the natural world of animal, vegetable, and mineral, and the built world of human artifice. Its scope covers three dimensions of perception: individual objects; settings such as home, city, and region; and global moods or feelings for the world. These three dimensions—objects, settings, and moods—may be isolated for study, but in lived experience they are inseparable. (1)

Emotional responses to a landscape constitute environmental memories of particular natural and cultural phenomena. Memory and emotions, or moods, are inseparable; environmental memory develops from a stratum of feelings held in the body. So, in addition to temporal, spatial and material memories aligned to the processes of thinking, the content of memory is a sensory and emotive scape, or an "emotional geography." Liz Bondi, Joyce Davidson and Mick Smith (2005, p. 3) further characterise the central aim of emotional geography as endeavouring "to understand emotion – experientially and conceptually – in terms of its socio-spatial mediation and articulation rather than as entirely interiorized subjective mental states." Transcending individual cognition and the construction of memory as the outcome of the rote processes of the brain, memory is influenced by society and space, culture and landscape, collective and personal proclivities. Reflecting Chawla's claim for the indivisibility of "objects, settings, and moods" as unified lived experience, memory is an anatomical topography, orchestrated by cultural and natural cues. As the interview process reveals, memory may be said to live, particularly when spoken and when embodied.

Research into the embodied, emotional and environmental aspects of memory is well-suited to the narratives of interviews. Memory narratives can emerge from the process of ethnography when interviewees are given space to reconstruct their sensory or emotional memories through a semi-structured or open-ended format (Brewer, 2000, pp. 63-66; Hammersley & Atkinson, 2007, pp. 97-120). Indeed, the exploration of memory is amenable to inscriptive forms, including interview transcripts, that record memory (Jones, 2005), but the movement between the incorporating practice of the interview, especially one in the field with the interviewee, and the inscribing practice of the transcripts exemplifies Hamilakis's (2010, p. 192) call to "collapse" Connerton's binarism between the two. Working with interviews and transcripts prompts reflexivity between incorporating and inscriptive practices of memory.

The connection between memory and ethnography is emphasised by anthropologist Johannes Fabian (2007, p. 132) who observes that ethnographic investigation inherently probes human memory: "Cultural knowledge, once articulated, is memory-mediated." In other words, the process of speaking during an interview is naturally an invocation of remembering. Fabian (2007, p. 132) asserts that "memory makes articulation possible whilst also coming between the person and the statement." Memory is often an unavoidable mediating formation in ethnobotanical interviewing in which usage or significance of plants is compiled from the recollections of the interviewee (for example, Martin, 2004). During an ethnobotanical interview, plant specimens, images or other cues can be used to "jog interviewee memory" about actual plants in the field and to access information about the cultural uses of flora (Hoffman & Gallaher, 2007, p. 203).

An interface with memory through the interview process prompts questions of how memories are expressed and what values inform their communication. The accounts of botanical memory produced during an interview could entail a series of visual images, a recounting of bodily sensations or a chronological recitation of events devoid of emotional context. But how does the interviewee feel about an experience of plants? Is there an emotional or sensory narrative behind the facts or the timeline of experiences? For the purpose of sensory plurality in the interview process, anthropologist Sarah Pink (2009, p. 2) points to the growing literature of multi-sensorial ethnography. The term "sensory ethnography" is defined as "a process of doing ethnography that accounts for how this multi-sensoriality is integral...to the lives of people who participate in our research." Reflecting Seremetakis's (1994) conceptualisation of memory, multi-sensorial ethnography assists researchers in understanding "the meanings and natures of the memories that research participants recount, enact, define or reflect on to researchers" (Pink, 2009, p. 38). The research discussed thus far collectively suggests that memory is more than a series of images recalled cognitively. Due to the heterogeneous ways in which human memory can be approached, botanical memory considers these multiple aspects. Crossing into embodiment, environment and emotions, botanical memory encompasses feelings and facts about plants, particularly with the use of ethnography to bring forth accounts of human engagements with the botanical world.

Botanical Memories Of Mourning: Biodiversity, Beauty and Grieving

As both collective remembrance and embodied subjectivity, memory intersects with emotions and place (Jones, 2005, p. 213). Mourning in particular can form an emotional node between people and places in response to what Porteous (1989, p. 230) terms "topocide" or the annihilation of place, and what Giblett (1996, p. 68) calls "aquaterracide," or the killing of wetlands. Glenn Albrecht (2010, p. 227) defines solastalgia as "the pain or sickness caused by the ongoing loss of solace and the sense of desolation connected to the present state of one's home and territory." While emotional in quality and tone, the statements of loss conveyed in the following interviews are not necessarily embodied evocations that point to a visceral absence in the felt worlds of the interviewees. Multi-sensoriality can be evident, but respondents also tend towards chronologies of despair or narratives of loss: a sequence of events set in temporal, rather than in embodied, space.

Themes of loss are evident in an interview with David James of Forrestdale, Western Australia. Born in the early 1950s, a passionate activist and self-trained botanist, David has lived near Forrestdale Lake all his life. Forrestdale Lake is located on the Swan Coastal Plain, shared by Perth, and protects numerous indigenous animals and plants (Giblett, 2006). Nearby Anstey-Keane Damplands is one of the most botanically significant places on the Swan Coastal Plain and more diverse than the popular Kings Park, adjacent to Perth Central Business District (Giblett & James, 2009) (see chapter 2). It lies at the northern tip of the Pinjarra Plains, a system of flat damplands—moist, shallow sinks—including the most suitable soil on the Swan Coastal Plain for pasture and development (Beard, 1979, p. 27). A member of multiple local conservation organisations, David expresses exasperation over the seemingly insurmountable pressures on the bush exerted by development in his area. Forrestdale Lake Nature Reserve and Anstey-Keane Damplands constitute "emotional geographies" for him, linked to conservation agendas in the rapidly suburbanised southern areas of Perth. David's involvement with different organisations dedicated to preserving native habitats is "a period of—shall we say—activism [with] organisations that are actually trying to preserve the environment." Activists, such as David, exemplify commitment to preserving a sense of place connected to the protection of the floristic character of a place.

During David's childhood in the 1950s, the bush around Perth seemed limitless and immune to modern suburban expansion: "We'd walk through bush to catch the school bus." His memories convey a perception of the flora as all-encompassing. The abundance of the bush intermingles with recollections of the encroachment of development and attendant emotion of powerlessness:

> In those days, the bush was everywhere. Nowadays, you realise how threatened it is, but in those days, it was common. We'd walk through bush to catch the school bus. We took it all for granted. Even in those days, people were destroying bushland but, because there was so much, as a kid, you just accepted it. (pers. comm., D. James, September 23, 2009)

David's recollections reveal how botanical memory can include information about the gradual transformation of the land by the juggernaut of contemporary progress. His childhood reverie has been displaced by an anxiety over the manifold threats to the local landscape, compelling him to align with conservation initiatives:

> What we took for granted is now threatened by housing, expansion of agriculture and roads. The piece of bush that we took for granted as kids is now being threatened and it's a bitter shame because now I spend more time trying to protect wildflowers than actually going out trying to enjoy them for what they are. (pers. comm., D. James, September 23, 2009)

David's interview exemplifies botanical memory as a reservoir of emotions about nature. Moreover, the memory narratives of amateur botanists and local conservationists like David can impart a sense for the scale of change. For example, the proliferation of exotic species on road verges is a distinct difference: "Years ago roads were narrow and roads with vegetation in good condition were quite normal." His recollections evidence the progressive incursion of exotic plants on the west side of Forrestdale Lake and the gradual disappearance of certain orchid species. These distinctive intrusions to the composition of the landscape near his home have occurred during his lifetime:

> Nowadays this side of the lake's pretty much weed infested but in those days the weeds weren't quite so bad. We used to get orchids growing alongside the road here, spider orchids and different species growing amongst the weeds. (pers. comm., D. James, September 23, 2009)

Slow-growing plants, such as banksias and the zamia palms, have been severely affected by development of bushland areas like Anstey-Keane Damplands and Forrestdale Lake (Giblett, 2006). According to his direct observation, the climax character of the bush has been permanently altered, despite claims about the regeneration of the bushland by replanting:

> After thirty or forty years, it looks quite natural, but believe me, if you went back before that, there was a lot of big stuff in there. But that won't be seen again in our lifetime because you need five-hundred years to grow big zamia palms or big banksias. (pers. comm., D. James, September 23, 2009)

Here, David evokes one of the distinguishing qualities of the South-West flora: its ancientness and slow-growing tendencies (for example, Breeden & Breeden, 2010). Certain plants require hundreds of years to reach a mature state, posing an amplified sense of loss reminiscent of the emotional debates surrounding the clearing of old-growth forests (for example, Humphries, 1998).

The emotions of mourning are further elicited in an interview with Nyoongar elder Noel Nannup (see chapter 7). From 1978 to 1989, Noel served as a National Parks ranger and became the first Aboriginal head ranger in Australia. Noel recalls the destruction of indigenous vegetation near Geraldton that he witnessed as a child. He expresses a "mixed" emotional state of despair over the clearing of a million acres a year and affection for the botanical heritage of his birth region:

> When I say mixed, I mean loving it and watching it get smashed to smithereens as they cleared a million acres a year during the 60s and the 70s. That's heartbreaking. And my father would say, 'This is Australia's greatest asset, its natural vegetation and look what we're doing to it' as we followed the bulldozers along around places like Dalwallinu, Wongan Hills, Ballidu, Talingeri, to 'round that country there. We were sad to watch it. (pers. comm., N. Nannup, July 21, 2010)

The loss of indigenous plants is more than an ecological abstraction, but has cultural and spiritual ramifications. Noel expresses being torn between affinity for the plants that are his totems and dismay over the onslaught against indigenous bushland, and hence against Aboriginal spirituality, as the two are consanguineous (for a discussion of Australian Aboriginal relationships to land, see Rose, 1992; M. Graham, 2008). He says:

> But I've always tried to balance that. How do you cope, when your dad's telling you that these [plants] are our totems and yet we're watching them get smashed to bits. (pers. comm., N. Nannup, July 21, 2010)

As with David James, Noel's memories track changes in the character of the South-West landscape. Near Geraldton, plant biodiversity occurred along railway lines and road verges before the introduction of herbicides and developments in transportation technologies that impacted species such as the more prominent everlastings:

> I used to ride a pushbike from Geraldton out along the railway lines towards Mullewa. In those days they didn't use a lot of herbicides and sprays, and all the wildflowers were still there, including big pink everlastings. You'd watch them form little plants that come up, grow, and flower in the railway line. They started to bring out an X-class diesel which didn't have spark arresters on it. The railways had to burn. You'd watch the burning and you'd know that the seeds of the everlastings were all getting burnt. And because they burnt them all the time, some of them coped and some didn't. (pers. comm., N. Nannup, July 21, 2010)

This statement points to the heterogeneous quality of botanical memory, comprising emotional impressions of plants linked to memories of childhood, communities, and rapid changes to each.

For Noel, everlastings also prompt associations with family and place. The common name "everlasting" denotes various species of the cosmopolitan Asteraceae family, such as the pom-pom head (*Cephalipterum drummondii*). Memories of pom-poms call to mind his father who taught Noel to snip the stems and preserve the flowers in hot wax:

> There used to be pom-poms, little round ones, just north of Three Springs, near Arrino, and also in Coalseam [Conservation Park] at Mingenew. There's still a lot of them there but they're little ones. I'd always cut them into bunches and put rubber bands on them. Dad always said, 'if you want them to last a long time, snip them off and dip them in hot wax'. That seals it. Then your flowers stay colourful for a long time. (pers. comm., N. Nannup, July 21, 2010)

Noel's tactile recollection of flowers near Mullewa affirms the mixed quality of the botanical memory of someone with a lifelong history in a place. His statements suggest that memories of plants can be concurrently of wonderment and loss over the tragedy of topocide. Moreover, such recollections can constitute "a mnemonics of the body" (Connerton, 1989, p. 74) in which inscriptive, material practices (the sealing of the stems of pom-poms in hot wax) are inseparable from incorporating practices of memory (the transmission of this local, haptic knowledge between Noel and his father). Whereas the colours of the pom-poms would last indefinitely, the wreath flower was notably evanescent and delicate. With a genus name celebrating the French botanist Jean Baptiste Leschenault de la Tour and a species name derived from Greek for "large flower," wreath flowers (*Leschenaultia macrantha*) are prostrate-growing plants endemic to the northern Wheatbelt areas near Mullewa and Perenjori, Western Australia. Noel associates the rarity of the wreath flower with the mining of Tallering Peak where he recalls that the flower grew:

> They were rare. I remember if you picked those, by the time you stood up after you picked one, it started to wilt. Talk about an amazing flower. Just pick it up and it's gone, it's finished. But they were up around Mullewa and Bullardoo Station. There's a big hill there. It was iron ore, sadly, and they loaded it onto trucks and railway cars and carted it away. Tallering Peak was a really important place. (pers. comm., N. Nannup, July 21, 2010)

Memory itself is ecosystemic. South-West plants instigate recollections of birds and the broader landscape, reflecting the concept of environmental memory posited by Chawla (1994, p. 1) as encompassing natural "objects," landscape "settings" and emotional "moods." In other words, memories of plants are rarely of plants only.

The notion of ecosystemic connectivity as integral to botanical memory and mourning is brought up by Don Williams, conservationist and proprietor of Hi Vallee Farm in Badgingarra, Western Australia. He implies ecologically integrated mourning, in which the loss of one species has larger consequences for the biotic system of which it is part, including other animals, insects and plants:

> No one can put a value on individual species. Some people will say if we lose it, it doesn't matter, more will evolve. It would appear that they won't evolve as quickly as we can wipe them out. What a lot of people forget is that if you lose a plant species, you could lose an animal or insect species. And a lot of the orchids have evolved around one individual insect which pollinates them. (pers. comm., D. Williams, August 28, 2009)

The exploration of plant memories unavoidably confronts mourning the loss of biodiverse habitats. All three interviewees have been life-long residents of the South-West, and have thereby engendered emotional attachments to plants through regular exposure to their ecologies and immersion in the depths of the botanical integrities of their respective places. All have been involved in conservation efforts. Although their responses shift between lament and celebration, their long-term perspectives on the South-West express a startling sense of the bush in transformation by social and environmental pressures of suburban development, pesticides, technology and mining. Noel Nannup and Don Williams suggest particularly the importance of "connectivity mourning" of habitat decline in relation to species loss, in distinction to Freudian mourning in which the isolated plant object becomes the singular focus of emotional transference (see chapter 11).

Interlude XVIII: Sheoak Reverie

A jolt of memory can occur at an unpredictable or mundane juncture, such as lagging up a hill in the rain in an underpowered car. Sheoaks reminded early Australian settlers of the treasured native oaks back home: a durable timber able to be crafted. Without question, Western sheoak (*Allocasuarina fraseriana*) gained part of its common name from the similarity of its grain to its English namesake. Yet, the origin of the prefix *she* has been the subject of some speculation amongst foresters and botanists. For example, some forest researchers note:

> 'She' may refer to the sex of female trees or may be a corruption of the word for the sound made as wind blows the branchlets of a casuarina tree. However, it is most likely that the prefix 'she' means inferior, i.e. with oak-like timber but not as good as in true English Oak. (Boland et al., 2006, p. 74)

Like the eucalypts, the sheoak was considered inferior in many ways to its northern counterparts. Sheoak timber was used in settlement years for shingles and yokes, but is now confined mostly to novelty crafts and cabinet-making (Corrick & Fuhrer, 2002, p. 39). "Sheoak Reverie" expresses the "fogging together" (l. 5) of memories of trees through this instance of botanical catachresis, verging on personal catharsis on the dark, wet road to Kalamunda:

Welshpool Road mounting
the Hills above Perth City
soused in eventide spawl

sheoaks rummy with radiata pine
all fogging together—
my mobile reverie cut short (ll. 1-6)

The poem is about the melding of personal and place histories, as the indigenous sheoaks themselves "rummy with radiata pine" (l. 4) introduced from California. A reflection on the incongruencies of naming conjures the interweavings of lore: the haunting sound of the wind through sheoak cladodes; the inferior oak not up to par with the demands of settlement; remarkably adapted indigenous species turned into the bane of colonists (to read "Sheoak Reverie" in full, see Ryan, *in press*).

Botanical Memories Of Celebration: Orchids, Love and Rapture

Whereas botanical memory can be characterised by mourning, it can also express the emotions of celebrating flowering beauty. Wildflower tourists such as Lyn Alcock, originally from the Eastern States of Australia, emphasise the beauty and awe-inspiring qualities of indigenous Western Australian plants. They do recognise the threats to botanical diversity, but more tangentially than David James, Noel Nannup and Don Williams. Of all the South-West plant species, Lyn is passionate about orchids and dryandras. When I asked her about the attractiveness of orchids, she relates to me her first memory of encountering wild-growing plants in the Murchison River area after having first grown them in her home:

> I think I had been given some *Cymbidium* bulbs at home and started growing orchids. Then, ten years ago on our first trip, [my husband and I] were travelling up north around the Murchison River area. I came across all these tiny little orchids about ten centimetres high. They were in a mass like the little snail orchids we saw this morning. For some reason they just grabbed me and I said 'Wow, these are amazing!' (pers. comm., L. Alcock, September 8, 2009)

For some flora enthusiasts, the promise of spotting rare or elusive iconic species can invigorate botanical passions. When I asked about which orchids she feels particularly drawn to, Lyn reflected on the characteristics of rarity and visual strikingness, while also communicating a sense of exaltation at locating a rare plant in flower. A beautiful and hard-to-find localised species, such as the Queen of Sheba (*Thelymitra variegata*) or the winter spider orchid (*Caladenia drummondii*), is a peak moment for an ardent wildflower aficionado and an experience which necessitates the mobility of walking:

> The Queen of Sheba is such a spectacular orchid. Just recently, we were up at Kalbarri and I saw a little orchid I thought I would never get to see. It's called the winter spider orchid. It's a very tiny orchid, which only occurs mid-winter. And because it's not very common I never thought I'd see it, but I was walking through the bush and there it was. (pers. comm., L. Alcock, September 8, 2009)

The critical perspective afforded by sensory scholarship makes us more aware of the centrality of embodied acts, such as walking, in the appreciation of wildflowers. Stoller (1989, p. 9) advocates attunement to "the sensual aspects of the field" so that visual experience is analysed in relation to multi-sensorial entanglements with those we study. Similarly, the uncommonness of the Western underground orchid

(*Rhizanthella gardneri*), with only two hundred and twenty recorded sightings, impacts the intensity of her recollection. As an orchid enthusiast, the underground orchid holds a prominent position in Lyn's memory, as well as the botanical imagination of the South-West. First identified by its sweet smell in 1928 by John Trott on his farm near Corrigin, Western Australia, the white leafless orchid is perhaps the rarest endemic plant species in the state and is known to occur in only two locations in the South-West: between Corrigin and Babakin, and northwest of Munglinup (A. Brown, et al., 2003). Lyn relates the delicate tactile memory of digging through the soil to unearth the extremely unusual, and very easily disturbed, subterranean species. The orchid requires a mutualistic relationship with certain South-West plants to survive. Lyn summarises:

> My most amazing sight has been the underground orchid. Because it only ever grows underground, you have to have sand for it to grow. You have to have a particular indigenous plant that it seems to grow with, and you of course have to have the fungus that stimulates this orchid. So you have to have all these three things. It grows very near the roots of that plant, but you could look under the roots of a thousand and you won't find one. It won't be the thousandth that it's under. Yes, I was taken, there were three of them all growing in a similar spot. And for me that was absolutely spectacular. It grows under leaf litter, and so you need to pull the leaf litter apart very, very carefully and then the flower is just sitting below the leaf litter, and then slowly excavate it, very carefully, very, very carefully. (pers. comm., L. Alcock, September 8, 2009)

Digging through the leaf litter and slowly excavating the flower of the underground orchid are "sense making rituals" (Waskul, et al., 2009) that invoke incorporating practices of memory—here between plants and people—and sensory recollection. The memory of the orchid is intrinsically bound up with the tactile and olfactory practice of exhuming it in order to momentarily see it. As collective or cultural memory (Boyer & Wertsch, 2009; Hua, 2009), the unearthing of the flower is practiced by anyone who wants to claim the experience or participate in exposing the flower. Other than digging through the leaf debris, there is no way to perceive the orchid. The action of pulling the leaf matter apart very carefully becomes part of a practice shared collectively by orchid enthusiasts, as part of the cultural network of memory surrounding the plant. Lyn (re)enacts this collective memory central to the wildflower tourism and conservation community.

Orchids furthermore play a significant role in the formation of sense of place for Ayleen Sands of the Stirling Range Retreat who recounts her first spring in the area through her experiences of learning the orchid species of her property. The mountainous Stirling Range National Park includes about 1,500 plant species, or about one-third of the flora in the South-West, including 87 endemic species dispersed throughout five main botanical communities (Keighery & Beard, 1993). Her sense of place has developed through the progression of time synchronised to the flowering of orchids:

> In May 1996 someone came to the office and asked me 'did you know your Hare Orchids (*Leporella fimbriata*) were out'? I said, 'ooh, Hare Orchids, have we got Hare Orchids here'? Off they took me. They showed me hundreds of orchid bed basal leaves of the little Hare Orchid, so I began to think, 'oh, this is quite an important place'. (pers. comm., A. Sands, September 8, 2009)

Like other self-trained botanists, Ayleen learned through the tutelage of more experienced local orchidologists. The local transmission of botanical knowledge can occur as a type of informal apprenticeship based on experiential and verbal guidance by mentors:

> In August the same year a lovely couple named Jack and Lena came. They were both in their late seventies. We introduced ourselves. 'My name's Lena, and my husband's name is Jack, and we've been coming here for thirty years'. And I said, 'you must love it'. 'It's the orchids we love', she said. (pers. comm., A. Sands, September 8, 2009)

As mentioned also by Lyn Alcock, the Queen of Sheba is a sought after orchid for its iridescent visual appeal. Ayleen recalls her ecstatic first encounter with "the Queen" (*Thelymitra variagata* and *T. speciosa*) as a ceremonious baptism into the natural world of the Stirling Range:

> A day or two later they both returned to the office with big smiles on their faces and said 'The Queen's out. You'll have to come and see the Queen'. And I asked, 'what's the Queen?' And they both were stunned. So off I went with them and I found the Queen. And so began my love of orchids and my passion to share them with visitors, whilst protecting their habitat. (pers. comm., A. Sands, September 8, 2009)

Ayleen's memories of the Queen of Sheba orchid are highly associated with interpersonal memories of Jack and Lena, demonstrating that botanical memory crosses into the natural and cultural worlds and that memories of plants are not merely of plants alone. A larger narrative of cultural memory contextualises the orchid flower in a network of social relations and approximates Hua's notion of "tangled memory" (2009, p. 139).

For Lyn Alcock and Ayleen Sands, the celebration of the South-West flora emanates from appreciation of visual beauty embedded in personal, embodied accounts of acquainting themselves with their places, as well as collective memories that are shared by a community of wildflower enthusiasts. Revealing human topographies of the plant world, botanical memory can consist of such storied engagements with plants. The interviewing of people in communities, therefore, is an integral approach for eliciting the emotional and multi-sensorial content of human memories, for tracing the landscapes of emotions and remembrance in the South-West. However, in this sample of interviewees, memories of celebrating wildflowers tend to characterise the experiences of interviewees whose sense of places are newly emerging and for whom the notion of topocide (Porteous, 1989, p. 230) can go unregistered because of affinity with legally protected reserves where pressures are diminished. For Ayleen and Lyn, orchid flowers are cause for celebration because of their rarity, colour and delicate structure, as well as community knowledge of orchids provided by senior figures like Jack and Lena. Botanical memories may consist of storied engagements reflecting emotional absorption or corporeal interaction with plants. Yet these memories can also be detached from concerns of habitat destruction or, in the case of the Stirling Range, alarming rates of plant disease. Whereas long-term residents might emphasise awareness of the threats to the viability of plant populations, tourists and other short-term visitors and more recent residents might not be exposed to these conservation realities. Further studies of botanical memory should focus on the dynamics between the closely related emotions of mourning and celebrating environments, and the differences and similarities between the responses of tourists, visitors and long-term residents of places.

Botanical Memories of Embodiment: Sucking Banksia Nectar

Botanical memory influences how people construct a sense of place through multi-sensorial interactions with the living landscape. However, sense experience can be rendered non-corporeal through the construction of remembrance as a procession of images rather than bodily and sensory memory (Casey, 2000; Connerton, 1989). Although vision is integral to the process of recollection, as the accounts of Lyn and Ayleen suggest, memories of plants can also be rich with sensation. In practice, Kevin Collins of Banksia Farm in Mount Barker, Western Australia, exemplifies the conceptual frameworks of sensory ethnography (Pink, 2009; Stoller, 1989, 1997), sensory memory (Seremetakis, 1994), and bodily memory (Connerton, 1989) as constitutive of botanical memory. His embodied approach to educating the public about the flora of the region uses incorporating practice as one strategy for engendering appreciation of plants. This section details an autoethnographic approach to botanical memory, tracing my generation of bodily memory through multi-sensorial interaction with indigenous South-West flora on Kevin's farm.

Kevin and his family bought the property that is now Banksia Farm in 1984 and began planting banksias a year later. By 1987, thirty species had been planted. In ensuing years, the family completed a collection of all seventy-six species while developing overnight facilities for visitors. His interest is a long-term fascination for propagating and cultivating banksia:

> We thought to ourselves, well we've got thirty species and there's only seventy-six, so let's keep going. It became an obsession to collect all known species and, despite earlier collecting for *The Banksia Atlas* [first published 1988] it drove the kids barmy, driving up and down. They later relished holiday trips to various parts of Australia. In fact we flew to Cairns [in Queensland] in 1991 to get the last species which grows on Hinchinbrook Island. New species discovered in 2000 and 2007 were subsequently added to complete the arboretum. (pers. comm., K. Collins, June 4, 2012)

Kevin draws from botanical science, Aboriginal bushtucker knowledges and actual bodily contact with seeds, flowers and leaves to foster appreciation not solely hinging on detached perception of flowers as aesthetic objects. For instance, the checkerboard symmetry of the banksia flower head consists of thousands of tiny flowers, which he encourages me to experience through my tactile sense: "So there's thousands of flowers in this banksia flower head or inflorescence and they're always in that beautiful checker board pattern. Touch some of the little buds, run your fingers up, then go down. They feel just like a file or rasp." Kevin's approach to wildflower tourism exemplifies the interactive and corporeal generation of botanical memory.

Half of my interview with Kevin occurred as a "mobile interview" (Hitchings & Jones, 2004, p. 8) as we walked amongst his collection of plantings. Participation in the learning act involved tasting the nectar of the banksia flower and eating its nutty seeds. The ritualistic aspects of plant encounters recur as Kevin advised me to pluck the nectareous flower bunches and suck out the sweet liquid, then put a flame to the seeds to roast them before eating:

> There are plenty of florets on the flower heads, so pull out a few of the recently opened ones, chew the extracted ends and then you can suck the sweet nectar like Aboriginal people did. Sometimes just squeezing the tube of the removed floret will cause small beads of nectar to appear. On some of the spherical flowered species, globules of nectar are visible at the base of open florets. Yummy. You can also eat the seeds of most species. They are delicious but not known as bush tucker. Usually we put a flame on them just to take the kernel off. They taste quite nutty. (pers. comm., K. Collins, June 4, 2012)

This passage expresses physical encounter involving gestural language; pull, chew, suck and eat are actions necessitating the closing of distance and the removal of the sensory distance associated with taxonomic knowledge. Kevin interfaces with plants with sensorial openness and abandonment. Although scientifically competent, his knowledge is partly imbued with the sensation of contact:

> This one smells quite sweet. You have to put your nose close to it. As kids we would part the open florets, poke our tongue in there and suck the delicious nectar from the flower. (pers. comm., K. Collins, September 9, 2009)

Recollections may be sense-rich, expressing the assertion by Seremetakis (1994, p. 9) that memory is a "culturally mediated material practice that is activated by embodied acts." Memory of sweetness crosses into smell and taste actively constituted by the diction "poking" and "sucking," implying an intimacy with his plantings. He further encourages me to "squeeze," "chew" and "suck" the flowers to draw out the sweet nectar.

Odours also shape a visitor's experience when, for instance, Kevin points out the smell of boronia, one of the characteristic fragrances of the native South-West vegetation: "This is boronia. One of the highly aromatic plants. Just have a smell of that." References to modern taxonomic nomenclature commingle with his treatment of ancient aromatic qualities of the Rutaceae family of which boronia is part:

> Boronias are in the Rutaceae family and they all have highly aromatic foliage. Not only are the leaves aromatic, but often the flowers are highly aromatic as well. Another family here is Myrtaceae which includes gum trees (eucalypts); coastal peppermints and tea trees (agonis); and bottlebrushes (melaleucas and callistemons). This family likewise has highly aromatic foliage. Each species smells slightly different and very nice. (pers. comm., K. Collins, June 4, 2012)

Kevin bridges seamlessly the technical language of scientific botany and the sensuous language of bodily experience in a manner similar to nineteenth century American writer and naturalist Henry David Thoreau (1993, 2000) and Australian place writers such as Edmund Banfield (1968). Sensory practice interweaves with his comprehension of banksia ecology:

> Feel how soft these banksia flower heads are. Put your face on one of these. Very few banksias are aromatic. The odd one smells musty or garlicky, attracting small marsupials. This soft fawn one smells sweet like tropical fruit or dessicated coconut and is pollinated by a moth visiting at night. See those tiny little pollen presenters. They're very close together so bees, ants, wasps and moths are possible pollinators. (pers. comm., K. Collins, June 4, 2012)

He hybridises scientific facts with embodied experience, emphasising material human and non-human interdependencies. Also featured at Banksia Farm, dryandras offer possibilities for multi-sensorial involvement with visitors, affirmed by the appellation "honey pot flowers:"

> Several of the dryandras species have flower heads with pistils folding inwards to form a circle and these are what we call 'honey pot flowers'. When those little loops open, nectar is present and if you get some rain the pot holds water making a nectar drink into which you can dip and then suck your finger. (pers. comm., K. Collins, June 4, 2012)

This mobile interview with Kevin Collins occurred at the crossroads of botanical science and human corporeality. As sites of memory expression, bodies mediate the natural and cultural worlds (Giblett, 2008). Smells, tastes, and textures undergird abstract scientific explanations that can seem meaningless or irrelevant to wildflower tourists. Incubated in the fibres of the body, memories of plants gestate ongoing re-enactments of sense experiences.

The Intricacies of Botanical Memory

This chapter has advocated a sensory and emotional approach to the study of botanical memory, just as Classen (1997, p. 410) outlines the timeliness of "a sensory approach to culture" through an anthropology of the senses. Memories of plants are inextricably tied to the complexities of the human faculty of remembrance and its sensory and emotional aspects. As an area of potential research into human-plant interactions, botanical memory represents a poignant component of environmental memory that is intimately related to the flora of diverse scales of place, from locales to regions. The beginning of this chapter outlined various memory frameworks including sensory memory (Seremetakis, 1994), body memory (Boyer & Wertsch, 2009; Connerton, 1989), collective memory (Hua, 2009), emotional geography (Bondi, et al., 2005) and environmental memory (Chawla, 1994) to show potential directions for botanical memory research and to establish the modes through which plant enthusiasts wend during the act of recollection. Through readings of excerpts from interview transcripts organised under the overlapping categories mourning, celebration, and embodiment, this chapter has argued that botanical memory can be multi-sensorial, emotional, and place-specific, but more typically alternates between these various mnemonic modes. Indeed, it is more germane to consider a plurality of botanical memories reflecting cultural and place values, rather than to conceptualise memory as a homogeneous discourse shared by people who interact with flora in a particular locale or region. Botanical memories reflect personal and collective proclivities, values and dispositions towards plants and places.

Different kinds of interviewees have been presented: wildflower tourists and proprietors involved in the consumption of wildflower tourism or the provision of tourism services (Lyn Alcock and Ayleen Sands); scientific botanists and horticulturalists who can control, within limits, the health and longevity of their plant collections (Kevin Collins); and long-term South-West residents turned conservationists who have witnessed the destruction of plant biodiversity first-hand (David James, Noel Nannup and Don Williams). As an educator, Kevin Collins could be considered a facilitator of botanical memory involving the senses. The experience he allows for visitors galvanises the multivalent capacity of memory for sensorial content, but his interview largely excludes feelings or "moods" that foster emotional attachments to flora. Lyn Alcock and Ayleen Sands, on the one hand, express celebratory emotions of flowers fixed firmly in their personal stories but missing the technical specialisation of Kevin Collins or the comfort he displays with sensory immersion in the botanical world. David James, Noel Nannup and Don Williams, on the other hand, refer to attachments to place through memories of flora reflecting a sense of mourning as plant populations vanish.

In a rapidly changing region such as the South-West of Western Australia, the study of human and plant interactions is a much-needed complement to botanical conservation research. Increasingly, the region's indigenous flora exists as remnant pockets within agricultural, pastoral, industrial or suburban terrains. As actual plants become less common, so do memories of those plants; memories which contain rich emotional and sensory narratives with the power to instil in others the emotions of appreciation, wonder and dismay. The botanist Stephen Hopper affirms this in saying

> When I grew up, having a patch of bush in the backyard or close enough within walking distance to the city was normal. Increasingly, as areas of bushland diminish and the city expands, that experience is on-offer less and less. So, I think this is where we have to be vigilant and continue to celebrate Western Australia, what's special about it. (pers. comm., S. Hopper, September 9, 2009)

An integral part of celebration and its obverse, mourning, is the faculty of memory in all its sensory and emotional diversity.

On a final note, botanical memory addresses the past, but raises questions about the future. On this point, Noel Nannup's memories of the destruction of the vegetation translate into a hopeful sense for the future of conservation. The replanting of Dreaming Trails to recreate interconnected habitats represents a movement towards the restoration of regional ecological networks intertwined with Nyoongar cosmologies (see chapter 7). Noel states:

> What we have to do as Aboriginal people is get everyone to understand our spirituality. That will allow us to get back onto country and follow the Dreaming Trails. As we follow, we teach our youngsters and particularly those that are lost, who don't know who they are, to get back to know who they are. As we follow the Trails, they're going to want to do something for their land. So we can get them to plant the Dreaming Trails. (pers. comm., N. Nannup, July 21, 2010)

Cultural, spiritual and botanical futures converge. Somatic and emotional memories of places can infuse a more integrated ethos of botanical conservation in the South-West. Community futures depend on plant futures because people and plants share a common biosphere. Memories of the past, therefore, lead the way forward.

References

Albrecht, G. (2010). Solastalgia and the creation of new ways of living. In S. Pilgrim & J. Pretty (Eds.), *Nature and culture: Rebuilding lost connections* (pp. 217-234). London, England: Earthscan.

Bachelard, G. (1971). *The poetics of reverie: Childhood, language and the cosmos.* Boston, MA: Beacon Press.

Banfield, E. (1968). *The confessions of a beachcomber: Scenes and incidents in the career of an unprofessional beachcomber in tropical Queensland.* Sydney, Australia: Angus & Robertson.

Beard, J. (1979). Phytogeographic regions. In J. Gentilli (Ed.), *Western landscapes* (pp. 107-121). Nedlands, Australia: University of Western Australia Press.

Beresford, Q., Bekle, H., Phillips, H., & Mulcock, J. (2001). *The salinity crisis: Landscape, communities and politics.* Crawley, Australia: University of Western Australia Press.

Boland, D. J., Brooker, M. H., Chippendale, G. M., Hall, N., Hyland, B. M., Johnston, R. D., et al. (2006). *Forest trees of Australia* (5 ed.). Collingwood, Australia: CSIRO Publishing.

Bondi, L., Davidson, J., & Smith, M. (2005). Introduction: Geography's "emotional turn". In J. Davidson, L. Bondi & M. Smith (Eds.), *Emotional geographies* (pp. 1-16). Aldershot, England: Ashgate.

Boyer, P., & Wertsch, J. (2009). *Memory in mind and culture.* Cambridge, England: Cambridge University Press.

Breeden, S., & Breeden, K. (2010). *Wildflower country: Discovering biodiversity in Australia's Southwest.* Fremantle, Australia: Fremantle Press.

Brewer, J. (2000). *Ethnography.* Buckingham, England: Open University Press.

Brown, A., Batty, A., Brundrett, M., & Dixon, K. (2003). Underground orchid (Rhizanthella gardneri) Interim Recovery Plan 2003-2008 Retrieved from http://www.environment.gov.au/biodiversity/threatened/publications/recovery/r-gardneri/background.html

Casey, E. S. (2000). *Remembering: A phenomenological study*. Bloomington, IN: Indiana University Press.

Chawla, L. (1994). *In the first country of places: Nature, poetry, and childhood memory*. Albany, NY: State University of New York Press.

Classen, C. (1997). Foundations for an anthropology of the senses. *International Social Science Journal*, 153, 401-412.

Connerton, P. (1989). *How societies remember*. Cambridge, England: Cambridge University Press.

Corrick, M. G., & Fuhrer, B. A. (2002). *Wildflowers of Southern Western Australia*. Noble Park, Australia: The Five Mile Press.

Daniels, S., & Cosgrove, D. (1988). Introduction: Iconography and landscape. In D. Cosgrove & S. Daniels (Eds.), *The iconography of landscape* (pp. 1-10). Cambridge, England.: Cambridge University Press.

Dewey, J. (1967). *John Dewey: The early works 1882-1898* (Vol. 2). (H. Schneider, Ed.). Carbondale, IL: Southern Illonois University Press.

Engen, T. (1982). *The perception of odors*. New York, NY: Academic Press.

Engen, T. (1991). *Odor sensation and memory*. New York, NY: Praager.

Fabian, J. (2007). *Memory against culture: Arguments and reminders*. Durham, NC: Duke University Press.

Giblett, R. (1996). *Postmodern wetlands: Culture, history, ecology*. Edinburgh, Scotland: Edinburgh University Press.

Giblett, R. (2006). *Forrestdale: People and place*. Bassendean, Australia: Access Press.

Giblett, R. (2008). *The body of nature and culture*. Basingstoke, England: Palgrave Macmillan.

Giblett, R., & James, D. (2009). Anstey-Keane Botanical Jewel. *Landscope 24*(4), 42-44.

Graham, M. (2008). Some thoughts about the philosophical underpinnings of Aboriginal worldviews. *Australian Humanities Review, 45*, unpaginated. Retrieved from www.australianhumanitiesreview.org/archive/Issue-November-2008/graham.html

Hamilakis, Y. (2010). Re-collecting the fragments: Archaeology as mnemonic practice. In K. Lillios & V. Tsamis (Eds.), *Material mnemonics: Everyday memory in prehistoric Europe* (pp. 188-199). Oxford, England: Oxbow Books.

Hammersley, M., & Atkinson, P. (2007). *Ethnography: Principles in practice* (3 ed.). New York, NY: Routledge.

Herz, R., & Cupchik, G. (1995). The emotional distinctiveness of odor-evoked memories. *Chemical Senses 20*(5), 517-528. doi: 10.1093/chemse/20.5.517

Hickman, E., & Ryan, J. (*in press*). *Two with nature: Ellen Hickman and John Ryan*. Fremantle, Australia: Fremantle Press.

Hitchings, R. (2003). People, plants and performance: On actor network theory and the material pleasures of the private garden. *Social & Cultural Geography, 4*(1), 99-113.

Hitchings, R., & Jones, V. (2004). Living with plants and the exploration of botanical encounter within human geographic research practice. *Ethics, Place & Environment, 7*(1 & 2), 3-18.

Hoffman, B., & Gallaher, T. (2007). Importance indices in ethnobotany. *Ethnobotany Research & Applications, 5*, 201-218.

Howes, D. (2003). *Sensual relations: Engaging the senses in culture and social theory*. Ann Arbor: The University of Michigan Press.

Hua, A. (2009). "What we all long for:" Memory, trauma and emotional geographies. In M. Smith, J. Davidson, L. Cameron & L. Bondi (Eds.), *Emotion, place, and culture* (pp. 135-148). Farnham, England: Ashgate.

Humphries, C. (1998). Nyoongar creation - tall trees and king parrot dreaming. In T. McCabe (Ed.), *Nyoongar views on logging old growth forests* (pp. 10-11). West Perth, Australia: Wilderness Society.

Jones, O. (2005). An ecology of emotion, memory, self and landscape. In J. Davidson, L. Bondi & M. Smith (Eds.), *Emotional geographies* (pp. 205-218). Aldershot, England: Ashgate.

Keighery, G., & Beard, J. (1993). Plant communities. In C. Thomson, G. Hall & G. Friend (Eds.), *Mountains of mystery: A natural history of the Stirling Range* (pp. 43-54). Como, Australia: Department of Conservation and Land Management.

Kurongkurl Katitjin. (2009, Kambarang). Nyoongar elder in residence at Kurongkurl Katitjin. *Our Place: Official Newsletter of Kurongkurl Katitjin* Retrieved from http://www.kk.ecu.edu.au/data/shared/documents/newsletter/kambarang_newsletter_09_final_we.pdf

Lawrence, D. H., & Skinner, M. (1924/2002). *The boy in the bush*. P. Eggert (Ed.). Cambridge, England: Cambridge University Press.

Lehrer, J. (2007). *Proust was a neuroscientist.*. New York, NY: Houghton Mifflin Company.

Lines, W. (1994). *An all consuming passion: Origins, modernity, and the Australian life of Georgiana Molloy*. St. Leonards, Australia: Allen & Unwin.

Lledo, P, Gheusi, G., & Vincent, J. (2005). Information processing in the mammalian olfactory system. *Physiological Reviews, 85*, 281-317. doi: 10.1152/physrev.00008.2004.

Martin, G. (2004). *Ethnobotany: A methods manual.* London, England: Earthscan.

Pink, S. (2009). *Doing sensory ethnography.* London, England: Sage.

Porteous, J. D. (1989). *Planned to death: The annihilation of a place called Howdendyke.* Manchester, England: Manchester University Press.

Proust, M. (1913/2008). *Swann's way* [Google Books]. (C.K.S. Moncrieff, Trans.) Retrieved from http://books.google.com

Rose, D. (1992). *Dingo makes us human: Life and land in an Aboriginal Australian culture.* Cambridge, England: Cambridge University Press.

Ryan, J. (*in press*). John Charles Ryan. In S.-P. Mitchell (Ed.), *Fremantle poets 3: Performance poets.*. Fremantle, Australia: Fremantle Press.

Seremetakis, C. N. (1994). The memory of the senses, Part I: Marks of the transitory. In C. N. Seremetakis (Ed.), *The senses stilled: Perception and memory as material culture in modernity* (pp. 1-18). Chicago, IL: University of Chicago Press.

Spooner, A. (2003). Hakea neurophylla Meisn. *FloraBase* Retrieved March 27, 2011, from http://florabase.calm.wa.gov.au/browse/profile/2186

Stoller, P. (1989). *The taste of ethnographic things: The senses in anthropology.* Philadelphia, PA: University of Pennsylvania Press.

Stoller, P. (1997). *Sensuous scholarship.* Philadelphia, PA: University of Pennsylvania Press.

Stratford, E. (1997). Memory work, geography and environmental studies: Some suggestions for teaching and research. *Australian Geographical Studies, 35*(2), 206-219.

Summers, L. (2011). Wildflower season: The development of native flora protection legislation in Western Australia, 1911-1975. In A. Gaynor & J. Davis (Eds.), *Studies in Western Australian History: Environmental Exchanges, 27,* 31-44.

Synnott, A. (1991). Puzzling over the senses: From Plato to Marx. In D. Howes (Ed.), *The varieties of sensory experience: A sourcebook in the anthropology of the senses* (pp. 61-76). Toronto, Canada: University of Toronto Press.

Thoreau, H. D. (1993). *Faith in a seed: The dispersion of seeds and other late natural history writings.* Washington, DC: Island Press.

Thoreau, H. D. (2000). *Wild fruits: Thoreau's rediscovered last manuscript.* New York, NY: W.W. Norton.

Trigger, D., & Mulcock, J. (2005). Forests as spiritually significant places: Nature, culture and "belonging" in Australia. *The Australian Journal of Anthropology, 16*(3), 306-320.

Urry, J. (2005). The place of emotions within place. In J. Davidson, M. Smith & L. Bondi (Eds.), *Emotional geographies* (pp. 77-86). Aldershot, England: Ashgate.

Vivienne, M. (1901). *Travels in Western Australia: Being a description of the various cities and towns, goldfields, and agricultural districts of that state.* London, England: William Heinemann.

Waskul, D., Vannini, P., & Wilson, J. (2009). The aroma of recollection: Olfaction, nostalgia, and the shaping of the sensuous self. *Senses & Society*, *4*(1), 5-22. doi: 10.2752/174589309X388546

Part IV

Botanical Languages

Part IV moves towards a more focused consideration of the interstices between plants and language, as prefigured by the discussion of Linnaean language in chapter 1 and the embedded poetics of scientific nomenclature in chapter 2. Chapter 9, "An Unlikely Marriage," revisits, in greater depth, the conceptual complementarity between Thoreau and Heidegger. Experience of the botanical world can be generated as an encounter between language and the senses in the philosophies of both writers. Chapter 10, "Poetic Ecologies of Flora," concretises the Thoreauvian-Heideggerean position by presenting an analysis of poetry composed by Western Australian writers about South-West plants. The chapter maintains that the microcosmic orientation of habitat poetry intensifies and particularises landscape poetry as a macroscopic mode of writing. Within habitat poetry, most notably of Western Australian poets Andrew Lansdown, Alec Choate and John Kinsella, we find the emergence of a distinctive poetics of multi-sensoriality, encouraging bodily intimacies between people and the plants of places.

9

Language

Abstract: This chapter will examine of the role of language—particularly language that expresses aesthetic meanings—within the diverse range of spoken and written resources addressed so far. The theorisation of language is a keystone towards conceptualising participatory and sensory relationships between people and plants. A comparative reading of the works of Thoreau and Heidegger provides a framework for approaching the spectrum of language sources garnered in *Green Sense*. Despite political differences, Thoreau and Heidegger shared a mutual conviction about the generative powers of language. Thoreau's literary practice partly involved immersion in places such as swamps and forests. Fittingly, Heidegger's explication of Rilke's concept of "the Open" mirrors the participatory aesthetics of Thoreau. Both thinkers looked towards the capacities of poetic thought to galvanise language's evolution

> *By the 25th of May this bud [of the sweet flag], before it has blossomed and while yet tender, is in condition to be eaten and would help to sustain a famished traveller. I often turn aside my boat to pluck it, passing through a dense bed of flags recently risen above the surface. The inmost tender leaf near the base of the plant is quite palatable, as children know. They love it as much as the musquash does.*
>
> Thoreau (1993, p. 184)

Commonalities between the works of Thoreau and Heidegger provide the foundation for theorising an approach to the widely dispersed sources of *Green Sense*. Heidegger (1971, p. 194) claims that "in its essence, language is neither expression nor an activity of man. Language speaks." An enigmatic but potentially productive phrase, "language speaks" attributes agency to words. The assertion dislodges language from categorisation, as either a conduit of information or a rote output of internal cognition. The statement corporealises language, as something that enables sensing bodies to experience. For Thoreau, language, habitats and bodies coalesce inseparably. Rather than distant speculation, his poetic interaction with plant life intimated that language underpins embodiment. This is because language, as a whole, is neither a passive medium of communication to be processed and received by the rationalistic self nor a mechanical manifold of grammatical structures to be analysed. In response to the increasing dissection offered by contemporaneous linguistic theories, Thoreau and Heidegger held the notion of language as a body in itself, one animated by sensuous bodies in contact. For each theorist, language was both bodily and a body. Their works evidence that sensuous encounters can be captured in language, but the metaphor of language *as a body* also runs prominently. The body of language can be encountered as a whole living phenomenon, rather than a dissected corpse.

Interlude XIX: Blood

Kino is the dark red resin produced by gum trees, such as redgum or marri (*Corymbia calophylla*). For D.H Lawrence visiting Perth in the 1920s, the sight of redgum kino evoked the melancholy of the inspirited Western Australian bush: "This tree seems to sweat blood. A hard dark blood of agony. It frightens me—all the bush out beyond stretching away over these hills frightens me, as if dark gods possessed the place" (Skinner, 1972, p. 112). The hardened marri blood was diabolical in Lawrence's eyes. However, in Nyoongar, marri denotes sustenance as "flesh or meat" (Moore, 1884/1978, p. 51) and its kino has a history of therapeutic use. Francis Armstrong (1836/1979, p. 199) noted the use of kino as an antiseptic for spear wounds, while Hassell (1975, p. 24) wrote of its ingestion for the alleviation of diarrhoea. In "Blood," marri kino instigates memories of seeing plant blood—the bloodroot flower of Western Australia and the bloodroot of North America—and human blood splattered on a Seattle street or issuing forth from the nose of a friend affected by altitude sickness. As a catalyst for recollection, the rejuvenative blood of marri is tasted in the poem as a nutriment, not shunned as a poison or bad omen.

Blood

once you've seen blood
you look for it everywhere,
the glowing dark enamel

seeping from chambers
where organs pulse—
blood impregnating blood

wave after wave
in the columnar light
of late afternoon, a marri

performs a plasma-letting
I dodge the feasting flies
to taste flecks of kino sugar

disintegrating on my tongue
imparting an acrid sting
agreeable as an antiseptic;

lineages inside my blood:
bloodroot spicing the bland roots
of King George Sound

bloodroot white under oaks,
the profusion of my blood
after a summer camp slashing

how it spilled like a springtide
or an open tap in my eyes,
I asked would it ever stop.

strange spangles of crimson
along a suburban Seattle street
after the drunken night ranting

of the neighbors, a terse friend's
nose ruptured from the altitude
of New England Green Mountains

his only ebullition of the day
my scalp sopping like fresh paint
as the surgeon excises a lesion

all the births I will never see,
including my own, but blood
is everywhere, though the body

dams it back in its remotest gorges
it gushes forth at improbable
moments of indifference—

blood *ortus* blood *nativitas* blood.

Gathering of Tongues: Aesthetic Language and Flora

> One September I gathered some of the peculiarly formed nuts of the witch hazel, which grow in pretty clusters, clothed, as it were, in close-fitting buckskin, amid the yellowing leaves, and laid them in my chamber ... Three nights afterward, I heard at midnight a snapping sound and the fall of some small body on the floor from time to time. In the morning I found that it was produced by the witch hazel nuts on my desk springing open and casting their hard and stony seeds across the chamber.
>
> Thoreau (1993, p. 184)

Rather than focusing on visual media, *Green Sense* has taken a linguistic turn in presenting the aesthetics of South-West plants. A language-based orientation has aimed to underscore the challenges for an embodied aesthetic theory of plants (see chapters 4 and 5) and the potential of participatory interaction towards multi-sensorial appreciation (see chapters 12 and 13). Multiple genres of language have been gathered, analysed and created. What do these genres point to collectively? What themes do they share? The sources have included ethnographic interview transcripts with botanists and wildflower tourists (see chapters 7 and 8); creative writing as poetic enquiry (see chapter 2, and throughout); the prose and poetry of South-West authors writing about plants (see chapter 10); Nyoongar words for plants such as *marri* and *kino* (see chapter 3); the marketing language of wildflower tourism web sites, brochures and organised tours (see chapter 7); the colonial journals of explorers and settlers as private language in the public domain (see chapters 3, 4 and 6); the managerial language of conservation governance organisations (see chapter 5); and plant names as part of the taxonomies of classical science (see chapters 1 and 5). These genres hybridise to create composite discourses of flora. For instance, interviewees, such as Kevin Collins, can switch adeptly between the technical nomenclature and participatory aesthetics (see chapter 8).

A comparative reading of Thoreau and Heidegger offers a coherent way to navigate these sources. Poetic language, for both thinkers, creates experience, rather than objectively accounting for it. Heidegger's claim that language speaks presents a critical principle. The documents examined have become living expressions of values, moods, inclinations, perceptual preferences and bodily experiences. Although the language can be a hundred years old, it breathes, pulses, sings and clamours. It is clearly animate. My engagement with it is not an act of resuscitation, but of intermingling between my senses and the living words. The sources can be full of complex views of the South-West plant life that I have come to know and feel as a contemporary reader and an American more acquainted with pines than banksias (see the Prologue).

From a distance, it might seem a counter-intuitive decision to examine written documents, rather than visual sources like illustrations or paintings, in response to questions of aesthetics. After all, aesthetics as it has been theorised since the eighteenth century denotes the pleasure of appearances, rather than the complex sensations of experience (see chapter 5). As such, the study of aesthetics should be limited to the products of visual culture (for example, see Mitchell, 2002, pp. 87-88). As Adorno (1984, pp. 112-113) argues, for the early nineteenth century German philosopher Hegel, aesthetics narrows to visual artefacts in contrast to natural beauty, which is "prosaic" and thus amenable to prose. However, unlike Adorno's bifurcation of written prose and graphic objects of art, in lieu of solely analysing pictorial renderings of flora, my approach highlights the shared ground between written and graphic sources: the word-image dialectic. Like photographs or paintings, aesthetic language can be a product of visual culture, if focusing on sight (Eaton, 1989, p. 42).

The word-image dialectic describes the dynamic between textual depictions, objects of art and visual language. It refers to words as pictures and images as diction in an interwoven aesthetics of visual culture that proves apposite to textual or graphic media alike. As such, aesthetic values entwine with visual writing to craft "aesthetic language" in which language emphasises spatial perception rather than multi-sensorial experience (see chapters 7 and 8). The concept of the "horticultural object" suggests a visual language of plant aesthetics (see chapter 4). As chapters 7 and 8 also suggest, authors compose images, construing broad-sweeping vistas or colourful flowers as artistic objects and scenes. Participatory modes of aesthetics, in contrast, provide no definitive statements of beauty, sublimity or picturesqueness because they more characteristically juxtapose opposite qualities, such as pleasure and pain. Several people have lamented to me at conferences that "South-West flora is prickly," after politely considering ideas about the embodied relationships between plants and people that have existed in the region for 50,000 years. I respond: "Yes, prickly as an adaptation, but also soft, nourishing, fragrant, colourful, medicinal, edible and primordially linked to the human perception of time and place."

In other words, a critical position on aesthetic language parallels the critique of ocularcentrism championed by Martin Jay (1993). Although, as Levin (1993, p. 186) argues, Heidegger was not a prominent critic of ocularcentrism but he took a "very critical position with regard to the everyday experience of seeing and the ocularcentric metaphysics which reflection on this experience has brought forth." In asserting that language speaks, Heidegger demonstrates a shift in his thinking away from "vision-generated discourse to a discourse formed by listening," or being attentive to the stimuli of one's surroundings (Levin, 1993, p. 186). The consequence of a focus on what is seen is the problematic prioritisation of surface beauty. Aesthetic language, in its ocularcentric manifestations, composes plants as artefactual scenes—paintings, photographs or landscapes—of diction. The aforementioned modes of language that have been gathered, however, suggest jointly a continuum of interaction between plants and people from the cognitive detachment of aesthetics to the sensory immanence of aesthesis (see chapter 13). Language can measure variation and change in the human experience of plant life. However, there is a cultural propensity to communicate about non-cultivated plants in surface-oriented terms or functional scientific language, rather than spiritual, cultural, medicinal, gustatory or multi-sensorial poetics. When we taste a plant in the field, can language express this? At what point does "vision-generated discourse" constrain a participatory aesthetics of flora?

When Language Speaks, Who Listens?

> Does he paint? he fain would write a poem;
> Does he write? he fain would paint a picture
>
> Robert Browning (1855/2008, p. 2590)

Heidegger's contention that language speaks has become my position on language. Indeed, the answer to the question of "who listens" provides a glimpse into the philosophy of language of Thoreau and Heidegger. Language can be conceptualised as immanent and autonomous: as a body. Both thinkers would concur that a poetic disposition towards the natural world enables a heightened ability to invoke the many senses. Rather than denoting specific forms for constructing words, poetic thought is a way of participating in sense-based meaning-making (Leggo, 2007). Through its speaking, language invigorates a receptivity—a listening—to the sensuous possibilities of the plant world. As intrinsically connected to experience rather than passively produced, language assumes an aliveness in the works of Thoreau. Critical reflexivity with words and an entrainment to the lingua of swamps, plants and forests were hallmarks of his interaction with the plant life of Concord, Massachusetts.

Thoreau theorised language in *Walden* (1854/1966) and the essay "Walking" (1862/2007). He later manifested these language concepts in the unfinished manuscripts on forest ecology, *Faith in a Seed* (1993) and *Wild Fruits* (2000). Nearly a century after Thoreau's death, Heidegger also took a revelatory turn towards poetic language within the conceptual frameworks of *Dasein*—the philosophical project of being or personhood—in *Poetry, Language, Thought* (1971) and *On the Way to Language* (1982). Although an uneasy, but potentially productive, relationship exists between their works, the characterisation of Thoreau as a "proto-Heideggerean" (Garber, 1991) is outside this scope. In his analysis of the connections between building, dwelling and cultivation, Garber (1991,

p. 146) observed Thoreau's "proto-Heideggerean recognition of the relations of being, dwelling, and location and what these have to do with being at home in the world." While the philosophy of Heidegger provided "some useful tools with which to understand the matter of housing in *Walden*" (Garber, 1991, p. 148), the idea of a proto-Heideggerean lineage between their philosophical thought warrants recognition as a possibility but deserves to be developed elsewhere.

In the essay, "Night and Day: Heidegger and Thoreau," by Stanley Cavell (2000), the yin-yang interplay between dark and light encapsulates the simultaneous tension and complementarity between their works. Cavell points to the concern that, despite synergistic theories of language, both theorists mark radically different politics. Thoreau practiced the constructs of language that Heidegger would later propound through an approach to the plant world that underpins the idea of cultural botany (see chapter 1) and botanic field aesthetics (see chapter 2). Cavell (2000, p. 46) characterises Thoreau as a thinker who "matches, I would say uncannily, so many of the philosophical configurations of Heidegger, while reversing his political sensibilities." The question of Heidegger's philosophy with reference to his Nazi allegiances has been a perennial debate among scholars of the history of philosophy (see, for example, Wolin, 1993). Whereas Heidegger can be seen as an example of a philosopher with problematic politics, Thoreau's paradigm of civil disobedience would later influence important social activists such as Mahatma Gandhi and Martin Luther King, Jr.

Recognising this framework of political difference, the potential complementarity between their linguistic philosophies merits deeper consideration. Both thinkers considered language an experience, rather than merely a vehicle for describing experience objectively. Moreover, both were preoccupied with the limitations of constructing the natural world visually. Translated as "configuration" or "enframement," the principle of *Ge-stell* underscores a concern for the increasing influence of imagistic thinking on the two-dimensional conceptualisation of the natural world (Heidegger, 1977). The implication of *Ge-stell* is the pictorialisation of living phenomena. Heidegger (1938/2009) contends that hierarchical power structures emanate from the transfiguration of the sensuous world into two-dimensional space:

> The fundamental event of modernity is the conquest of the world as picture. The word 'picture' now means the formation of presenting production. Within this formation, the human fights for the position in which he can be that being that provides the measure for all beings, and draws up the guidelines. (221)

Ham and Senior (1997, p. 2) suggest that *Ge-stell* connotes "the exploitive, instrumental representation at the core of technology, the 'enframement' of nature as a vast reservoir of energy to be stockpiled and spent." Heidegger's project of language is to resist the depiction of the world in structural terms, essential to the hegemony of "vision-generated discourse" (Levin, 1993, p. 186).

In its manifold forms, language has been conceived of in Western thought as a technical medium. Conceivably, Thoreau and Heidegger could have been responding to their respective intellectual currents that propounded linguistic reductionism. In 1798, French philosopher Étienne Bonnot de Condillac (cited in Whewell, 1840, p. 462) claimed that "every language is an analytic method and every analytic method is a language." The early twentieth century structuralist Ferdinand de Saussure (1916/1986) characterised language as a system:

> A language is a system of signs expressing ideas, and hence comparable to writing, the deaf-and-dumb alphabet, symbolic rites, forms of politeness, military signals, and so on. It is simply the most important of such systems. (15)

Saussure (1916/1986, p. 15) proposed *semiology* as the scientific method of analysing language or "a science *which studies the role of signs as part of social life.* It would form part of social psychology, and hence of general psychology [italics in original]." According to de Condillac and Saussure, language is conceptually produced in the mind and systematically issued, as an assemblage of signs, through the actions of the mouth, eyes and hand. Such linguistic theory presupposes that language mediates the unspoken inner life and the perceptible outer world through the output of the language products: speech, prose or transcripts.

Both Thoreau and Heidegger conceived of language as a living body, instead of a manifold of parts comprising a machination. Engagement with the body of language yields something considerably different to interaction with its anatomised constituents. For Thoreau, physical immersion in swamps, forests and mountains inherently dispelled the illusion of distance between the subjective self and the objective other that upholds the objectification of language. The sounds, smells, tastes and sensations of the plant world were sensed directly through his sense organs and skin. However, Thoreau's idea of language as immanence contradicts the construction of language as a mechanical discourse. Similarly, Heidegger (1971, p. 191) contested the narrow conceptualisation of language as the "audible utterance of inner emotions, as human activity, as a representation by image and by concept." Heidegger (1971, p. 194) problematises the assumption that language is "the expression, produced by men, of their feelings and the world view that guides them." Thoreauvian and Heideggerean language is more than a technical conglomeration of syllables and sentence structures yielding a mode of communication.

Heidegger (1971, p. 191) maintained that the reduction of language into products is part of a scientific epistemology that maintains the primacy of "grammar and logic, philosophy of language and linguistics." Not only an anatomy, language is a physiology through which ideas, emotions, sensations and values are continually gestated. Language can thus be approached as a living body. While having the capacity for palpability, poetic language can also entail slippage of certainty and other desirable tensions with what it expresses. Hence, Heidegger challenges the assumptions of technical linguistics for an animated sense for language as the fabric of experience itself. Heidegger (1982, p. 57) suggests the essential human ability to "undergo an experience with language." Rather than describing experience from a distance, language generates experience internally, and so becomes experience. Although subtle, this distinction is central to Heideggerean phenomenology. When language speaks, experience is given shape through a physiology of interaction between words and flesh.

But do all forms of language speak in the same way? Who listens? For Heidegger and Thoreau, poetic sensibility enables a participation in language in a manner comparable to two living bodies engaging in sensory exchange. Both philosophers classify language according to its somatic affect, rather than its visible structure as verse or prose. In fact, Heidegger (1971) recognised no taxonomic distinction between poetry and prose:

> Poetry proper is never merely a higher mode (*melos*) of everyday language. It is rather the reverse: everyday language is a forgotten and therefore used-up poem, from which there hardly resounds a call any longer ... Pure prose is never 'prosaic'. It is as poetic and hence as rare as poetry. (205)

The prose writings of Thoreau retrospectively typify Heidegger's thesis that poetry is not necessarily a higher mode of language structured in line breaks. Thoreau's writings are quintessential examples of poetic prose that reveals embodied interactions with plants and habitat consciousness (see chapter 1). Thoreau was a poet of plants for whom language was an essential way to know non-human life.

Language, Vision and Flora: Towards "the Open"

> By the 'Open', therefore, I do not mean sky, air, and space; they, too, are 'object' and thus 'opaque' and closed to the man who observes and judges. The animal, the flower, presumably is all that, without accounting to itself, and therefore has before itself and above itself that indescribably open freedom which perhaps has its (extremely fleeting) equivalents among us only in those first moments of love.
>
> Rainer Maria Rilke (cited in Heidegger, 1971, p. 105)

How does the redefinition of language as a body relate vision and flora to the spectrum of language sources in this project? The concept of "the Open" is instructive here. Thoreau and Heidegger sought to galvanise sense through poetic language. Correspondingly, Rilke distinguishes between being "*in* the world" and standing "*before it*" [italics in original], which results from the assertion of human reason that "observes and judges." "Sky, air, and space," of which "the animal, the flower" are part, are closed off to us through the exclusive attention to seeing and evaluating. Rilke offers the suspension of observing and judging towards sensuous participation. Whereas the diction of aesthetics tends to visually frame and thereby block us off from plants, sensuous participation enraptures us in "sky, air, and space...the animal, the flower."

In contrast, category-driven language, such as taxonomy, enframes plant life with prescriptive aesthetic values or scientific assertions. From such a perspective, a jarrah tree is beautiful or ugly—and taxonomically *Eucalyptus marginata*—but not indeterminate, mutable or mysterious. Categorical language conceptualises the world as external. In contrast to science, poetic writing can offer the manifold of a tree at a particular moment expressed through bodily meaning without epistemological closure. Heidegger (1934/2009, p. 12) explicates that "if we, therefore, assign language to a philosophy of language, then we are immediately already seized by a certain determinate conception." Yet, through a reconfiguration of language, determinate conceptions can be softened and made sensual and bodily.

In the interviews highlighted in chapter 8, some examples of language correlate flora closely to sight as the language of visual aesthetics: plants are as they are seen. Other language mixes scientific analysis, emotion, sensoriality and colloquialism. For instance, the ethnographic language from interviews can be divided into the technical language of botanical nomenclature and the sensuous language of human memory. In speaking of her affinity for orchids and her love of wildflower photography, Lyn Alcock shifts from emotional and sense-rich language to technical expositions of plant hybrids, biogeographical distribution and species rarity. Part of the excitement of the South-West flora, for Lyn, is the opportunity to be part of an evolving science of plants:

> I have found several hybrids. For me, that's what's exciting, finding something different. And because I spend so much of my time looking at wildflowers, I have more of an opportunity to come across those things. And not just in orchids, I mean, in other flowers too you come across hybrids. And the other thing I look for is just maybe I'll be photographing a plant that is rare and endangered or that hasn't been found before in that area. (pers. comm., L. Alcock, September 8, 2009)

Interviews prove aesthetically multilingual as respondents oscillate between seeing flora, to sensuously and emotionally engaging, to considering plants the objects of scientific discourse. The perspectives of most speakers and writers are ostensibly syncretic, demonstrating cross-currents between science and aesthetics.

A lens through which this vast scope of language has been read is for its visual content because language is correlated intensely to sight. Jay (1993, p. 395) characterises the relationship between images and words as "the complicated interlacing of language and vision." Such linkage has a biological, rather than cultural, rationale in which "unlike the other senses of smell, touch or taste, there seems to be a close, if complicated, relationship between sight and language, both of which come into their own at approximately the same moment of maturation" (Jay, 1993, p. 8). Authors like Marcel Proust (1913/2008), however, depart from the visual crafting of prose towards sense immanence. The unravelling of the "complicated interlacing" requires a conscious bodily effort. As etymology reminds us, the word *language* derives from *lingua*, for the tongue. Language is something tasted, smelled, touched and embodied in the gestures of limbs, facial expressions and repose of the chest. Taste occurs on the tongue as the transmitter of audible language and the receptor for sensory inputs.

The trouble with ocularcentric language is that landscapes, which were once embodied spaces where people could nourish themselves from wild flora, are constructed as "visual taxonomies" (Lenman, 2003/2006, p. 93). Surface-oriented depictions of the plant world abound as illustrations, photos and botanical desiderata, but also as the language of appearances. As visual appreciation only, aesthetics reiterates categoricalism. However, "the Open" illuminates multi-sensoriality in space. For Heidegger, language reaches into the Open when liberated from rectilinear structuring. Thoreau and Heidegger argue compatibly that poetic writing broadens the narrow experience of sight towards bodily contact. The Open prompts a distinction between the externalisation of the world and the participation in experience through language. Heidegger (1971) maintains that the Open:

> Means something that does not block off. It does not block off because it does not set bounds. It does not set bounds because it is in itself without all bounds. The Open is the great whole of all that is unbounded. (104)

As closed systems, the beautiful, picturesque and sublime set bounds on perception by compartmentalising experience (see chapter 5). The Open, in contrast, entails gesture and contact between sentient bodies in which living beings are drawn together centripetally and centrifugally. Heidegger (1971) argues that the Open is a state of wholeness that:

> Lets the beings ventured into the pure draft draw as they are drawn, so that they variously draw on one another and draw together without encountering any bounds...They do not dissolve into void nothingness, but they redeem themselves into the whole of the Open. (104)

As "the whole of all that is not objective" (Heidegger, 1971, p. 104), the Open broadens the botanical field constructed by scientific thought or aesthetic evaluation.

I suggest that Heidegger's sense of the Open was practiced by Thoreau *avant la lettre* as habitat immersion. Thoreau's interaction with local plants dismantled perceptual distance, while also testing his physical limits as a form of body-engaged judgement. For instance, although appearing glossy and attractive, a berry can be toxic, nauseating or too tough to chew. Thoreau drew consistently on the interplay between corporeal interaction and distant speculation vis-á-vis the local flora of Concord. Thoreau (2000, p. 66) writes: "I *eat* the high blueberry, but I am also interested in the rich-looking, glossy-black chokeberries, which nobody eats and which bend down the bushes on every side—sweetish berries, with a dry and so choking taste [italics in original]." Although the berries reflect desirable forms and colours, they were not pleasing to the local palate. As a mode of botanical enquiry, Thoreau's body intervened by tasting the uncelebrated chokeberries and making his own conclusions.

As implied by this excerpt, language gestates aesthetics, while potentiating aesthesis. As contact between the bodies of people and plants, aesthesis takes place in the Open, reconstituted as the botanic field. Language reveals gradations of interaction, the ways in which we announce the natural world in images, and the ways in which bodily sensation interrupts visual systematising. Heidegger (1971) writes that the natural world is more than a collection of seen phenomena:

> In what the senses of sight, hearing, and touch convey, in the sensations of color, sound, roughness, hardness, things move us bodily, in the literal meaning of the word. The thing is the *aistheton*, that which is perceptible by sensations in the senses belonging to sensibility. (25)

Extending Heidegger, participatory experience in the Open depends on language. Heidegger (1971) argues that words in themselves constitute activities and cannot be separated from what they denote:

> When we go to the well, when we go through the woods, we are always already going through the word 'well', through the word 'woods', even if we do not speak the words and do not think of anything relating to language. (129)

Entering into the Open, for Heidegger (1971, p. 94), necessitated a "dialogue with poetry," one which becomes "an unscientific violation" of analytical scholarship. But as a consilience of opposites, poetic thought and analytical thought can be brought "into the most extreme discord and so to establish their concord" (Heidegger, 1971, p. 96). Such reconciling tension between discord and concord occurs in the centripetal and cetrifugal Open, rather than in the two cultures divisiveness (see chapter 1). Thoreau's practice of the Open in Concord entailed syncretism between botanical science and somatic enquiry and between mimesis and methexis, both conceptualised as dialogic. Poetic thought brings us bodily into the Open where the flower already is at home (see chapter 10). The Open is the field of contact in which multi-sensoriality promotes bodily aesthetics, such as tasting wild berries in conjunction with viewing the fruits as beautiful objects. Embodied language is the catalyst for participation in the field of plants (see chapters 2 and 3).

Interlude XX: Wheatbelt Pneuma

With a genus name celebrating the French botanist Jean Baptiste Leschenault de la Tour and a species name derived from the Greek for "large flower," wreath flowers (*Leschenaultia macrantha*) are prostrate-growing plants endemic to the northern Wheatbelt areas near Mullewa and Perenjori. The flowers appear alongside railway lines and, like everlastings, are iconic of the northern wildflower areas of the South-West. They are part of the soul of the place, its non-material essence:

> they listen to the old stories;
> when pneuma fused breath and heat,
> the woven flower of *Leschenaultia macrantha*
> flexed a green heart girded by blood fire
> : asking the secret earth to sing (ll. 26-30)

The Stoic philosophers of ancient Greece conceived of pneuma as breath, formed of the elements fire and air. Pneuma is the "sustaining cause" or animating principle of all living bodies; in plants, pneuma was known as *phusis* or literally "nature" (Baltzly, 2008, para. 7-8). "Wheatbelt Pneuma" celebrates the genius loci of the northern inner wheat-growing areas through an inspirited interpretation of a popular flower clinging to the wild margins (to read "Wheatbelt Pneuma" in full, see Hickman & Ryan, *in press*).

Language, Senses and Flora: Botanical Metaphor

> What flavor can be more agreeable to our palates than that of this little fruit, which thus, as it were, exudes from the earth at the very beginning of the summer, without any care of ours? What beautiful and palatable bread! I make haste to pluck and eat this first fruit of the year, though they are green on the underside, somewhat acid as yet, and a little gritty from lying so low. I taste a little strawberry-flavored earth with them. I get enough to redden my fingers and lips at least.
>
> Thoreau (1993, p. 188)

Thoreau approached plants through multi-sensoriality and language. Botanical metaphor is one expression of the triadic interrelatedness between language, flora and the senses. Thoreau resisted the standardisation of language in which the derivation of words is buried behind the sheen of popular diction. In a letter from 1857, Thoreau (cited in West, 1984, p. 747) lamented the sterilisation of lived experience by stock means of communication: "How shall we account for our pursuits, if they are original? We get the language to describe our various lives out of a common mint." Prefiguring Heidegger, Thoreau witnessed the systemisation of language, especially as the theories of the French Enlightenment began to influence American literature (West, 1984). In contrast to the structuring of language by semiotics, Thoreau (cited in West, 1984, p. 749) advocated universal language of which human verbal expression is only a part: "In all the dissertations on language, men forget the language that is, that is really universal, the inexpressible meaning that is in all things, everywhere."

So began Thoreau's life-long pursuit of language-sense immanence through the diction "in all things," including the sounds, tastes and aromas of plants. The chapter "Sounds" in *Walden* opens with "we are in danger of forgetting the language which all things and events speak without metaphor, which alone is copious and standard. Much is published, but little printed" (Thoreau, 1854/1966, p. 75). Thoreauvian language is not the rote transmission of internal states to the external world, but a mode of broadening the bearers of language to include swamps and forests. In lieu of semiotic coding, the language of all things speaks immanently. Biologist and author Gary Nabhan (1993) observes that:

> Thoreau became obsessed with 'learning the language of these fields' in the years just prior to his death. For him, the forest was a book waiting to be read: one simply had to devote sufficient time to grasp its grammar, to learn the rhythms of its syntax. (xvi)

For Thoreau, the forest was also a body poised to be sensed. The experience of a forest became inseparable from language and sensation. From a perspective that continuously cross-bred linguistics and natural science with poetry, the flowering of a plant, the groan of a bog and the taste of a wild apple were kinds of language to be felt by Thoreau (see, for example, Sellers, 1999). His urgent task was the tracing of the language of all things into poetic prose: to etch his bodily experience of the sounds, smells, tastes, textures and images of habitats into written words. His subsequent literary projects embodied this belief that language is not only in the parlour or the lecture hall as "audible utterance," but endemic to the swamp and the woodland. Language speaks from all corners. Hence, nature was brought into cultural domains. His task would admit neither intellectual distance nor the comfort of visual speculation; it necessitated physical engagement and immediacy.

A concern with the integrity of words was at the heart of Thoreau's project. He read the philosophies of the French Enlightenment, including Antoine Court de Gebelin's assertion that each vowel relates to a distinctive sense (West, 1984, p. 764). Multi-sensorial language was critical to Thoreau's philosophy of the written word and the natural world. In "Walking," Thoreau (1862/2007) propounds a theory of language reflecting Heidegger's claim that language speaks sensuously: The poet nails "words to their primitive senses, as farmers drive down stakes in the spring, which the frost has heaved." This extract conveys the idea that language is a shared faculty of human expression and the natural elements. Thoreau wove literary expression with ecological processes as one of the first examples of American nature writing (Murray, 1995). The final sentence is the most concise statement of Thoreau's relationship to the "primitive senses." The poet hammers words back to their origins, through to their roots, after language has been jostled out of place by popularisation, commercialisation, globalisation and parlour-speak. Thoreau sought a primeval bodily language. The relationship between a farmer and the land symbolises the pragmatic physical conjunction between corporeality and words.

Botanical metaphor is one mode through which Thoreau and Heidegger mutually theorise the meeting of language, senses and flora. A botanical metaphor can be defined as the use of an anatomical part or ecological process of a plant as a trope for a cultural process or concept. One of the more prevalent botanical metaphors in postmodernism is the rhizome, which "has no beginning or end; it is always in the middle, between things, interbeing, *intermezzo*. The tree is filiation, but the rhizome is alliance, uniquely alliance" (Deleuze & Guattari, 1987, p. 27). The rhizome counters the dominant growth-oriented, hierarchical metaphors of modernity by "burrowing through substance, fragmenting into simultaneous sprouts, moving with a certain stealth, powerful in its dispersion ... [destabilising] the conventions of origins and endings" (Kaplan, 1996/2000, p. 87).

Before postmodernity, Thoreau made use of botanical metaphor. After theorising the purpose of language as "nailing words to their primitive senses," Thoreau goes on to use floristic figures of speech to propound linguistic ideas.

The "poet who could impress the winds and streams into his service" would also be one:

> Who derived his words as often as he used them—transplanted them to his page with earth adhering to their roots; whose words were so true and fresh and natural that they would appear to expand like the buds at the approach of spring, though they lay half smothered between two musty leaves in a library—aye, to bloom and bear fruit there, after their kind, annually, for the faithful reader, in sympathy with surrounding Nature. (Thoreau, 1862/2007, p. 29)

This excerpt consists of three major botanical tropes: "with earth adhering to their roots;" "like the buds at the approach of spring;" and "to bloom and bear fruit there." Though "smothered," language speaks through certain kinds of literature as it blooms and bears fruit on the dusty page. This language is in "sympathy with surrounding Nature" that is closely attuned to the smells, sounds, tastes, textures and sights of flora. Such language takes us out of the strictures of the parlour and into the botanic field where the senses create bodily posture rather than visual prospect, where the "bodily eye" Thoreau speaks of becomes a hexis of visceral engagement. Although the medium of the page goes stale, poetic language blooms and fruits with the earth still at its roots.

In the same way, Heidegger's work deploys plant metaphors in theorising language. Plants epitomise Heideggerean phenomenology as *physis* in which their existence is intrinsically a process of standing forth and going back, revealing and concealing. A process-oriented nature of plants contrasts with *Ge-stell* as the enframing of a moment in the life cycle of a plant. Feenberg (2005) argues that a plant symbolises:

> Rootedness in the earth from out of which it emerges. It stands forth from the earth by going back into the earth, sinking its roots into its source. This double movement—standing forth and going back—characterizes the specific motility of living things. (30)

Hölderlin's elegy "Bread and Wine" uses the metaphor of a flower: "Now for it words like flowers leaping alive he must find" (cited in Heidegger, 1982, p. 100). The poetic thinker finds "words like flowers" just as he or she brings "words to their primitive senses." Hölderlin's phrase was taken up by Heidegger (1982, p. 99) to characterise language as "the flower of the mouth. In language the earth blossoms toward the bloom of sky." Thus, Heideggerean language engages the Earth and thereby blossoms as flowers do.

In his final analysis, language defines the Open, as "language is the house of Being" (Heidegger, 1982, p. 63). Both Thoreau and Heidegger maintain an uncanny concurrence that language immanence depends on the oscillations of experience.

Conceivably, the plant is a core metaphor for Heidegger's phenomenology. The flower is an essential trope for expressing immanence as the blooming of the senses. Poetic thinking reflects the *physis* of the living plant "standing forth and going back" as a process since "all reflective thinking is poetic, and all poetry in turn is a kind of thinking" (Heidegger, 1982, p. 136). As with Thoreau, concealment—dehiscence, the returning of plant energy to the root after wildflower season—and emergence—flowering, the boom of floral forms projecting towards the bloom of the sky—are etched in the grain of language.

Towards a Language of Plants and Sensation

> The gray blueberry bushes, venerable as oaks—why is not their fruit poisonous? It has the wildest flavor of any of the huckleberry tribe that I pluck. It is like eating a poisonous berry which your nature makes harmless. I derive some of the same pleasure from it as if I were eating the arum berries and musquash root with impunity, as if I were a Mithridates among the berries.
>
> Thoreau (1993, p. 199)

Despite political misalignments, Thoreau and Heidegger reflect a mutual understanding of language as a body. Thoreau practiced Heidegger's aphorism that language speaks *avant la lettre* through immersion in berry patches and woodlands. Correspondingly, Heidegger theorised the experience of enmeshing with the Open. Not only did Thoreau match "the philosophical configurations of Heidegger" (Cavell, 2000, p. 46), he put into earthly practice Heidegger's configurations. Additionally, both philosophers theorise plants as central metaphors for their respective ontological aims. The language emphasis of *Green Sense* befits Thoreau and Heidegger. The philosophy of multi-sensorial interaction with the botanical world gestates in a current of language that is organic and transformational, that is responsive and breathing, which houses the senses and restores sensuality to the human experience of plants. My theoretical premise hinges on the interrogation of the aesthetics and aesthesis of language. Is the language taxonomic, visual, multi-sensorial, palpable, tactile, evocative, sensuous, emotional or a hybrid of qualities? Is the language monolithic or heterogeneous? Does it take into account multiple voices, such as the diverse narratives of plants (see chapter 3)?

Language has been integral to questions of plant aesthetics because it reveals gradations of values. A language of sight undergirds what is perceived directly by the eyes as a striking scene and what is recreated in the mind when the *scene* is no longer *seen*, but expressed in words. At the threshold of seeing and sensation, *aesthesis* breaks through the distance of culturally constructed binaries between plants and people. Heidegger (1971) affirms that the movement from a "covetous vision of things" to corporeality is ultimately language-based:

> The hard thing consists not only in the difficulty of forming the work of language, but in the difficulty of going over from the saying work of the still covetous vision of things, from the work of the eyes, to the 'work of the heart'. (136)

Sensation entails an intimacy with plants that has been excised by classical scientific and visual language: "Intimacy obtains only where the intimate—world and thing—divides itself cleanly and remains separated. In the midst of the two, in the between of world and thing, in their *inter*, division prevails: a *dif-ference*" (Heidegger, 1971, p. 199). The hybrid languages of flora are characterised by an unassailable flux between incorporation and objectification. A heightened level of appreciation encompasses the senses, synchronicity and diachrony, as Heidegger says, the vision of objects and the work of the heart.

References

Adorno, T. W. (1984). The beauty of nature (C. Lenhardt, Trans.). In G. Adorno & R. Tiedemann (Eds.), *Aesthetic theory* (pp. 91-115). London, England: Routledge and Kegan Paul.

Baltzly, D. (2008). Stoicism. In E. N. Zalta (Ed.), *The Stanford encyclopedia of philosophy* (Spring 2008 ed.) Retrieved from http://plato.stanford.edu/archives/spr2008/entries/stoicism/.

Browning, R. (1855/2008). One more word. In C. Warner (Ed.), *A library of the world's best literature* (Vol. VI: Brantome-Bulwer, pp. 2589-2592). New York, NY: Cosimo Publications.

Cavell, S. (2000). Night and day: Heidegger and Thoreau. In J. E. Faulconer & M. A. Wrathall (Eds.), *Appropriating Heidegger* (pp. 30-49). Cambridge, England: Cambridge University Press.

Deleuze, G., & Guattari, F. (1987). *A thousand plateaus*. London, England: Continuum.

Eaton, M. (1989). *Aesthetics and the good life*. Cranbury, NJ: Associated University Presses.

Feenberg, A. (2005). *Heidegger and Marcuse: The catastrophe and redemption of history*. New York, NY: Routledge.

Garber, F. (1991). *Thoreau's fable of inscribing*. Princeton, NJ: Princeton University Press.

Ham, J., & Senior, M. (1997). Introduction. In J. Ham & M. Senior (Eds.), *Animal acts: Configuring the human in Western history* (pp. 1-8). New York, NY: Routledge.

Hassell, E. (1975). *My dusky friends*. Dalkeith, Australia: C.W. Hassell.

Heidegger, M. (1934/2009). *Logic as the question concerning the essence of language* (W. Gregory & Y. Unna, Trans.). Albany, NY: State University of New York Press.

Heidegger, M. (1938/2009). The age of the world picture (J. Veith, Trans.). In G. Figal (Ed.), *The Heidegger reader* (pp. 207-223). Bloomington, IN: Indiana University Press.

Heidegger, M. (1971). *Poetry, language, thought* (A. Hofstadter, Trans.). New York, NY: HarperCollins.

Heidegger, M. (1977). The question concerning technology. In D. Krell (Ed.), *Martin Heidegger: Basic writings* (pp. 284-317). New York, NY: Harper & Row.

Heidegger, M. (1982). *On the way to language* (P. Hertz, Trans.). San Francisco, CA: Harper & Row.

Hickman, E., & Ryan, J. (*in press*). *Two with nature: Ellen Hickman and John Ryan.* Fremantle, Australia: Fremantle Press.

Jay, M. (1993). *Downcast eyes: The denigration of vision in twentieth-century French thought.* Berkeley, CA: University of California Press.

Kaplan, C. (1996/2000). *Questions of travel: Postmodern discourses of displacement.* Durham, NC: Duke University Press.

Leggo, C. (2007). Astonishing silence: Knowing in poetry. In J. G. Knowles & A. Cole (Eds.), *Handbook of the arts in qualitative research: Perspectives, methodologies, examples, and issues* (pp. 165-174). Thousand Oaks, CA: Sage.

Lenman, R. (2003/2006). British photographers and tourism in the nineteenth century: Three case studies. In D. Crouch & N. Lubbren (Eds.), *Visual culture and tourism* (pp. 91-108). Oxford, England: Berg.

Levin, D. (1993). Decline and fall: Ocularcentrism in Heidegger's reading of the history of metaphysics. In D. Levin (Ed.), *Modernity and the hegemony of vision* (pp. 186-217). Los Angeles: University of California Press.

Mitchell, W. (2002). Showing seeing: A critique of visual culture. In N. Mirzoeff (Ed.), *The visual culture reader* (2 ed., pp. 86-101). London, England: Routledge.

Moore, G. F. (1884/1978). *Diary of ten years eventful life of an early settler in Western Australia,* and also *A descriptive vocabulary of the language of the Aborigines.* Nedlands, Australia: University of Western Australia Press.

Murray, J. (1995). *Nature writing handbook: A creative guide.* San Francisco, CA: Sierra Club Books.

Nabhan, G. (1993). Foreword: Learning the language of fields and forests. In B. Dean (Ed.), *Faith in a seed: The first publication of Thoreau's last manu*script (pp. xi-xvii). Washington, DC: Island Press.

Proust, M. (1913/2008). *Swann's way* [Google Books]. (C.K.S. Moncrieff, Trans.) Retrieved from http://books.google.com

Saussure, F. d. (1916/1986). *Course in general linguistics* [Google Books] Retrieved from http://books.google.com

Sellers, C. (1999). Thoreau's body: Towards an embodied environmental history. *Environmental History*, 4, 487.

Skinner, M. L. (1972). *The fifth sparrow: An autobiography*. Sydney, Australia: Sydney University Press.

Thoreau, H. D. (1854/1966). *Walden; and Civil disobedience: Authoritative texts, background, reviews, and essays in criticism*. (T. Owen, Ed.). New York, NY: Norton.

Thoreau, H. D. (1862/2007). *Walking*. LaVergne, TN: Filiquarian Publishing.

Thoreau, H. D. (1993). *Faith in a seed: The dispersion of seeds and other late natural history writings*. Washington, DC: Island Press.

Thoreau, H. D. (2000). *Wild fruits: Thoreau's rediscovered last manuscript*. New York, NY: W.W. Norton.

West, M. (1984). Thoreau and the language theories of the French Enlightenment. *ELH: English Literary History*, *51*(4), 747-770.

Whewell, W. (1840). *The philosophy of the inductive sciences: Founded upon their history* (Vol. 2). London, England: John W. Parker.

Wolin, R. (1993). "OVER THE LINE:" Reflections on Heidegger and National Socialism. In R. Wolin (Ed.), *The Heidegger controversy* (pp. 1-22). Cambridge, MA: MIT Press.

10

Poetry

Abstract: Habitat poetry represents the lives of plants, animals and the features of the natural world within their ecological networks. Commonly detailing physical contact with nature, habitat poetry narrates moments in which the senses engage with ecological processes. Additionally, habitat awareness in poetry tends to convey a sense of grappling with scientific discourses. These characterisations of habitat poetry will be articulated in the context of the South-West. The works of poets Alec Choate (1915-2010), Andrew Lansdown (1954-), and John Kinsella (1963-) use sensory language to express something about nature and convey the dynamics between science and poetry. The concept of habitat provides an interpretative framework for reading Choate, Lansdown and Kinsella. The three could be described not only as landscape poets but more precisely as habitat poets, a distinction pursued in this chapter.

> *The witching hour is noon in the gum forest.*
> *Foliage builds like a layering splash: ground water*
> *drily upheld in edge-on, wax-rolled, gall-puckered*
> *leaves upon leaves. The shoal life of parrots up there.*
>
> Les Murray (2007, p. 31, ll. 17-20)

Poeticising Flora

For centuries, writers across the globe have poeticised flora. The plant-poet nexus surfaces in the literature of most cultures. While some poets are conversant botanists, many are appreciative botanisers. As the quintessence of the poet-botanist, polymath Johann Wolfgang von Goethe (1790/2009) formulated botanical theories through literary discourse, ultimately resulting in hybrid treatises such as *The Metamorphosis of Plants*. Pablo Neruda (1995) wrote numerous odes to wildflowers and forests, including "To Flowers Along the Coast" and "To the Petrified Forest." In Europe, poet-botanists from William Wordsworth, Erasmus Darwin and George Crabbe to John Ruskin and D.H. Lawrence considered plants in their poetry (Mahood, 2008). In the contemporary collection *Terrain Seed Scarcity*, Peter Larkin (2001) exerts phenomenological consciousness on trees in "The Willow" (p. 187) and "The Leafless Tree" (p. 187).

The introductory excerpt from Les Murray indicates that the global practice of poeticising trees, shrubs and herbs is also evident in the Australian canon. Allusions to flora, set within a dialogue with science, occur commonly in the works of Judith Wright (1994) in "Phaius Orchid" (p. 88-89), "Nameless Flower" (p. 130) and "Sun Orchid" (p. 414). The writings of Wright and Murray suggest that poetic attention to flora facilitates interaction with the natural world as a complex living habitat: a poetic ecology of plants. Yet, the tradition of writing about plants is highly varied. For instance, published in 1910 in the evanescent literary journal *The Leeuwin*, "The Pillars of the West" by Willem Siebenhaar (cited in 2005, p. 282) elegises the destruction of the karri forests of the South-West. Siebenhaar's verses typify the depiction of trees as aesthetic objects (see chapter 5). The poem centres on the sublime visual impact of the forest, rather than its ecological importance as a habitat for animals, plants and people.

Through numerous examples from authors such as Siebenhaar, Murray and South-West poets, this chapter will argue that ways of writing about plants inherit aesthetic modes, ranging from inspection of pleasing visual features to embodied immersions. Some poetry attends to the static surfaces of individual plants or landscapes. Other poetry employs sensory acuity as a way to get beyond appearances alone and elicit the physical connections between humans, plants and animals. Some forms of poetry express a sensorily nuanced and interconnected—rather than visually privileged and immutable—aesthetics. Through language, poetry about plants can facilitate embodied investigation by tracing the corporeal entanglements between botanical and cultural worlds. Poetry about flora will often not be exclusively about plants—as a reductionist analogue of scientific botany—but rather about the cultural, historical, personal and ecological workings of flora and culture. Hence, in contrast to landscape poetry, *habitat poetry* embeds plants in ecological and cultural contexts. Describing subject-to-subject interaction, habitat poetry elicits bodily senses in relation to ecology, while often exposing a grappling with—scepticism of or faith in—empirical methods for understanding flora.

This chapter draws attention to the works of writers who poeticise South-West flora. The selected poets include Andrew Lansdown (1979), Alec Choate (1978, 1986, 1995) and John Kinsella (1997, 2005). Some poems trace multi-sensorial experience of flora, whereas others reveal how the science of plants interfaces with poetry and language. Lansdown, Choate and Kinsella poeticise plants in their habitats. Three themes characterise their works, sometimes within one poem itself, but more typically across a range of examples: engagement with science; the restoration of linguistic heterogeneity to the plant-people exchange; and the use of multi-sensoriality for exploring plants. Firstly, the poets tend to engage botanical knowledge and terminology—especially scientific names—as an acknowledgement of the technical discourses entered into when writing about plants. The relationship between a writer, a plant and the paradigm of science permeates some works as poets address the technical construction of nature (see Jolly, 2010). It could be claimed that science is unavoidable; a plant is demarcated visually from the rest of the strata of life through its assignment to the kingdom Plantae (see, for example, P. Clarke, 2008) (see chapter 1). Additionally, the term *plant* is a premise of taxonomic classification (Thomas, 1983, p. 65). For many writers, poetry initiates a dialogical space with science through the use of its specialised terminology.

Furthermore, by poeticising flora, writers restore non-specialist language, which can be co-opted by objectivist nomenclature. This is regionally invoked by Choate and Kinsella, for whom poetry redeems the usurpation of common and Aboriginal names. To state it differently, while taxonomy has expropriated language by systematising naming, habitat poetry, as one of its defining characteristics, restores linguistic heterogeneity: the West Australian Christmas Tree is known scientifically as *Nuytsia floribunda* and in Nyoongar as *mudja* (see chapter 3). Such poetry shows plants transgressing the technical language surrounding them. Lastly, multi-sensoriality is central to the restoration of language. The selected poetry traces sense-rich interactions with plants as habitat poetry, rather than speculative distances as landscape poetry. Beyond the scope of this discussion are the ethical responses of poetry to the destruction of plant life, elicited by Siebenhaar's "The Pillars of the West" and further politicised by Kinsella.

Landscape Poetry and Habitat Poetry

Poetry has the capacity to track the fluid qualities of plants and their processes. Poeticisation of plants is vital. Unlike animals whose movements tend to announce their presences, plants can more readily be depicted as static objects or as two-dimensional surfaces (see chapters 4 and 5). *The Lure of the Golden West* by Thomas Groser (1927) exemplifies the tendency of prose writers to represent flowering plants through a visual paradigm:

> There is scarcely a more lovely picture imaginable than a West Australian Bush in the Springtime. Pink is perhaps the prevailing colour—certainly where the 'everlasting' predominates. But flowers of every other colour of the rainbow are there—white daisies; pale blue leschenautia [*sic*]; red, blue and cream orchids; scarlet and yellow kangaroo paws growing to three feet in height on long slender stems; purple and mauve heather; golden buttercups and wattle—to mention but a few. The rich green undergrowth of Spring-time, and the evergreen and flowering eucalyptus trees, form a rich setting for this glowing pageantry of colour. (216)

The passage pictorialises flora through emphasis on colour and form. A "glowing pageantry of colour" echoes conventional wildflower tourism speak (see chapter 7). The excerpt constructs a scene implying the fixity of the flowers and their isolation from, for instance, their pollinators as well as other creatures and landforms co-occupying their space. Like Lindley's *Sketch*, Groser's text represents plants as objects of visual art. Hitchings and Jones (2004, p. 11) plainly note that "vegetation is something passive in contemporary understanding: to be in a vegetative state is to be without mind. Yet the root of the word 'vegetative' is associated with activity and enlivened animation." As *Green Sense* has argued, aestheticisation is the process by which the perception of passivity takes root. Visual focus picks out frames in the life cycle of plants, while privileging particular conventions of beauty (see chapter 5).

Encoding visual values, *landscape* is both a contentious term and a problematic way of perceiving nature. Elliott (1967, p. xi) in his seminal work *The Landscape of Australian Poetry* defines *landscape* as "the visible scene about us, the subject-matter of descriptive picture-making" and admits that, for synonyms, "I have sometimes employed the term *topography*, or referred to the *vista*, signifying what is seen by the eye [italics in original]." Although critical of "colonial topographical obsession," Elliott reiterates the imagistic focus of landscape poetry. He seeks "a natural revaluation of the environmental image," perhaps encompassing metaphorical or mental, but nevertheless based acutely in sight. Similarly, the geographer Denis Cosgrove (1998, p. 13) configures landscape as human perception of space. He defines the term as "not merely the world we see, it is a construction, a composition of that world. Landscape is a way of seeing the world." A way of seeing implies modes of viewership predicated on conventional gravitations of language: beautiful flowers, sublime forests or picturesque heath lands, for instance. Extending Cosgrove's notion, it could be said that landscape poetry is a way of seeing the world, a perception mediated by words and informed by conventions of aesthetics.

Interlude XXI: On the Desirability of Wildflowers Whilst Dog Walking

"On the Desirability" is about the inadequacies of poetry to express the sense of smell (see figure 23). In the novel *Perfume*, the protagonist Grenouille, born with hyperosmia or an acute sense of smell, begins to doubt the capacity of language for communicating his perceptual experience: "All the grotesque incongruities between the richness of the world perceivable by smell and the poverty of language were enough for the lad Grenouille to doubt that language made any sense at all" (Süskind, 1985, p. 27). In this poem, smell comes about at the convergence of domestic spheres and ecological contexts through walking a Rottweiler. The planners of the

Perth suburb Lesmurdie had foresight. The bush pops up everywhere in town. As the poem alludes, Lesmurdie's distinctive quality is a mixture of suburbanisation (a Telstra tower, sidewalks, a tennis ball) and indigenous plants (fuchsia grevillea and prickly moses) flowering amidst the infrastructures of civilisation. Prickly moses (*Acacia pulchella*) is an endemic South-West plant, widespread in the Perth Hills. Its vernacular name could be a corruption of "prickly mimosa," which it resembles in flower and foliage. Prickly moses is a member of the legume family and hence produces long slender cinnamon-coloured pods. Lindley (1840, p. xv) noted that *Acacia pulchella* is "peculiar to the Colony." Red-tailed black cockatoos cracking nuts in the jarrah and marri tree tops is an auditory outcome of the town's bushy conservation character. Cultural and botanical domains seem to merge together through the entanglement—or is it the juggernaut?—of being pulled along by the massive animal.

Figure 23 Wattle in Bloom in Dalwallinu Shire, Western Australia. *Image courtesy of the author.*

On the Desirability of Wildflowers Whilst Dogwalking

my Rottweiler Axle
suspires and strains at the lead
raring steadily like a sled dog
neck barreled and rippling,

musculature, a sleek black
as the hide of a gorilla, loping
along fickle Lesmurdie sidewalks,

overarched by wattle spoondrift
and the tentative reds of eucalypts
the sprawling coven of the winter hills
fuchsia grevillea and prickly moses—

his injections are succinic
episodes of micturus interruptus
into each cosmetic cranny of lawn,

each errant tuft of trans-hemispheric
grass issued an amber ordinance
a canine-decreed cease and desist
he knows the punctilio of the town—

in the enclaves of bush, a Telstra tower;
underneath the aureole of flowering acacia,
a radiance tempered by solstice mists,

a daily rumpus with a fenced-in dog
perturbs his mild-mannered countenance
sends him into twists of baritone
barking , shaking bowel-deep excitement,

quietened by the red-tailed cockatoo
wrestling jarrah nuts, splitting stone-hard
hulls with beaks like iron instruments—

my companion sees none of this (or perhaps)
has seen it all along, feigns being impressed
instead mesmerised by sand specks
studding a tennis ball I pitch in perfect,

slow camber over the sated scrubland
over low-dwelling insect sucking rosettes
above purple splendour herbal ringlets—

he tears down the track after it
snaps the nuclear green orb into jaws
in mid-stride, sniffs for dog sign language,
leaves a vileness that consumes the air—

> nose safe in wattle churnings of sweetness,
> it occurs to me, there are too few adjectives
> for the tragedies of smell, well, none that admit.

The distinction between landscape poetry and habitat poetry enables a more careful analysis of the nuances between people and plants. Does the written work stand back distantly and paint a "lovely picture," or does it enter into close contact with plants in places of entanglement through tasting, touching and smelling? Extending Giblett's (2011) distinction between landscape writing and nature writing, I differentiate landscape poetry from habitat poetry. Giblett points out a critical difference between writing that constructs a landscape through sight more-or-less exclusively—propounding a hierarchy of the senses and judgements of visual value—and writing that traces multi-sensorial contact and encounter:

> Landscape writing that aestheticises the static surfaces of nature can be contrasted with nature writing that celebrates its dynamic depths. I define 'nature writing' as the creative, written tracing of the bodily and sensory enjoyment of both the processes and places of nature. (26)

Through the creative intent of the poet or not, poetry that focuses on the surfaces of plants can be understood as landscape writing. In contrast, direct sensory writing about flora—where people make contact with plants—is a genre of nature writing or more precisely *habitat poetry*.

Aesthetic values filter into poetic representations. The perceptual legacy of a poem, considered along with its literary or historical context, reflects the decisions of writers to adopt ways of seeing and sensing. As one example among many, the sublime spectatorship evident in William Wordsworth's *The Prelude* forges asymmetries of the natural world in which "the superiority of mountains over other types of country comes down to a matter of sight, appearance and feeling" (Giblett, 2011, p. 69). Landscape poetry inherits the hierarchies of the beautiful, sublime and picturesque as points from which to view the surfaces of the natural world and form value judgements (Giblett, 2011, pp. 59-60) (see chapter 5). Rather than a interdependent and mutable living habitat, carefully attended to in all its sensuous particularity, *landscape* implies an externalised and artefactual scene. Landscape as concept makes possible the notion of a natural environment as separate from human culture.

Nature writing engages the processes of nature through attentive sensuous acts (J. Murray, 1995, p. 10). In contrast to landscape poetry, habitat poetry—as a genre of nature writing—attends to broad environmental processes and scales. Most importantly, in habitat poetry allusions to one plant or species mingle with allusions to other plants, animals, rocks, landforms and cultural features that constitute the living land. Landscape poetry is aligned to aesthetic traditions; habitat poetry refers to ecological and bodily interdependencies. Habitat poetry is a form of process-oriented nature writing; landscape poetry constructs visual products such as picturesque scenes. Conveying ecological associations, habitat poetry attunes to ecological rhythms as its essential context. It is also possible to discern poetry that attends to isolated plants or species. This poetic form replicates the reductionism and specialisation of botany, zoology and geology as branches of knowledge. *Species poetry* is wholly centred on a single plant through a classificatory way of seeing nature. In species poems, an individual plant—such as the Underground Orchid or the West Australian Christmas Tree—is given full attention. Species poetry can express sensory interactions with a plant. However, this form of poetry risks marginalising the representation of habitat connections. Admittedly, this way of rethinking poetry—landscape, habitat, species—is taxonomic. However, it is not intended to promulgate a hierarchy but rather emphasise a continuum between modes of representing plants across which many poets write.

All in all, the analysis of botanical poetry reveals that some works afford multi-sensorial representation through a habitat mode. Other works prioritise the surface features of plants through a landscape mode. Invariably, botanical poetry exists between landscape and habitat modes, thereby mixing vision and the multiple senses. In many of the selected works by South-West poets, curiosity initiates embodied investigation: smelling, touching, tasting, listening carefully and looking closely. As such, visual experience is tethered to sensory acts. Hence, landscape poetry shapeshifts into habitat poetry through the sensuous participation of the poet in what otherwise would be distant scenes or objects. Oscillation between seeing and sense acts characterise many plant poems; an unfolding vista is counterpoised by the immediacy of touch or the envelopment of smell. For Thoreau, swamps afforded bodily immersion in habitats, rather than Wordsworthian speculation about panoramas (Giblett, 1996). Indeed, Thoreau wrote habitats rather than landscapes; his sensory exposure to the natural world inhibited any pretence of distance (see chapter 1).

A comparative reading of Siebenhaar's "Pillars of the West" and Murray's "The Gum Forest" is illustrative of the critical differences between landscape and habitat poetry. Both poems deal with gum forests, but in strikingly divergent ways. "The Pillars of the West" (cited in Plate, 2005, p. 282) valorises the sublime upward thrusting form of the karri forest: "Uprise the Pillars of the West,/ Like kings in armoured splendour dressed" (p. 282, ll. 7-8). The morose sentiment of the poem echoes the characterisation of the Australian bush by Marcus Clarke (1876/1993) as "weird melancholy." Siebenhaar's poem refers to "the great and lonely bush" (p. 282, l. 9); "a hush/ That almost fills the heart with dread" (ll. 12-13); and a "sombre land" (l. 14). The smaller trees and birds that live beneath the canopy of the great forest comprise the dreadful underworld of the sublimated pillars: "And giant blackboys ghostlike stand" (l. 11) and "A shriek of savage fear is heard/ From yonder brightest emerald bird" (ll. 15-16). Not only is sight valorised, but priapic ("lofty") vegetative forms dominate the feminised and interconnected understory or "the tangle:"

> Out of the tangle here below
> Like lofty titan-lilies rise
> In slender grace the karri kings;
> Their silver grey cuirasses glow
> In the dim light of darkened skies,
> And from their cloud-kissed helmet springs
> The mazy verdure of each plume,
> So dense that all beneath is gloom. (ll. 17-24)

Siebenhaar presents the architecture of the scene. The dense gloom nullifies the smells, sounds, tastes and tactile sensations of the karri forest floor beneath the sublime reaches of the trees. Moreover, the poem uses numerous battle metaphors, such as "silver grey cuirasses" and "cloud-kissed helmet." The karri kings preside over an Earth that is hostile and melancholic. "Pillars of the West" could be read as a lamentation of the destruction of the sublimity of the karri forest. However, the subtext shows the formation of aesthetic hierarchies of flora and the indoctrination of writers of Siebenhaar's era in a perception of the Australian bush as macabre and unwelcoming. The bush was not a familiar place, a home terrain (see chapter 6). "The ploughman's son will know at best/ Where stood the Pillars of the West" (ll. 39-40), but who will memorialise the feminised understory where occur microscopic interconnections of entangled roots and soil? The minutiae of gesture and contact—an intrigue for the constant dynamism in a forest ecosystem—go unregistered by the poet's fixation on scenic monumentality.

Whereas the karri forest is a foreboding place for Siebenhaar in the early twentieth century, the gum forest is explicitly home for Murray later in the century. His poem "The Gum Forest" opens with the couplet "After the last gapped wire on a post,/ homecoming for me, to enter the gum forest" (p. 31, ll. 1-2). Like Siebenhaar, Murray deploys battle tropes to convey the imagery of the gum forest: "This old slow battlefield: parings of armour,/ cracked collars, elbows, scattered on the ground" (ll. 3-4). Yet, Murray refrains from pushing such war metaphors too far and instead turns to the processes unfolding microcosmically before him in a tone of wonderment: "New trees step out of old: lemon and ochre/ splitting out of grey everywhere, in the gum forest" (ll. 5-6). Further along, Murray comments on the interdependencies of the gum forest: "Flooded-gums on creek ground, each tall because of each./ Now a blackbutt in bloom is showering with bees/ but warm blood sleeps in the middle of the day" (ll. 14-16). Rather than sombre and hostile, the gum forest is a mysterious place of reverie, reverence and rejuvenation: "Delight to me, though, at the water-smuggling creeks,/ health to me, too, under banksia candles and combs" (ll. 24-25).

Landscape poetry prioritises sight, whereas habitat poetry encourages sense immersion. Murray's poem takes place on the floor of the gum forest, whereas Siebenhaar's verses are located in the sublimated canopy. "The Gum Forest" is exemplary of habitat poetry; seeing is linked to the ecology and the physical experience of the forest as a place of sensory openness: "I go my way, looking back sometimes, looking round me;/ singed oils clear my mind, and pouring sound high up" (ll. 28-29). Murray's perception is ecosystemic and sensory, leading to appreciation of forest processes, not merely its aesthetic products (see chapter 5). The poem closes with a question and an answer: "Why have I denied the passions of my time? To see/ lightning strike upward out of the gum forest" (ll. 30-31). In lieu of disdain or horror, mystery and reverence are the resounding notes: "You can never reach the heart of the gum forest" (l. 10). The sublimated reaches of tree crowns—"the mazy verdure of each plume"—figure only marginally in Murray's reading. For Siebenhaar, the canopy subordinates the Earth beneath, but for Murray the floor is a nexus of ecological events. As a mode of seeing, landscape entails acts of sight that reiterate traditions of visual appreciation. In contrast, sensory entanglements permeate habitat poetry and broaden "a way of seeing the world." Habitat poetry is a way of representing acts of becoming—changing, growing and processing through the seasons—as Murray's poem aptly suggests; it fosters contemplation, attentiveness and inquisitiveness in nature.

Reading Lansdown, Choate and Kinsella as Habitat Poets

Andrew Lansdown (1954–), Alec Choate (1915-2010) and John Kinsella (1963–) write with consistent attention to plants. With focused awareness on ecological processes and the sensory experience of the poet, their works are examples of habitat poetry. Their poems engage botanical knowledge through references to scientific names and the use of specialised terminology. Plants change in these poems; the reader's awareness of change occurs incrementally rather than through a set of visual instances. Often, the works include non-visual experiences—although not always—as a way of representing flora through the multiple senses. "A Few Weeks Later I Returned To Find" by Lansdown (1979, p. 8) sets in verse an awareness of plants in-becoming. In the poem, visuality is contextualised by the habitat processes of the balga flower, made haptic through interaction between the plant and the poet. Lansdown reminds us that investigation of flora is not restricted to the sciences; it can take the form of a patient opening of the human senses to the plant (see chapter 13). For Lansdown, enquiry blends science and sensuousness. With adeptness, "A Few Weeks Later" weaves the representation of physical encounter with perception of the well-formed features and pleasing symmetries of the flower. Moments of sensory contact with the balga are preceded by a kind of taxonomic itemising:

> Centred in the stamens,
> the shorter stylus – surrounded,
> and at times, 'covered
> by a glistening glob of transparent nectar
> which, in turn, was caught in the cup-pit
> of the six guardian stamens. (ll. 11-16)

After an anatomical inventory noting the stylus and stamens, the poet turns towards embodied modes of engagement.

At this point, the poem moves up and down the balga stalk: it constructs not a single moment of perception, but a moving awareness of the *Xanthorrhea* as a changeable plant. Morphological descriptions of the kind found in any scientific guidebook are enlivened by bodily analogues such as "tiny yellow vulvas of pollen" (l. 10). Inquisitiveness as a mode of openness culminates when:

> I thought each flower had mysteriously
> caught last night's dew,
> so I put my tongue to it
> (Descartes would not have approved) to see:
> it was a powerful, honey-thick
> nectar. The odour was a heavy
> sweetness. I wiped the pollen from my nose. (ll. 17-23)

Lansdown's poem rather self-consciously brushes up with scientific authority: "Descartes would not have approved" (l. 20). Here, sense acts deconstruct the scientific derivation of knowledge. Sensuous, and even scandalous, moments of intimacy with the balga provoke sensory entanglements: "So I put my tongue to it" (l. 19). The senses form into a sensorium; the whole sensing body participates in, rather than distantly views, the microscopic and macroscopic features of balga. Vision—"to see" (l. 20)—joins with taste—"a powerful, honey-thick/ nectar" (ll. 21-22)—and smell—"The odour was a heavy/ sweetness" (ll. 22-23). The poet responds physically: "I wiped the pollen from my nose" (l. 23).

Contrary tactile qualities typify the South-West flora (see chapters 1 and 2). For instance, dryandras bear down-soft flowers guarded by stiff thorny foliage. The balga offers a comparable range of sensation. After summoning a colloquial name "blackboy" as a direct address of the plant, the poem reveals the tactile metamorphosis of the flower:

Blackboy,
the compact, coarse sandpaper
of your flower-spear
has turned to softness. (ll. 24-27)

The core human awareness of the plant is its mutability; the senses change in response. The balga bears the capacity for self-generated evolution from coarseness or even armour—its "flower-spear"—to softness. The title, "A Few Weeks Later I Returned to Find," intimates that awareness results from the diligence—or happenstance—of revisiting plants throughout the year. The balga tree is poised in-becoming, not frozen in a state of perception as a conglomeration of aesthetic features. The perception of the balga comprises the evolution of the plant in relation to a the body of the poet. Awareness of the habitat of which the balga is part and of the balga *as* a habitat is evident in the last stanza:

Later
the seeds will come;
then the parrots.
I see the sharp, triangular lengths
of your deep-green leaves
shimmer in anticipation. (ll. 28-33)

Not a stand-alone image or icon, the balga is situated ecologically with seeds and parrots imminent. "Shimmer in anticipation" suggests the sensory possibilities of the plant with the poet poised to engage its materiality.

Plants also figure conspicuously in the works of poet Alec Choate. Born in Hertfordshire, England in 1915, Choate spent most of his life in Western Australia and much of his poetry deals with the desert regions of the state where he worked as a surveyor (Choate, 1995). "Nuytsia floribunda" and "Poverty Bush" exemplify the connection of his poetry to science and naming. "Nuytsia floribunda" refers to the endemic West Australian Christmas Tree (Choate, 1986, p. 16) (see chapter 3). The movement of science into poetry is reflected in the allusion to the taxonomic name of the plant, rather than its multiple common names, such as Cabbage Tree. The poem represents Nuytsia in its habitat. Still, Choate refrains from technical jargon like *hemiparasite* or *haustoria* used scientifically to denote the parts of Nuytsia (see, for example, Hopper, 2010). Choate begins by explaining the symbolic meaning of the intriguing plant Nuytsia for the Western Australian summer:

This tree could only find root
in a land whose heart belongs
to its summer, no other

season, and where the summer
feels bound to repay that heart
with an emblem of grandeur. (ll. 1-6)

The middle stanzas trace the progression of the seasons and the anticipation of the arrival of the flower as the "emblem of grandeur:"

Autumn and winter linger
to make ready as tinder
all that is dark in their time,

the dark of life waste and mould
and rain when blind among roots,
the spleen of night's halting hours

or simply the crape of clouds.
Spring is the easing of limbs,
the young leaves winking for warmth. (ll. 7-15)

In later stanzas, the connection between fire and flower reflects the colonial trope of Nuytsia as a "tree on fire," as explored in chapter 3:

But when the summer returns,
the culminating summer,
it breathes upon the tinder,

and from the gauntlets of green
of this chosen tree in its
thousands, lights torch after torch

of amber that floods to gold,
its own heart naked as fire
the land's heart naked as flower. (ll. 16-24)

"Tinder," "torch after torch," "heart naked as fire" and "heart naked as flower" suggest the metabolic processes of the plant. As homophones, fire and flower refer to regenerative processes of the land (see Hallam, 1975). Flower colour crystallises as the season progresses: "Amber that floods to gold" (l. 22). Rather than an eschatological end point, the golden blossom is a culmination—a crescendo—of a network of interactions between "waste," "mould," "rain" and "roots." Nuytsia in bloom is positioned as part of a broader habitat. By the end, fire is no longer figurative, but a material reality into which the flower projects upward. The effect is not the sublimation of the tree, in Siebenhaar's terms, but the relation between flower and fire in literal and figurative ways:

> while from skyline to skyline
> the haze is a secret's veil
> that a fierce trust has shredded. (ll. 25-27)

The haze of fire reflects the time-worn processes of the land. The fierce trust that shreds the haze is the uplifted golden canopy of the Nuytsia, entwined with the tacit agreements between organisms that ensure regeneration. In this sense, the poet represents ecology as a pact between humans, plants, animals, landforms and elements of fire, wind, water and earth. Choate's nuanced reading of the flower within its habitat contrasts distinctively with a surface-oriented reading of the flower in isolation, as a beautiful object (see chapter 5).

In "Poverty Bush," Choate (1995, pp. 94-95) comments on the historical misperceptions inscribed in plant names vis-á-vis the desert plant *Eremophila alternifolia.* Poverty Bush is concentrated on the arid eastern edge of the South-West Botanical Province (Spooner, 1999). The etymology of *Eremophila* is from the Greek *phila* for "to love" and *eremo* for "lonely places" or "desert" (Coleman, 2008). The opening two tercets connect the plant to a love for dry places. However, the love is tainted by a poverty of popular perception, as indicated by its common name:

> The desert cries out for love,
>   and no shrub answers
> With more heart to return it
>
> than this whose vast sisterhood
>   is far from the name
> and seeming of poverty. (ll. 1-6)

Although Poverty Bush has adapted to desert country, the same soils in which it prospers inhibited the spread of colonial European plants. Choate historicises the common name as a misnomer applied by European settlers:

> Poverty of mind rather
>   was theirs who crammed mouths
> on the pastoral reaches

and who when the ground feed died
        so named the shrub, it
being no browser's standby. (ll. 7-12)

The tough, spiky foliage of the bush thwarts grazing by animals. The act of granting common names preceded knowledge of the plant's ecological adaptivity. Names reveal the prerogative of settlement to consume and convert land. However, the metaphorical aspects of scientific names reveal the affinity of Poverty Bush for desert conditions. The technical name conveys that the plant has adapted to the same conditions that repelled early settlers. For Choate, the act of taxonomic naming entails the messianic arrival of botanists who redeem the plant from cultural misconstruction as a wasteland species:

But someone came, someone saw
        it lacquer its leaves
against the wind's rainless lips,

scatter its seeds and trust roots
        to the rust-red sand,
saw how it decored itself

in wool, a ripple of scales,
        a mantle of hair,
or posed sepals as petals,

its means, and its miracles,
        for coming to terms
with skies and their gaze of stone.

Someone came. The name blossomed.
        "*Eremophila*,"
he said, "or Desert Loving." (ll. 13-27)

Perhaps the "someone" was Robert Brown who first identified the species. The poeticising of Poverty Bush reveals familiarity with verbed processes: "Lacquer its leaves," "scatter its seeds," "trust roots," "decored itself/ in wool" and particularly "posed sepals as petals." The habitat is austere: "The wind's rainless lips" evokes the parched earth. The final tercet suggests that, if comprehended, the poetic meanings of names can animate scientific understandings with cultural symbolisms (see chapter 2).

Poetic deliberation on plants also figures significantly into the works of John Kinsella (1963–), a prominent Western Australian writer and literary critic. His poetry often displays an adept visual awareness of landscapes. However, the slippage between visual construction of landscape as image and bodily involvement with landscape as moving sensation are evident in "Everlastings" (1997, pp. 29-30). The poem opens with bodily analogues to describe the picking of wildflowers:

> A couple pick flowers
> while their child lies
> cribbed in the dry rustle
> of stalks & petals. (ll. 1-4).

Although outlawed in 1935 (Summers, 2011), the harvesting of wildflowers is a haptic—though not necessarily an environmentally conscious—mode of engaging flora. In the poem, the plant habitat is protective and nurturing; flower stalks and petals cradle a small child like a crib. Anatomical tropes connect people and plants through a common corporeality: the everlastings are "broken-necked." Similarly, "Paperbarks" (Kinsella, 1997, p. 174) brings people and plants into contact as "skins peel and flake/ about the grasping roots" (ll. 6-7). Skin imagery describes the exfoliated tree bark, a distinguishing visual characteristic: "Absorbent skins will not extinguish when voice/ falls and memory lingers, for these are ghosts" (ll. 9-10).

Although not a corporeal poem, "Everlastings" employs bodily imagery to assemble a scene that eventually ripples to life with habitat experiences. The "painting" (l. 17) breathes through the movements of the breeze enlivening "rippling waters" of everlastings:

> The breeze stirs the feeling
> deep inside the painting,
> the sun flickers
> & is passed over
> by pumice clouds, bunches
> hanging rigid in the shade,
> mock-glorious in their brilliance. (ll. 16-22)

The sensations of landscape mix with images of flowers. Through lyrical assertion, the painting is made fluid like water pouring into an empty vessel. The visual features of the everlastings are perceptible as "white pink rose scarlet" (l. 9). Yet the colours are synaesthesiac and full of motion: "Cool almost ice their swayings" (l. 10). Allusions such as "Bees, laden" (l. 12), "the sun flickers/ & is passed over/ by pumice clouds" (l. 18-20) reveal habitat consciousness tinged with a visual aesthetics. In a similar vein, the poem "The Bottlebrush Flowers" (Kinsella, 1997, pp. 174-175) deconstructs the colloquial descriptors "bristling firelick"

(l. 11) and "a spiral of Southern Lights" (ll. 11-12) through reference to the habitat relations of the tree: "I've also seen/ honey-eaters bob upside down/ and unpick its light in seconds" (ll. 12-14).

"Exposing the *Rhizanthella gardneri* Orchid" (Kinsella, 1997, p. 227) exemplifies Kinsella's habitat poetry. The poem's sensory language is linked to an exposition of orchid ecology. The progression is sense-based and earth-anchored. The poem's title indicates the orchid's taxonomic name, but the plant also has the common name "Underground Orchid." The opening stanza alludes to the mutualistic adaptations of the orchid. The flower is seldom exposed to sunlight and must survive through particular dependencies on surrounding flora (Brown, Batty, Brundrett, & Dixon, 2003). As chapter 8 discussed, the orchid relies on other indigenous plants:

> Above the roots
> of a Broom Honey Myrtle
> the beak of an orchid
> tastes the acrid air.
> Its mouth sweet with flowers.
> Termites roaming the pollen. (ll. 1-6)

The diction is gustatory as "the beak of an orchid/ tastes the acrid air" (l. 3-4) and "its mouth sweet with flowers" (l. 5). The orchid anatomy—its beak—is observed along with its processes, including termites spreading pollen. The second stanza further represents the ecology of the Underground Orchid. Scientific terminology outlines a relationship between orchid, host tree and fungus:

> Saprophyte,
> and guest-host
> to a root-invading fungus,
> its liaisons go unnoticed
> as the scrub is peeled back,
> and are only half-revealed
> with the lifting
> of the surface. (ll. 7-14)

In fact, the Underground Orchid is not a true saprophyte deriving nutrients from dead things, but instead thrives on a symbiotic fungus formed between the orchid and the myrtle. The unnoticed "liaisons" (l. 10) are the unseen mutualisms occurring under the soil. The ritualistic sequence of actions required to unearth the orchid parallels Lyn Alcock's reflections in chapter 8. Processes linger beneath the "surface:"

Excavated,
its leaves unfold
and termites roam the pollen,
its dark heart
reddening
with exposure. (ll. 15-20)

Motion infuses images: "Its leaves unfold" (l. 16). The flower reddens; termites distribute pollen as its only known pollinators. In the final analysis, the poem mediates the above-ground act of unearthing and the normal concealed state of the orchid. The writing of the poem seems to take place at the edge between surface and depth, dispelling any preoccupations with vistas or broad-sweeping expanses, with landscapes.

Interlude XXII: A Colony of Royal Hakea

During the spring of 2009, I went on a bushwalk sponsored by the Friends of the Fitzgerald River National Park, an important local conservation group dedicated to protecting the global biodiversity heritage site, the Fitzgerald Biosphere. "A Colony of Royal Hakea" recounts fragments of the walk, in which our group encountered a pod of Royal Hakea (*Hakea victoria*), named in honour of Queen Victoria and known less commonly, though more evocatively, as Lantern Hakea (see figure 24):

erratic columns of cabbage heads,
crowned with russet splotches
as if gnawed on by locusts (ll. 17-19)

The practice of walking invokes a tradition of foot travel with botanising for enjoyment as one of its central motivations (see chapter 12). Movement on foot conjures images of the old explorers like John Eyre (1845/1964) who reported disparagingly on the present-day Fitzgerald River area (see chapter 6). Along with the Qualup Bell, Royal Hakea is an endemic species, identified by botanist James Drummond in 1847 en route to West Mount Barren. Drummond (1848, p. 2) praised the rich greenness and the "blood-red colour" of the bracts of Royal Hakea: "To this, the most splendid vegetable production which I have ever beheld, in a wild or cultivated state, I have given the name of our gracious Queen, Hakea Victoria. It will soon be cultivated in every garden of note in Europe, and in many other countries" (to read "A Colony of Royal Hakea" in full, see Ryan, *in press*).

Figure 24 Royal Hakea at the Fitzgerald River National Park. *Image courtesy of the author.*

Celebration of Landscapes and Appreciation of Habitats

How can poetry represent habitat relations? To begin with, South-West poets who engage plants multi-sensorially often evidence a dialogue with scientific discourses. The poems about flora offered by Kinsella, Choate and Lansdown reveal critical interactions with the terms of science. Interpenetration between science and the senses is a trademark of a habitat poem. Corporeal diction vis-á-vis plants deconstructs claims to objective knowledge by representing the human body as part of the scene. As both Thoreau and Linnaeus remind us, most enquiries into nature begin in language (see chapters 1 and 2). A habitat comprises earth, science and language in a matrix of relations; but habitat is also a mode of interaction. As such, habitat is a way of engaging with plants, just as "landscape is a way of seeing the world" (Cosgrove, 1998, p. 13). In "Borrow Pit," John Kinsella (2005, p. 142) ends with:

> I borrow words
> from before I could speak, the tones of wandoo and mallee,
> intricacies of roots, and palettes of gravel
> that stare us in the face, trunks horizontal, parallel
> to the rippling undersurface, those winning ways. (ll. 87-91)

Attending to the complexities of the unseen as "the rippling undersurface" and "intricacies of roots," habitat poetry connects language to Earth. Siebenhaar's sublimated canopy codifies the sublime and beautiful. In contrast, Kinsella elicits the intersection of habitat thinking, regional plants and poetic language. It could be argued that habitat poetry inherits Earth whereas landscape poetry reiterates aesthetics. The selected works of South-West poets demonstrate that habitat poetry focuses on the tangible intersections between plants, people and place. As habitat, poetry engages science, restores linguistic heterogeneity to flora and fosters multi-sensorial relations. Each of the poems presented accentuates some of these themes. For example, Choate's poetry attends to science and taxonomic language, with "Poverty Bush" a strong commentary on the idiosyncrasies of naming. However, few of his poems introduce the sensations of the writer on the ground. Lansdown's "A Few Weeks Later I Return to Find," in contrast, exhibits a critique of scientific methods along with the poet's experience of the plant but lacks the nomenclatural interests of Kinsella and Choate.

Augmenting scientific knowledges, sensuous language is accessible to a wide audience of appreciators, not only specialists. In *Wildflower Country*, a collection of poetry about the plants of the Fitzgerald River National Park, Jenny de Garis (cited in Southern scribes, 2007) mentions on the back cover "the tastes and perfumes that belong only here." In the collection, short poems such as "Xanthorrhhoea sp." (p. 22) by Gai Downes and "Dampiera sacculata" (p. 25) by Mary Smith draw attention to the local flora. Poetry has the capacity to instil appreciation for plants by shaping the ways in which people regard flora and by distilling ecology into a palatable and palpable form. As a complement to visual representations, habitat poetry opens up possibilities for curious sensuous interaction: touching the balga stalk or unearthing the orchid. The adaptive successes of many South-West plants can thus be brought to public light without the narrowing effect of specialist language.

References

Brown, A., Batty, A., Brundrett, M., & Dixon, K. (2003). Underground orchid (*Rhizanthella gardneri*) Interim Recovery Plan 2003-2008 Retrieved from http://www.environment.gov.au/biodiversity/threatened/publications/recovery/r-gardneri/background.html

Choate, A. (1978). *Gifts upon the water*. Fremantle, Australia: Fremantle Arts Centre Press.

Choate, A. (1986). *A marking of fire*. Fremantle, Australia: Fremantle Arts Centre Press.

Choate, A. (1995). *Mind in need of a desert.* Fremantle, Australia: Fremantle Arts Centre Press.

Clarke, M. (1876/1993). Preface to Adam Lindsey Gordon's Sea spray and smoke drift. In M. Ackland (Ed.), *The penguin book of 19th century Australian literature* (pp. 43-46). Ringwood, Australia: Penguin Books.

Clarke, P. (2008). *Aboriginal plant collectors: Botanists and Australian Aboriginal people in the nineteenth century.* Dural Delivery Centre, Australia: Rosenberg Publishing.

Coleman, H. R. (2008). Eremophila R. Br. *FloraBase* Retrieved from http://florabase.calm.wa.gov.au/browse/profile/22098

Cosgrove, D. (1998). *Social formation and symbolic landscape.* Madison, WI: University of Wisconsin Press.

Drummond, J. (1848). Notice of Mr. Drummond's discovery of three remarkable plants in South-West Australia. *Companion to Curtis's Botanical Magazine, 74,* 1-3 Retrieved from http://books.google.com

Elliott, B. (1967). *The landscape of Australian poetry.* Melbourne, Australia: F.W. Cheshire.

Giblett, R. (1996). *Postmodern wetlands: Culture, history, ecology.* Edinburgh, Scotland: Edinburgh University Press.

Giblett, R. (2011). *People and places of nature and culture.* Bristol, England: Intellect.

Goethe, J. W. V. (1790/2009). *The metamorphosis of plants.* Cambridge, MA: MIT Press.

Groser, T. (1927). *The lure of the golden West: Experiences and adventures in a bush brotherhood of Western Australia: Early problems and conquests: All about group settlements of the West: The land of sunshine and opportunity.* London, England: Alexander-Ouseley.

Hallam, S. J. (1975). *Fire and hearth: A study of Aboriginal usage and European usurpation in south-western Australia.* Canberra, Australia: Australian Institute of Aboriginal Studies.

Hewitt, D. (2001). *Halfway up the mountain.* Fremantle, Australia: Fremantle Arts Centre Press.

Hickman, E., & Ryan, J. (*in press*). *Two with nature: Ellen Hickman and John Ryan.* Fremantle, Australia: Fremantle Press.

Hitchings, R., & Jones, V. (2004). Living with plants and the exploration of botanical encounter within human geographic research practice. *Ethics, Place & Environment, 7*(1 & 2), 3-18.

Hopper, S. (2010). Nuytsia floribunda. *Curtis's Botanical Magazine, 26*(4), 333-368. doi: 10.1111/j.1467-8748.2009.01671.x

Jolly, E. (2010). *Challenging the divide: Approaches to science and poetry.* Adelaide, Australia: Lythrum.

Kinsella, J. (1997). *Poems 1980-1994*. Fremantle, Australia: Fremantle Arts Centre Press.

Kinsella, J. (2005). *The new arcadia*. New York, NY: Norton.

Lansdown, A. (1979). *Homecoming*. Fremantle, Australia: Fremantle Arts Centre Press.

Larkin, P. (2001). *Terrain seed scarcity: Poems from a decade*. Cambridge, England: Salt.

Lindley, J. (1840). *A sketch of the vegetation of the Swan River colony*. London, England: James Ridgway.

Mahood, M. (2008). *The poet as botanist*. Cambridge, England: Cambridge University Press.

Murray, J. (1995). *Nature writing handbook: A creative guide*. San Francisco, CA: Sierra Club Books.

Murray, L. (2007). *Selected poems*. Melbourne, Australia: Black Inc.

Neruda, P. (1995). *Neruda's garden: An anthology of odes* (M. Jacketti, Trans.). Pittsburgh, PA: Latin American Literary Review Press.

Plate, C. (2005). *Restless spirits: The life and times of a wandering artist*. Sydney, Australia: Pan Macmillan.

Ryan, J. (*in press*). John Charles. In S.-P. Mitchell (Ed.), *Fremantle poets 3: Performance poets*. Fremantle, Australia: Fremantle Press.

Southern Scribes. (2007). *Wildflower country*. Ravensthorpe, Australia: Southern Scribes.

Spooner, A. (1999). Eremophila alternifolia R.Br. Poverty Bush. *FloraBase* Retrieved from http://florabase.calm.wa.gov.au/browse/profile/7180

Summers, L. (2011). Wildflower season: The development of native flora protection legislation in Western Australia, 1911-1975. In A. Gaynor & J. Davis (Eds.), *Studies in Western Australian History: Environmental Exchanges*, *27*, 31-44.

Süskind, P. (1985). *Perfume: The story of a murderer*. Camberwell, Australia: Penguin Books.

Thomas, K. (1983). *Man and the natural world: Changing attitudes in England 1500-1800*. New York, NY: Penguin Books.

Wright, J. (1994). *Judith Wright: Collected poems 1942-1985*. Sydney, Australia: Angus & Robertson.

Part V

Botanical Futures

Having analysed the multitude of synergies between plants and language in the preceding part, Part V, "Botanical Futures," moves towards forward-looking prospects and possibilities for cultural alliances with flora. This section will continue to focus on historical and contemporary events in the South-West in further developing concepts of flora. Chapter 11, "Mourning," contextualises the region's flora in the global crisis of biodiversity loss. This chapter maintains that the diminishment of plant diversity entails sensorial deprivations and that the mourning of botanical loss can be given direction through the premise of connectivity to habitats. Chapter 12, "Walking," conceptualises walking as an integral *habitus* of embodied connectivity that can bring about multi-sensorial relationships to flora. Chapter 13, "Aesthesis," posits a theory of gesture, contact and sensation with respect to plants. This chapter presents several cultural practices that entail somatic engagements between indigenous flora and post-colonial societies in South-West places.

11

Mourning

Abstract: Botanical extinction is not only an environmental crisis to be solved by biological, political or legal measures. As a complete extinction or as a gradual decline in occurrence, the disappearance of a species has emotional and aesthetic consequences that entail loss and absence. Not only are the flesh, bone and blood of the organism and its collective members gone, so are its colours, sounds, smells, behaviours and habitat relations. With every extinction, our sensuous worlds are impoverished. An uncanny silence of the landscapes make the loss more profound. The mosaic of life—of which we and plants are part—is irrevocably altered. Extinction is ecological and eschatological, requiring both science and poetics as palliative measures. Through the tragedy of near extinction or complete extinction, our connectedness to the world and to other beings, paradoxically, can be reconfirmed. This requires a conceptual framework for loss and actual modalities for mourning plant extinction.

A "Species" Is Bodiless

The expansion and growth of human populations and the demands of technological development have placed global pressures on wild plant communities. Accelerating declines in habitat and the pollution of air and water have led to an extinction crisis unprecedented in the history of three billion years of life on our planet. Biodiverse places, such as the South-West, are particularly susceptible to the pressures and transformations ecological systems are undergoing worldwide. The diverse flora of this corner of Australia offers a poignant study of the pandemic of biological extinction. Public appreciation combined with scientific recognition of South-West species offer possibilities for conserving plants, alleviating the pressures leading to their disappearance and enhancing opportunities for human engagement with non-human life (Hopper, 1993, 1998, 2004; Nikulinsky & Hopper, 2005).

Extinction involves broad temporal and spatial scales. Its monumental finality can be hard to comprehend. As Hopper (1993, p. 89) remarks cogently, "the loss of biodiversity is irreversible. Extinction, the death of birth, is final." The finality of loss, as in any death, should involve rituals of mourning. However, Shiv Visvanathan

(1996, p. 311) observes that "science has no mourning rituals." Thus, societies struggle to rationalise why extinctions matter. Part of the difficulty in coming to terms with extinction lies in the concept of the species itself. A *species* is part of the "discourse of nature" (Foucault, 1966/2002, p. 171). The taxonomic construct of *species* enables the classificatory schema of science, emphasising the reproductive structures of fruits, flowers and seeds as well as hierarchical relationships between organisms (see chapter 1). A species is an idealised form that generalises how an organism should appear—on the ground and in the field—without consideration of individual herbs and trees, or the behavioural dynamics between lives in an ecosystem. In other words, its abstractedness makes absent the body, flesh, pulse and sensorium of the plant it signifies. Beyond its scientific appellation, an organism is a sentient life, with multiple relations and a sensorial distinctiveness of its own. The individual plant slips between categorical abstractions, such as taxonomies, through breathing, sensing, consuming and all the various volitions that express it as being alive.

The disembodiment of *species* compounds the scale and severity of the extinction of species. The way through mourning extinction hence confronts two layers of absence. The first layer is the reification of the species itself, while the second layer is the loss of botanical life in actuality. A species cannot be touched, smelled, seen, heard or loved. Taxonomy functions in the first place because it generalises; it posits the physical absence of the plant enmeshed in a habitat. Classificatory naming succeeds as the primary method of natural science because it interrupts the connectivity of living plants. The natural world is reduced to a series of signifiers. The disjunction between species as signifier and organism as signified raises several problems for human mourning. How do we mourn the loss of a species when a species does not actually exist but rather points to the botanical body *in absentia*? How can we even approach the question of why extinctions matter when a species is formulaic, substituting images for actual living plants and abnegating human sensoriality? To come to terms with extinction as an ongoing contemporary possibility, we move to embrace—figuratively and actually—plants as organisms with corporeality. In order for extinction to matter, we need to engage the power of our senses and emotions, and contact the life cycles and seasonal nuances of plants to make meaningful such a monumental and hard-to-grasp occurrence as extinction.

How do landscapes of extinction look, smell, taste, sound and feel? This question calls forth the host of absences—the sensory deprivations—that extinctions precipitate. With attention to mourning the loss of South-West flora, there is a need to reconceptualise plants as bodily presences rather than taxonomic abstractions; to engage poetic depiction and bodily appreciation in the comprehension of extinction; and to theorise connectivity as a guiding principle of mourning the loss of biodiversity. This chapter will consider several artistic responses to the decline of South-West plants along with the importance of the arts in making extinction a felt reality. What is the role of the arts in creating botanical memorials of loss? Painter Adolph Plate depicted South-West flora in the early twentieth century, writer Barbara York Main in the mid-twentieth century and artist Gregory Pryor in recent years. Botanical memorials occur as artists and writers record or respond to South-West landscapes on the brink of change. Further, engagement of more-than-human life requires frameworks for mourning that revision the concept of the mourned object propounded by Freud (1917/2009). Freudian mourning dichotomises the subject and object, the mourner and the lost one. It replicates what Foucault terms the "discourse of nature." Sensuously rich experience of plants, however, concretises absence. The "connectivity ontology", theorised by ecological humanities scholars such as Deborah Bird Rose, potentiates sensory experience and aesthetic engagement in mourning, allowing the way for intersubjectivity between the mourned one and the mourner.

Interlude XXIII: Under the Wattle Scrub, Coalseam Park

This villanelle is dedicated to Frank Cook, an American herbalist, wild foods forager and gatherer of traditional plant medicine knowledge worldwide who passed away in 2009 in the middle of his life. I wrote notes for the poem at Coalseam Conservation Park near Mingenew, WA, while sitting in the shade of a wattle overlooking an expanse of yellow everlastings in the Irwin River valley:

> Under the wattle scrub, shimmering lawn
> With the lissom Irwin River grasping,
> I hold still watching the swift yellow dawn (ll. 13-15)

The common name "everlasting" denotes various species of the cosmopolitan Asteraceae family, such as the pom-pom head (*Cephalipterum drummondii*). In the spring of 2009, Coalseam was regarded as having the best display of everlastings in the state. At the verge of the *kwongan* and the arid desert, the park is one of the most botanically diverse areas in the northern Wheatbelt. In addition to its botanical interest, the first coal deposit in the state was discovered in 1846 in present-day Coalseam by the Gregory brothers who also explored Lesueur National Park. Pom-poms are discussed in chapter 8 by Noel Nannup in the context of botanical memory. The flowers, known for carrying their colour well after being clipped, symbolise immortality: a *pneuma* infusing a bodily existence that eventually ends. Frank's legacy is now being carried on by the non-profit organisation Plants and Healers International: www.plantsandhealers.com (to read "Under the Wattle Scrub" in full, see Hickman & Ryan, *in press*).

The Sixth Great Extinction

Extinction threatens biodiversity. Scientists estimate that 50% of the planet's species might disappear by the end of the twenty first century (Leakey & Lewin, 1996, p. 233). Van Dooren (2010, p. 271) observes that "we are living in the midst of the Earth's sixth great extinction event, the first to be caused by a single species, our own. At present, species are dying more quickly than we can count them—let alone conserve them." Whereas geogenic extinction is an evolutionary event that has given rise to the biodiversity we now protect, anthropogenic extinction is an outcome of modern technology-driven overconsumption. *Extinction* here refers to the global disappearance of a plant species along with its economic, ecological, ethical and aesthetic value. Extinction also points to the processes through which a plant reaches extermination or critical endangerment. The factors catalysing extinction—habitat loss, introduced exotic plants, overgrazing by feral animals and the effects of diseases—impact species differently (Hopper, 1993, p. 89). Local extinctions are of tantamount significance to global extinctions; a local or regional extinction is the complete disappearance of a plant from a locality, although the same species may be thriving elsewhere. Yet, the concise status of a plant as globally or locally extinct, endangered or threatened recapitulates the semantics of federal conservation protocols. Whether total or partial, the extinct, endangered and threatened situations of plants are absences.

The question "Why do extinctions matter" necessitates the companion question of "Why does biodiversity matter?" If extinction is absence, biodiversity is the proliferation of species over time. Biodiversity is a measure of biological variety expressed commonly as the number of species in a region as small in scale as a forest ecosystem or wetland or as large in scale as the globe (Lovejoy, 1997). Biologist Edward O. Wilson (1997) defines biodiversity as everything living that exists:

> All hereditarily based variation at all levels of organization, from the genes within a single local population or species, to the species composing all or part of a local community, and finally to the communities themselves that compose the living parts of the multifarious ecosystems of the world. (1)

Writers identify four reasons why biodiversity is important: economics, ecology, ethics and aesthetics. Economics refers to the preservation of the genetic pool of crop plants on which human civilisation depends. The transfer of genes between wild and domesticated plants is a necessary strategy to enhance crop yields and maintain disease resistance over time (Lovejoy, 1997, p. 8). The economic argument for biodiversity preservation deploys metaphors like "genetic storehouse" and "unread library" to describe the value of wild plants. The ecological rationale for protecting biodiversity observes that the planet is (or was) a balanced system, and that the extinction of species causes overt or unforeseen ecological consequences. Both the economic and ecological arguments value plants as safekeepers of genetic information, providers of food, fibre and medicine, and agents of balance in the biosphere. The ethical rationale asserts that human societies have the responsibility to protect biodiversity for the intrinsic sake of the plants themselves and for the benefit of future human generations. While these reasons have been argued for persuasively in conservation literature, the fourth value of aesthetics is generally positioned within environmental philosophy (see chapter 5).

Roger Lewin and renowned paleoanthropologist Richard Leakey characterise the "sixth great extinction" as the current human-driven mass disappearance of species. Although extinctions have been common throughout biological history, our present extinction is unusual for the variety of species impacted (Leakey & Lewin, 1996, p. 253). For Lewin and Leakey, three motivations for valuing and preserving biodiversity include aesthetic, economic and ecological considerations. Aesthetic value is the least tangible, yet the most crucial for tapping "deep into what it is to be human" (Leakey & Lewin, 1996, p. 127). They emphasise the value of an integrated perspective of nature: "And when we understand the Earth's biota in holistic terms...we come to see ourselves as part of that whole, not as a privileged species that can exploit it with impunity" (Leakey & Lewin, 1996, p. 253). Lewin and Leakey argue beyond economic and ecological justifications for biodiversity safeguarding. An ethical responsibility underscores biodiversity conservation, especially when the variables precipitating extinction are within human control. Aesthetic value is the third suggested by Lewin and Leakey, but the least articulated. It signifies that the extinction of a species is the impoverishing of the sensuous world.

Human impacts on the environment have become so significant that scientists, such as Nobel Prize-winning atmospheric chemist Paul Crutzen, have coined the term *Anthropocene* to denote the current epoch in which human societies have become a "global geophysical force" (Steffen, Crutzen, & McNeill, 2007, p. 614). Beginning around 1800, the Anthropocene corresponded to the rise of industrialisation and is marked by a drastic increase in fossil fuel consumption (Steffen, et al., 2007, p. 614). The authors plainly state that "enormous immediate challenges confront humanity over the next few decades as it attempts to pass through a bottleneck of continued population growth, excessive resource use and environmental deterioration" (Steffen, et al., 2007, p. 21). While concerns expressed by the scientific community focus on impacts on humanity, the farther-reaching consequences of resource exhaustion and land development extend to all life. Extinction is the most pronounced and irreversible of the effects of environmental decline. This has been evident in Australia where, according to Hopper (1993, p. 89), "colonisation has resulted in the extinction of 20 mammals and 100 flowering plants in the short space of two centuries. More than 3,000 Australian plant species are now rare or threatened, including 16% of our eucalypts."

In Australia and throughout the world, the rapid increase of human populations means greater competition for resources between people and plants. All too often, endangered flora cling to marginal habitats. Most rare and threatened plants survive as severely fragmented pockets. Leakey and Lewin (1996, p. 253) observe that "the most vulnerable species are those whose geographical distribution is limited." With limited geographical distribution, the flora of the South-West is particularly vulnerable. The greatest impact on the region's ecology has been the clearing of its vegetation for agriculture. In the Wheatbelt, the eradication of plants has led to the salinisation of the topsoil, or "the salinity crisis" (Beresford, Bekle, Phillips, & Mulcock, 2001). Other influences on plant populations include diseases, mining and introduced exotic species. Presently, only about one-third of the original vegetation found at the time of European settlement in the 1830s remains and almost 89% of the *Eucalyptus* woodlands, including many of the large trees, have been lost (Conservation International, 2008).

Landscapes of Absence

The extinction of plants is an aesthetic issue. It changes irreversibly the experience of places. Between 2008 and 2011, I surveyed sites of high botanical diversity in the South-West through bodily praxis (see chapter 2). One of my ancillary purposes was to clarify and expand upon Lewin and Leakey's third value of biodiversity in relation to flora. Lewin and Leakey argue that aesthetic value is as imperilled as economic and ecological value. Recognising the marginal status of many South-West plants, I set out to articulate aesthetics within a conceptual framework allowing bodily and emotional entanglement. While the field sites are botanically rich, they also reveal greater absences within the broader environment. Populations of flora cling to remnant habitats, as Hopper (1993, p. 89) observes. The Stirling Range provides an unparalleled vantage point for observing the island effect in which indigenous flora cling to montane pockets. The incursions of settlement register sensuously. The effect is one of pressure: roads and vehicles; fenced pastoral lands; and domesticated animals and feral exotics engulfing botanical reserves. Aesthetic experience of plants is entwined with the awareness of industrialised transformation of the land (see "Three Peaks Triptych," Interlude XI). Settlement pressures around the periphery of the reserve demarcate the montane flora as an island community. The flora is also severely threatened by the soil-borne fungus *Phytophthora cinnamomi*. Wills and Kinnear (1993, p. 135) identify *Phytophthora* dieback as one of the most serious threats to biodiversity conservation (see chapter 12).

Beyond the Stirling Range, the vast inner plateau of the Wheatbelt is a landscape of absence in which well-adapted plant communities have been eradicated by agricultural expansion, introduced flora and fauna, and federal programs sanctioning the destruction of indigenous taxa. The "million acres a year" scheme, mentioned by Noel Nannup in chapter 8, was a federally endorsed program to clear the bush for agricultural colonialism. Now removed of 90% of its pre-settlement composition, the Wheatbelt was once an extraordinarily biodiverse place. Travelling through the grasslands in 1854, surveyor Robert Austin (1856) observed "extensive casuarina (swamp oak) plains" (p. 235), "dense eucalyptus and acacia thickets" (p. 236), "dense scrub and gum-trees" (p. 237) and that "melaleuca (tea-tree), wattles, and cypress-trees present formidable thickets" (p. 238). Early European settlers, however, transformed the land for food and fibre cultivation. The gum trees presented a formidable obstacle to settlement by colonists for whom an agricultural agenda meant transfiguring the bush.

In an article entitled "Picturesque Western Australia," Adolph Plate (cited in Plate, 2005, p. 260), a German-born artist who immigrated to Western Australia in 1907, extols the natural virtues of the South-West: "There is the picturesque scenery of the South-West coast, the peaceful inlets, sheltered from the storm-swept seas, and the great forests of gigantic karri trees, that, were they in America, would form a pilgrimage for thousands of tourists." The landscape, however, was on the cusp of dramatic change. In 1913, Plate painted a South-West bush scene with a ring-barked gum in the foreground (Plate, 2005). The encircling gash in the tree reveals a complex story of Western Australian settlement through the eradication of eucalypts. Seddon (1997, p. 76) argues that an assumption, and ultimately an error of judgement, originated in the European worldview "that the bigger the trees, the better the soil, which led to heartbreaking attempts at settlement in the karri country in Western Australia." On the practice of ring-barking, the ABC Radio (cited in Plate, 2005) program, aptly entitled *Hindsight*, recalled:

> The tree died and then the branches fell and we spent years and years picking up those branches ... Dead trees kept dropping branches on fences...one of the worst risks was of bushfires ... There were hundreds and hundreds of acres of stark, ring-barked forests. (266)

Plate's watercolour of the ringbarked gum signifies the complexities of importing European values, especially the picturesque, to Australia (see chapter 5). The smoothness of the gum tree's trunk is dissected by the bark ring and scar tissue. The spindly gums of the background identify this tree as the last of its girth: an embattled giant. The ring is a gash embodying, in its grotesqueness, the history of its eradication. The wound and the roughened scar tissue are tragic distortions along the sleek, upward soaring trunk of the tree. The gash is tactile; the sensations are felt; and the hardened scar tissue symbolises the destructive widespread rationalisation of the practice (to view "Untitled [WA Bush Scene with Ringbarked Gum]," see Plate, 2005). Plate's selection of the ring-barked tree as subject matter marks a departure from the traditional pleasing picturesque scene and inverts the idealism of a pure, circumstantial or virgin forest devoid of a complex human history. Here, the history of the destruction of the karri forests is written into the tree's body—the flesh of its cambium. The scene is unlike some of Plate's other landscape paintings that depict the landscape as *terra nullius* without human inhabitation, lacking a history of usage and belonging to no one (Lindqvist, 2007). The ruination of these ancient "pillars of the west" represents the large-scale changes brought upon the Wheatbelt.

Plate wrote about the South-West for literary periodicals, including his ephemeral journal, *The Leeuwin: A Westralian Magazine* (Plate, 2005, pp. 276-286). He vacillates in his writings between conflicting values. Plate expresses scorn at the wanton destruction of the karri forests by farming practices, antagonism towards the trees themselves as hindrances to the conversion of the land and interest in them as fiscal objects. Writing under the pseudonym "Boree," an Aboriginal word for a species of acacia wood or a man who has chosen to live independently from society, he describes the parklike banks of the Blackwood River as a feast for the eyes (cited in Plate, 2005, p. 160). In the *Lloyd Guide to Australasia*, however, Plate reflects utilitarian values, advocating the conversion of the land from "unproductiveness into wealth" (Nord-Deutscher & Plate, 1906, p. 59). As a painter, Plate's invocation of the picturesque conflicts with the settler prerogative to transform the grasslands of the Wheatbelt into cultivated crops. Plate reflects other mixed attitudes of some early South-West writers such as Fred Davis in *The Lure of the Golden West* whose initial applauding of scenic value is followed by an expression of hope that the scene will become subjugated (Groser, 1927, p. 245).

Botanical Memorials

How do the literary and visual arts memorialise landscapes of absence? In an epoch of great pressures on wild plants, the arts can produce botanical memorials to reaffirm the aesthetic value of biodiversity, as noted by Lewin and Leakey. In the form of paintings, poetry, memoirs or performances, memorials elegise habitats in decline. Memorials record sensory experience or invoke an emotional expression of the imperilled status of the natural world. Two contrasting botanical memorials from the South-West address the importance of arts-based engagement with flora. Creative intersections are needed as much as science-based methodologies for tapping "deep into what it is to be human" through multi-sensorial experience

(Leakey & Lewin, 1996, p. 127). The first example of a botanical memorial is the 2005 exhibition *Black Solander* by Gregory Pryor. The second example is Barbara York Main's account of a tract of Wheatland bush, *Between Wodjil and Tor* (Main, 1967).

Main's first-person account is a naturalist's record through the seasons without explicit commentary on the social, political or technological factors driving the changes. An expert on trapdoor spiders, Main recreates the nineteenth century nature writing tradition of Thoreau, for whom careful observation and corporeal experience of flora were recorded in detail as journal reflections (see chapters 1 and 3). The lineage between Main and Thoreau is explicitly acknowledged in the "Preface" in which Main quotes Thoreau's qualification of wilderness as "the raw material of all our civilization." In comparison to Main and her Thoreauvian influence, Gregory Pryor's work explicitly comments on the precarious status of plants. Pryor takes an anti-aesthetic, or even *ascetic*, position that critiques the conventions of form and colour invariably influencing the rendering of flowers.

The installation work of Pryor (2005a,b) exemplifies the role of the visual arts in engaging extinction. *Black Solander* is a memorial to over 10,000 Western Australian plants rendered on rectangles of black sugar paper to create the ambience of a tomb within the gallery at Perth Institute of Contemporary Arts (PICA) in Western Australia. Shadowy profile portraits of plants substitute for idealised renderings of flowers to highlight the imminence and urgency of species extinction (to view images in "Black Solander," see Pryor, 2005b). Exhibition curator and art historian John Barrett-Lennard (2005, p. 3) comments in his introduction to the exhibition, "Outline and Absence," that "repeated in their thousands upon thousands, [these images] stand in for the living - the total flora of Western Australia - and for the dead, the disappeared, the erased - the plants that are no more." Pryor interrupts and extends the taxonomic *modus operandi*, starting with Charles Gardner whose specimens formed the initial collections of the Western Australian Herbarium.

Pryor's inversion of the aesthetic of beauty is made apparent through the shutting out of the sun and the modelling of the gallery as crypt. Paradoxically, the darkness of the exhibition summons, through the absence of the sun, the luminosity of the South-West environment, a factor of extreme solar exposure to which flora are well-adapted. The visitor enters into the immediacy, symbolised by the mausoleum within the gallery enclosure. Physical immersion and sensory deprivation, along with an empathic feeling for the flora, creates an exchange between the viewer and the subject matter. The effect is dialogical in which the viewer contemplates human impacts on flora and the severity and scale of extinction. The exclusion of light and colour invoke a melancholic viscerality. Impending, but hidden still, is a feeling for—a waiting for—the sensory pleasure that is yet to come: pleasures of form, colour, harmony, balance, sound and touch that could be at stake. *Black*

Solander inverts the disengaged viewership model that positions a subject at a comfortable and appreciative distance to a plant object (see chapter 5). The plants are neither aestheticised idealisations nor dissected schemata. Pryor (2005a, para. 9) upturns the paradigm of beauty and the "loaded aesthetic appeal of the flower." The ghostly adumbrations of flora are huddled together *en masse* seeking refuge from the dissolution of their sustenance: uninterrupted habitat, clean water and air, as well as mutually beneficial relationships.

In contrast, Barbara York Main's *Between Wodjil and Tor* is both a scientific and literary account of the bushland flora with only minor commentary on the status of indigenous plants. Rather than invert the "loaded aesthetic appeal" of the flower as Pryor does, Main memorialises the experiential features of the flora with scientific veracity. In the "Preface," Main (1967) comments on the value of her dual scientific and literary focus: "But perhaps it is the juxtaposition of wilderness and the man-made orderliness of farmlands, which is becoming the situation now in our countryside, that will provoke a realization of the inherent interest and value—both scientific and aesthetic—in the natural bushland." Through the increasing absence of intact bushland in the Wheatbelt and the juxtaposing of indigenous vestiges to pastoral lands, as discussed of the Stirling Range, the scientific and aesthetic value of the flora becomes more apparent. Absence invokes ecological consciousness, in Main's terms. Main captures with unwavering precision and attention the pleasurable features of plants through her immersion in the Wheatbelt. Main (1967) writes of the botanical community, the mallees:

> Dominant, in great shaggy clumps, stood the mallees. These mallees, the Burracoppin mallees (*Eucalyptus burracoppinensis*) sprawled low, many of the boughs even lying on the ground, with their crowns of tough, grey-green leaves sweeping and resting on the yellow sand, each clump forming a great, shaggy, untidy tent. Everything about them indicated age. The twisted trunks of each clump were top-heavy with old dead branches. (7)

Main evokes corporeal engagement with flora. Bodily descriptors such as *sprawled*, *lying*, and *sweeping* denote active posture and reflect back to the reader the author's tacit participation in the habitat. Main (1967) writes of the seasonal, cyclical quality of the narrative: "The recurrent theme of the book is the annual rhythm – the changes wrought by the seasonal cycle - with a particular landscape, its dormant plants and a selected group of animals." The descriptions of the undisturbed vegetation serve as a record of a landscape that has lost nearly 90% of its flora.

John Kinsella (2008, pp. 64-65) characterises Main as a hybrid author bridging scientific zoological writing and literary reflection with "acute local observation of place—flora and fauna, habitat and specificities, but also a sense of the mediating self, of the author being part of the process of identification and belonging." Main's involvement in the process of belonging to the parcel of bushland, however, seems restrained. The scientific quality of the prose overpowers a literary voice. Yet, a narrative perspective comes through multi-sensorial diction. The palette of colours—"grey-green leaves" and "yellow sand"—mixes with tactile images of decay and gnarly age. Main's approach is one of naturalistic realism, depicting both the beautiful and grotesque features of the mallee vegetation through an emphasis on their mutability. As with the rendered images of *Black Solander*, plants are not solely depicted through idealised forms and colour but balanced by the grotesque: shaggy clumps, untidy tent, twisted trunks and old dead branches.

Unlike Pryor, however, Main's botanical memorial is less critical of the effects of anthropogenic change. In this sense, Main's memorial is less an elegy to extinct species and departs markedly from Pryor's metatext. She writes of her greatest hope that the landscape will retain its "partially cultivated mosaic structure in which the *uncultivated* interstices will continue to inspire [italics in original]" (Main, 1967, Preface). The juxtapositions of remnant bush and pastureland inspire, rather than disorient. Kinsella (2008, p. 176) argues that, while Main's writing reflects the cyclical sense of place, it neglects the impacts of European agricultural and pastoral practices on those cycles: "So there is a classical Beethovian sense to the place, but not to the place as cleared and rearranged by European farmers." In other words, the work glosses over the magnitude of environmental change in the Wheatbelt. Beresford, Bekle, Phillips and Mulcock (2001) comment that:

> The Wheatbelt has a little-known, but unenviable, reputation: in no other region in the world is there thought to be an area as large which has been cleared of its natural vegetation in so short a period - most of it in the sixty years between the 1920s and the early 1980s. (11)

The vegetation evolved deep-penetrating root systems that has helped to keep water tables at a depth from the surface (Beresford, et al., 2001, p. 2). The clearing of indigenous species disrupts habitat equilibrium, brings salt levels to the topsoil and inhibits long-term agricultural viability. *Between Wodjil and Tor* makes no ostensible commentary on the broiling crisis.

Interlude XXIV: Notes on a Dying Chook

Poetry has the capacity to express the senses. However, the act of writing can also entail engagement with the sensuousness of place and the particularities of mourning. Leggo (2006) comments:

> I have always focused a great deal of attention on the senses, especially the well-known senses of sight, smell, hearing, taste, and touch, but I am now learning to attend also to the sense of proprioception, the sense whereby we know our physical location, our point of being, as opposed to our point of view. (143)

I suggest that, in addition to our point of being, a bodily sense of place—or proprioception in Leggo's terms—is also a process of becoming. Part of being and becoming is an awareness of bodily passage. "Notes on a Dying Chook" concludes with the wandoo forest as a place of salvation, a freedom from the cold restraint of the steel cage, a place older than animal lives. A Nyoongar term denoting the resin of the tree, *wandoo* refers to *Eucalyptus wandoo*, the white gum tree endemic to the South-West (Boland et al., 2006, p. 348). Through close attention to sight, sound and bodily motion, the poem traces the decline of a chook. The passing of animals, forests and flowers have been symbolically and actually connected to human death (see chapter 8). Yet, flora also can represent eternal life, as the example of the everlasting flower indicates (see "Under the Wattle Scrub," Interlude XXIII). By the end of the poem, wandering off into the eternal wandoo forest is the denial of corporeal decline in the eyes of a human mourner, aware of his own interrelated proprioceptive teleology.

Notes on a Dying Chook

an amoeba of auburn hues
pontificating on one leg

meditatively in a t'ai chi
posture

lashless eyes tightened to slits
red rubber viscera under-jaw

hung like a windless
flag

when the patio door creaked
her orange moons and absolute

black opals flashed full
embouchure

her shape particularised,
dashed underfoot hoping for

the salmon-coloured ceramic
floor and

when denied entry sputtered like
a dervish on dinosaur feet

mohawk comb a flabby
mess

of raw meat (bird-sign of dehydration),
lapped the swamp under potted

plants, accepted no freshwater
almsgivings

the hangman was silent with his deed
lest dogs would smell the knell

and, impassioned, unearth the
entombed

but somehow, when the chirruping
in the steel cage ceased, I convinced

myself she had simply wandered off
into the wandoo forest.

From Subject-Centred Mourning to Connectivity Mourning

The prose of Barbara York Main and the visual art of Gregory Pryor are modalities for memorialising species and expressing grief. Although science may lack mourning rituals, as argued by Shiv Visvanathan, the arts and humanities can fill a need. The absence of sensory and emotional attachment to plants constitutes the objective foundations of botanical science. Arguably, we turn towards artistic and literary interpretations of mourning because emotional expressions conflict with objectivity. Perhaps inadvertently, Plate's watercolour and Main's bushland account memorialise Wheatbelt landscapes in the midst of change. Since then, these places have been transformed even more intensively. More self-consciously than Main's prose or Plate's watercolours, Pryor's renderings are simultaneously elegies to species lost and expressions of grief.

Why does science lack mourning rituals? Although a detailed enquiry is beyond the present scope, it is nevertheless critical to revisit the Western construction of nature (see chapter 1). This conceptualisation has influenced how science and the public approach the mourning of plant loss. Foucault describes scientific botany as part of a discourse of nature that objectivises the natural world into institutionalised modes of seeing, saying and doing (Giblett, 2011, p. 1). Through a discourse of nature, as Foucault (1966/2002) argues:

> Nature is posited only through the grid of denominations and - though without such names it would remain mute and invisible - it glimmers far off beyond them, continuously present on the far side of this grid, which nevertheless presents it to our knowledge and renders it visible only when wholly spanned by language. (174)

The discourse sets up a binary between humans and plants (see chapter 5). This dichotomy, which is also at the core of disinterested aesthetics, can underscore the commodification, marginalisation or plain destruction of flora. Connectivity between people and plants is excised out of the discourse for an empiricism of an externalised nature.

Taxonomic science encodes the visible grid of denominations. In the discourse of taxonomy, the idealisation of a species undergirds the abstraction of extinction itself. A species is a generalisation, a classificatory act. A species neither exists phenomenologically nor is experienced by human sensory faculties. While the concept of species is functionally valuable, it is insufficient as an aesthetic, sensory or emotional medium. A species cannot be grieved. Only an individual organism or a community of organisms in a place can be grieved. Yet, plants, if regarded as objects of grieving in Freud's terms, risk being objectified through the discourse of nature and the power relations it structures. The problematic conceptualisation that infers an object of mourning necessitates an alternate framework for grieving lost species. As Giblett (1996, p. 242) comments on the loss of wetlands worldwide, "the world should mourn its wetlands which gave it life and nourished it." If science and taxonomy have no rituals for dealing with loss, absence and mourning, to where do we turn?

As applied to plant loss, Freudian mourning extends Foucault's discourse by formulating nature as an object of a subject's mourning. Freudian mourning is "regularly the reaction to the loss of a loved person, or to the loss of some abstraction which has taken the place of one, such as one's country, liberty, an ideal, and so on" (Freud, 1917/2009, p. 19). One mourns the embodied resonances that issue from an idea, such as liberty, not the abstraction itself. The mourning of a species is the mourning of the image, scent, feel, sound or taste of an individual or community. Freud's "Mourning and Melancholia" suggests that mourning ceases when the mourner severs emotional linkages to the lost object-cathexis. This severance frees the discrete ego to focus on a new object, and thus create a substitution for the absence of the lost one. Freud (1917/2009, p. 246) differentiates between mourning as the outward extension of grief and melancholia as a pathological inward projection of grieving emotions through egotism: "In mourning it is the world which has become poor and empty; in melancholia it is the ego itself." Mourning entails the loss of an object, whereas melancholia is the loss of the ego. Giblett (1996, p. 177) rephrases this distinction as "in mourning the world is experienced as loss whereas in melancholia the ego is experienced as loss."

Freud's mourning and melancholia circumscribe the traditionally unified subject in which his or her allegiance is to personal desire rather than to the objects of desire (Clewell, 2002, p. 1). Mourning depends on the dynamic between the mourning subject and the mourned object, whereas melancholia compresses grieving emotions into solipsistic self-identity in which the world vanishes. For the mourning of plants, what is lacking in Freudian mourning and melancholia is an expression of human interrelationship to the world beyond the pathologies of the unified subject. Clewell (2002, p. 1) stresses that "Freud's mourning theory has been criticized for assuming a model of subjectivity based on a strongly bounded form of individuation." In botanical terms, the extinct plant is mourned, not out of the intrinsic loss of the plant itself, but rather as a reflection of the subject's inevitable decline. Freud's position is strongly subject-centred. The "bounded form of individuation" reflects a discourse that requires object-subject juxtapositions to produce transferences. Freudian mourning constrains the evolution of transpersonal mourning that is ecological and based in connectivity thinking. Rather than subject-centred mourning, a connectivity ontology offers a framework for mourning lost species through bodily empathy and ecologically networked sensoriality. As experiential complexity diminishes along with botanical biodiversity, regional responses to plant loss require artistic or literary responses to galvanise transpersonal mourning.

The ecological humanities provide a useful conceptual structure for approaching the mass and scale of the issue of species extinction, for grounding the abstractedness of the species concept itself and for engendering emotional and sensory responses to loss (see chapter 1). As an alternative to Freudian individuation, the ecological humanities provide a way through the conceptual frameworks that place the human ego at the centre of being and thus reaffirm the dualities that lie at the heart of human-induced species extinction. The ecological humanities theorise that organisms are intermeshed and interrelated, rather than purely separate and discrete as the discourse of nature purports. An object of mourning is untenable in the context of networked relations. And in reality the mourning of a lost plant is the mourning of the host of ecological and sensuous losses that disappear along with it. Describing her concept of "ecological existentialism," Deborah Bird Rose (2011, pp. 2-3) observes "from certainty the shift is to uncertainty. From atomism the shift is to connectivity. The West has reached these big shifts through the working of its own intellectual and social history." The "connectivity ontology" of the ecological humanities provides a perspective grounded in intersubjective relationality. The mourning of the self and the mourning of the lost occur as intrinsic connection between the two. Simply put, one mourns the loss of self, the loss of the other and the loss of the connectivity. Mourning must take on a distinctly mosaic quality that recognises causality between the self, the other and the environment.

Rose and Robin (2004, para. 3) stress that "an ontology of connectivity entails mutual causality: organism and environment modify each other. Relations between organism and environment are recursive." The same factors that bear upon the extinction of a species are necessarily related to human lives, as we share the same networks:

> The imperative of learning to think about and with connectivity can be operationalised as an imperative to enlarge the boundaries of thought and to enlarge thinking itself—to enhance our ability to think in dialogue and, perhaps, in empathy with others. (Rose & Robin, 2004, para. 7)

A connectivity ontology suggests the shared mourning of the consequences of extinction rather than the subject-centred mourning of the lost object or the inversion of mourning to melancholia. We mourn the absence of the web of relations. Most importantly, extinction starts to matter because people and plants share common ecological circumstances specific to being embodied (requiring water, nutrients and air) and emplaced (requiring relational connections developed over time in a prescribed region). Detachment from extinction, as a media issue, conceptual abstraction or subject-driven emotional platform, is counterpoised through interconnection. As sensuously related to experience, extinction can be made real, imminent, sensible and a matter of survival for all. Rose and Robin (2004, para. 16) call for embodied spatiality as part of a connectivity ontology: "Permeability [opening] persons not only to place, but to the substance and history of the place." Through bodily interactivity in a place, artistic engagements with the natural world, as expressed in *Between Wodjil and Tor* and *Black Solander*, ensure that extinction can continue to resonate transpersonal connections to nature. Consequently, pathways for mourning begin to open.

Mourning the Loss of Flora

Why do extinctions matter? The visual and the literary arts provide sensory-rich means of creating botanical memorials to the loss of plant biodiversity. Artists working in the South-West have taken the threatened flora as subject matter.

Taxonomic science is detached intrinsically from complex human experience. Hence, a framework for mourning is constrained by the discourse of nature echoed in Freud's theory of mourning and melancholia. Scientific discourse governs knowledge-driven interaction with the natural world. Rather than looking towards science, this chapter has posited the importance of bodily and emotional engagement with plants as a way to approach the mourning of life lost. In addition to ecological and economic consequences, extinctions impoverish our sensuous worlds. A more tenable ecological conceptualisation of mourning considers connectivity between beings, rather than demarcated subjectivity, as a principle for exploring the deep channels of grief over the loss of botanical heritage.

References

Austin, R. (1856). Report by Assistant Surveyor Robert Austin, of an expedition to explore the interior of Western Australia. *Journal of the Royal Geographic Society of London, 26*, 235-274. doi: 10.2307/1798360

Barrett-Lennard, J. (2005). Outline and absence. In *Black solander, Gregory Pryor*. Perth, Australia: PICA Press.

Beresford, Q., Bekle, H., Phillips, H., & Mulcock, J. (2001). *The salinity crisis: Landscape, communities and politics*. Crawley, Australia: University of Western Australia Press.

Boland, D. J., Brooker, M. H., Chippendale, G. M., Hall, N., Hyland, B. M., Johnston, R. D., et al. (2006). *Forest trees of Australia* (5 ed.). Collingwood, Australia: CSIRO Publishing.

Clewell, T. (2002). Mourning beyond melancholia: Freud's psychoanalysis of loss. *Journal of the American Psychoanalytic Association, 52*(1), 43-67 Retrieved from http://www.apsa.org/portals/1/docs/japa/521/clewell.pdf

Conservation International. (2008). Biological diversity in Southwest Australia. K. Caley (Ed.) *Encyclopedia of Earth* Retrieved from http://www.eoearth.org/article/Biological_diversity_in_Southwest_Australia

Foucault, M. (1966/2002). *The order of things: An archaeology of the human sciences*. London, England: Routledge.

Freud, S. (1917/2009). Mourning and melancholia. In L. G. Fiorini, T. Bokanowski & S. Lewkowicz (Eds.), *On Freud's* Mourning and melancholia (pp. 19-34). London, England: Karnac Books.

Giblett, R. (1996). *Postmodern wetlands: Culture, history, ecology*. Edinburgh, Scotland: Edinburgh University Press.

Giblett, R. (2011). *People and places of nature and culture*. Bristol, England: Intellect.

Groser, T. (1927). *The lure of the golden West: Experiences and adventures in a bush brotherhood of Western Australia: Early problems and conquests: All about group settlements of the West: The land of sunshine and opportunity*. London, England: Alexander-Ouseley.

Hickman, E., & Ryan, J. (*in press*). *Two with nature: Ellen Hickman and John Ryan*. Fremantle, Australia: Fremantle Press.

Hopper, S. (1993). *Kangaroo paws and catspaws: A natural history and field guide*. Como, Australia: Department of Conservation and Land Management.

Hopper, S. (1998). An Australian perspective on plant conservation biology practice. In P. Fiedler & P. Kareiva (Eds.), *Conservation biology for the coming decade* (2 ed., pp. 255-278). New York, NY: Chapman Hall.

Hopper, S. (2004). Southwestern Australia, Cinderella of the world's temperate floristic regions. *Curtis's Botanical Magazine, 21*(2), 132-180.

Kinsella, J. (2008). *Contrary rhetoric: Lectures on landscape and language*. G. Phillips & A. Taylor (Eds.). Fremantle, Australia: Fremantle Press.

Leakey, R., & Lewin, R. (1996). The *Sixth Extinction: Biodiversity and its survival.* London, England: Orion Books.

Leggo, C. (2006). Attending to winter: A poetics of research. In W. Ashton & D. Denton (Eds.), *Spirituality, ethnography, and teaching: Stories from within* (pp. 140-155). New York, NY: Peter Lang.

Lindqvist, S. (2007). *Terra nullius. A journey through no one's land* (S. Death, Trans.). London, England: Granta.

Lovejoy, T. (1997). Biodiversity: What is it? In M. Reaka-Kudla, D. Wilson & E. Wilson (Eds.), *Biodiversity II: Understanding and protecting our biological resources* (pp. 7-14). Washington, DC: Joseph Henry Press.

Main, B. Y. (1967). *Between wodjil and tor*. Melbourne, Australia: Jacaranda.

Nikulinsky, P., & Hopper, S. (2005). *Soul of the desert*. Fremantle, Australia: Fremantle Arts Centre Press.

Nord-Deutscher, L., & Plate, A. G. (1906). *The Lloyd guide to Australasia*. London, England: Edward Stanford.

Plate, C. (2005). *Restless spirits: The life and times of a wandering artist*. Sydney, Australia: Pan Macmillan.

Pryor, G. (2005a). Black death: Species extinction in WA. *Artlink, 25*(4) Retrieved from http://www.artlink.com.au/articles.cfm?id=2227

Pryor, G. (2005b). *Black solander, Gregory Pryor*. Perth, Australia: PICA Press.

Rose, D., & Robin, L. (2004). The ecological humanities in action: An invitation. *Australian Humanities Review, 31-32* Retrieved from http://www.australianhumanitiesreview.org/archive/Issue-April-2004/rose.html

Seddon, G. (1997). *Landprints: Reflections on place and landscape*. Cambridge, England: Cambridge University Press.

Steffen, W., Crutzen, P., & McNeill, J. (2007). The Anthropocene: Are humans now overwhelming the great forces of nature? *Ambio, 36*(8) Retrieved from http://www.allenpress.com/pdf/ambi-36-08-06_614..621.pdf

Van Dooren, T. (2010). Pain of extinction: The death of a vulture. *Cultural Studies Review, 16*(2), 271-289 Retrieved from http://epress.lib.uts.edu.au/ojs/index.php/csrj/article/viewFile/1702/1834

Visvanathan, S. (1996). Footnotes to Vavilov: An essay on gene diversity. In F. Apffel-Marglin & S. Marglin (Eds.), *Decolonizing knowledge: From development to dialogue* (pp. 306-339). Oxford, England: Clarendon Press.

Wills, R., & Kinnear, J. (1993). Threats to the Stirling Range. In C. Thomson, G. Hall & G. Friend (Eds.), *Mountains of mystery: A natural history of the Stirling R*ange (pp. 135-141). Como, Australia: Department of Conservation and Land Management.

Wilson, E. O. (1997). Introduction. In M. Reaka-Kudla, D. Wilson & E. Wilson (Eds.), *Biodiversity II: Understanding and protecting our biological resources.* Washington, D.C.: Joseph Henry Press.

12

Walking

Abstract: This chapter maintains that walking and botanising represent an embodied convergence between people and places. Throughout South-West Australian history, cultural realities have been shaped by the dynamic intersection of foot travel, botanical life and bodily interactions with the land. Following Mauss, walking, rather than a mundane action, is a "body technique," a learnt cultural and natural *habitus*. An emerging literature conceptualises bipedality as sensuous engagement with place. The theoretical complexities of walking will be grounded through reference to the Bibbulmun Track, a long-distance walking track linking Perth and Albany, Western Australia.

> *The bushwalker may justly claim that this pastime is one of the very few that develops both the mind and the body. It takes him far away from the hustle and bustle of the modern city and he may tread in places where no man has trod before. He learns to appreciate the strange, peaceful charm of the bush, and realises that man and his civilisation form only a small part of a wonderful creation.*
>
> The Federation of Australian Bushwalking Clubs, 1939
> (cited in Baker, 2010, p. 29)

The Botanising Stroll: People Embrace the Bush

Since the Romantic era, walking has shifted from an obligatory activity, connected to ways of making a living in the countryside, to a recreational pursuit of life quality (Amato, 2004, p. 102). The preceding quotation is from the FABC, one of the earliest confederations of independent bushwalking organisations in Australia. It describes how, as a therapeutic get away from the city, bushwalking remedies the effects of urban stress, the "hustle and bustle." The peripatetic urban dweller assumes the persona of an early explorer who charted *terra nullius* "where no man has trod before," or at least no European. Most significantly, the recreational bushwalker, typically an upper class male, undergoes a transformation of values.

With the walker immersed in the land, "civilisation" looms less largely, constituting "only a small part of a wonderful creation." Liberated from affinities of taste gestated elsewhere, the walker "learns to appreciate the strange, peaceful charm of the bush." In an era of post-Federation nation-building and identity-making, bushwalking would be an illuminating foray into the character of the Australian bush, one that is not discomfiting "weird melancholy," as Clarke (1876/1993, p. 45) lamented, but a peaceful, though strange, experience. Hence, a walk would not be an invocation of nationhood, at once personally rejuvenating, culturally instructive and environmentally integrative. Though initially encountering alienation, the walker's perceptions would become refined. He or she would learn, as Clarke did, to read a lingua: "He learns the language of the barren and uncouth and can read the hieroglyphs of haggard gum-trees" (Clarke, 1876/1993, p. 46).

This chapter will analyse the convergence of these three closely related, but often segregated, elements of bipedality: the *habitus* of bushwalking; the bush through which the walker ambulates; and the appreciation of flora. The bush in consideration is the South-West flora. Expressing corporeality, walking will be characterised as participatory engagement that collapses the divide between an appreciator and a picturesque, externalised scene. By forging a field for multi-sensorial awareness of plants, walking complements the spectatorship of wildflower tourism (see chapter 7). Thus, there is noteworthy historical convergence between walking (in the bush rather than in the metropole), knowledge of plants (scientific or experiential) and a depth of bodily engagement with nature (corporeality, or embodied sense of place). In Australia, the history of bipedality is closely interwoven with flora, including contemporary wildflower tourism, colonial and post-colonial plant collecting and botanical surveying, and Aboriginal bush tucker and fibre or medicine craft (see chapters 3 and 13).

The trope of the botanising *flâneur* used by Charles Baudelaire and Walter Benjamin, the detached male connoisseur "of the new human nature of the metropolis," symbolises visual speculation as well as the subordination of the senses (Clark, 2001, pp. 18-19). Although culturally refined, the *flâneur* is socially insensible, assuming a position of cautious disinterestedness and never attaining communion with "female fauna" (Clark, 2001, p. 18). In the terms of George Gissing's 1889 novel *The Nether World*, the cosmopolitan *flâneur* may also disparage the smell of the female flora as the "rank, evilly-fostered growth" of the festering city (cited in Giblett, 2009, p. 39). Observing the city habitat through a lens, rather than participating in it through the bodily senses, the *flâneur* gazes timidly over the metropolis, but fails to gain union with it. In the form of bipedal movement, embodiment entails contact, not of a botanising *flâneur* peripherally sensing but of multi-sensorial immersion in place.

The gradations of meaning between bushwalking, embodiment and botanical appreciation merit attention. First of all, *bushwalking* derives from Australian usage to indicate what is known as hiking, tramping, trailing or trekking in other countries. The word *bush* was used by one of the earliest hiking clubs in Sydney, the Waratah Walking Club, formed in the 1920s, which then changed its name to

"Bush Walkers" and later to "Sydney Bush Walkers" (SBW) (Caffin, 2002, para. 3). Popularised in the 1890s by Henry Lawson in "The Bush Undertaker," *bush* refers to the Australian landscape, its particular assemblage of eucalypts and shrubs, somewhat parallel to the terms *woods* or *outdoors*. In 1873, the Victorian novelist Anthony Trollope (1873/1967) observed the vernacular invocation of the term and advised its careful application:

> Readers who desire to understand anything of Australian life should become acquainted with the technical meaning of the word 'bush'. The bush is gum-tree forest, with which so great a part of Australia is covered, that folk who follow a country life are invariably said to live in the bush. Squatters who look after their own runs always live in the bush, even though their sheep are pastured on the plains. (263)

The bush can be apprehended pictorially through the eyes and viscerally through the senses (see chapter 13). As the third element of the triad, embodiment points to the ways in which "human and extrahuman realities are apprehended through the body" (Sellers, 1999, p. 487). Embodiment recognises the consanguinity between the bodies of botanical nature and those of culturally habituated bushwalkers. As a method of participatory corporeality, walking traces human exertion into *terra firma*, while simultaneously being shaped by the physicality of habitats in which one acts (see chapter 2).

A *botanising stroller* could be defined as a local amateur, hobby botanist or *flâneur* who explores plant diversity within a region but whose activities fall outside the auspices of tourism. The word *botanising* denotes a specific form of recreational walking, a speculative Romantic fusion of mobile philosophy and hobby science in which attention to a plant becomes a space for more fully experiencing the self in nature. Motivated by a curiosity about the bush, botanising can also be scientific, conservation-oriented, educational or even allied to land development. For instance, annual summer excursions to survey the vegetation of the Bogong High Plains was initiated in the 1940s by members of the Botany School at the University of Melbourne. These exploratory botanisings were triggered by concerns over the costly impacts that soil erosion presented to the Kiewa hydro-electric scheme (Gillbank, 2007, p. 26). Whether or not a walker reverts to *flâneurie* depends on touching, tasting, chewing, spitting out and listening. Hence, either pleasurable or perfunctory, walking is integral to botanical appreciation. Some of the richest botanical reserves in the South-West, and elsewhere in Australia, are traversed by recreational walking tracks. However, some forms of walking are more conducive to physical engagement. Kay and Moxham (1996, p. 174) tell us that "recreational walking is so diverse and dynamic that it merits careful classification of its many different forms."

During the Romantic era, the perception of walking shifted. No longer pastoral drudgery and increasingly an aristocratic recreational pursuit, walking gained new aesthetic connotations. Edensor (2001, p. 83) observes the emergence of "a set of interlinked reflexive conventions, aesthetic imperatives and practical endeavours which produced a distinctive relationship between the walking body and nature." Included within the changing human relationship to nature was the recovery of historical values related to the continuity of bodily movement interrupted by technological locomotion. The body-in-becoming of the walker, moving through a picturesque setting, contrasted to the body-in-being of the pastoralist, whose activities were fixed spatially to an agricultural matrix of labour. Thus, the botanising stroll enabled an educated gentry to cultivate meaning with respect to flora without engaging in the visceral demands of farming, woodcutting, medicinal herb gathering or preparing wild-crafted foods. In this new sense of bipedality, walking transformed the perception of the land into an act akin to reading—with nature as decipherable as poetry or taxonomic grids.

Interlude XXV: Sunday Zamia Swagger

Leggo speaks of attending poetically to the world through the body senses. He advocates that researchers balance quantitative inclinations with feeling, that we "focus on our senses, come to our senses, awaken our senses, live within common *sense(s)*...a poetics of research means being present and open sensuously to the whole earth [italics in original]" (Leggo, 2006, p. 148). Like Thoreau, Leggo describes engagement with the world through a bodily eye, through corporeal participation within it: "What I want is to revel in what I am seeing, to see with the whole body, so that my body is rendered alive, is written in the poems" (Leggo, 2006, p. 151). Poetry is a method for dwelling sensuously. As an example of a poetic narrative of openness to plants, "Sunday Zamia Swagger" plays on zamia staggers, a toxic shock known to develop in cattle grazing on the zamia palm (see chapter 3). I was intrigued by the dual notions of a plant as a poison to be shunned or destroyed by colonists, but a nutriment to be consumed and fostered by Aboriginal peoples as *djiridji*:

> Grey observed "violent fits of vomiting"
> Vlamingh "no distinction between death and us"—
> Savouring its bready fruits, unsoaked like hazels,
> Cattle staggered at the poison of the New World
> Encased in the sweet flesh of a nut (ll. 20–24)

All morning, I had been reading accounts of poisoning through the ingestion of unprocessed by-yu nuts. I wanted to get close to the villainous species. The result was not only a confrontation with the cultural history of a plant, but a sense for "the way to better things" (l. 13) through these essential expressions of gestural curiosity and sense openness (to read "Sunday Zamia Swagger" in full, see Ryan, *in press*).

Techniques of Walking: The *Habitus* of Bipedal Movement

> Isn't it really quite extraordinary to see that, since man took his first steps, no one has asked himself why he walks, how he walks, if he has ever walked, if he could walk better, what he achieves in walking...questions that are tied to all the philosophical, psychological, and political systems which preoccupy the world?
>
> Honoré de Balzac (cited in Amato, 2004, p. 1)

The past decade has witnessed a resurgence of interest in histories of walking in Australia (Harper, 2007), New Zealand (Ross, 2008), North America and Europe (Amato, 2004; Solnit, 2001), as well as the cultural significance of walking (Edensor, 2001; Giblett, 2008a). The history of walking is increasingly looked upon as intrinsic to other cultural formations in which the walking body is a nexus of biological and cultural influences. Giblett (2008a, p. 14) observes the unified natural and cultural dimensions of bipedal ambulation: "Walking in the sense of moving the body through space propelled by the body is natural but the way in which we walk is learnt and so cultural." In a speed-oriented society, bipedality is furthermore theorised as an antidote to the ill-effects and disembodied rapidity of vehicular travel, cutting across space and time and making the peripatetic body obsolete. In *Wanderlust*, Solnit (2001, p. 12) alludes to the politics of walking: "In this context, walking is a subversive detour, the scenic route through a half-abandoned landscape of ideas and experiences."

In the essay "Techniques of the Body," the French sociologist Marcel Mauss defined walking as a *habitus* after observing the influence of American walking styles, transmitted through Hollywood cinema, on the gait of French nurses (Mauss, 1935/2006). Mauss theorised the actions of the body as a congregation of acquired social, psychological and biological influences. Further evidenced through a socially esteemed, loose-hipped female Maori style of walking called *onioni*, Mauss posited that a specific technique, style or habit of walking is learnt as a cultural *habitus*. Techniques of walking—gait, speed, accent, cadence, tendency to pause along the way or continue to the end, talk with others or reflect alone, compete or cooperate with self or companions, or smell, taste, touch and listen—are culturally conditioned: learnt, re-learnt or unlearnt. Mauss (1935/2006, p. 82) defines a technique as "an action which is *effective* and *traditional*...It has to be *effective* and *traditional*. There is no technique and no transmission in the absence of tradition [italics in original]." The human body is the "first and most natural instrument. Or more accurately, not to speak of instruments, man's first and most natural technical object" (Mauss, 1935/2006, p. 83). By characterising it as a technique, Mauss implies that walking requires the development of effective practices through tradition, or the cultural transmission of styles and affinities. Not inherited as a purely anatomical action, bipedality is a cultural *habitus*, acquired like a spoken accent. The body is a technical, rather than technicised, site of cultural and biological processes, not a rote mechanical assembly of parts in a Vesalius *écorché* rendering (see chapter 4).

The synonyms for walking suggest the sophistication of techniques and the many styles of bipedality. Mauss (1935/2006) responds to what he saw as a prevalent tendency in sociology to classify walking (as well as swimming, standing and squatting) under the rubric of mundane "miscellaneous" activities with little technical differentiation and with no bearing on social preferences or cultural systems. The techniques of walking are instead highly varied: "The *habitus* of the body being upright while walking, breathing, rhythm of the walk, swinging the fists, the elbows, progression with the trunk in advance of the body or by advancing either side of the body alternately" (Mauss, 1935/2006, p. 89). Walking is remarkably diverse in style and setting. Shambling or shuffling is the slow dragging of the feet, whereas sauntering and strolling are carefree and enjoyable. Historian Joseph Amato (2004, p. 6) notes the variety of descriptors employed to discern between different forms of walking: "Passing people are said to slink, slither, stalk, shuffle, slog, trudge, hike, stroll, strut, swagger, promenade, gallivant, jaunt, mosey, wander, peregrinate, amble, or saunter." Promenading is aristocratic strutting for the purpose of being seen. Wading is walking through water. Prowling is stealthy searching. Ambulating is restless seeking.

Ambling can be free of fixed goals or destinations, in contrast to perambulating which intimates walking a perimeter or boundary to assert ownership. Richard Mabey (2010, p. 34) defines pootling as "walking for its own sake," "just roaming aimlessly" and "inquisitive walking." Similarly, Thoreau (1862/2007) sauntered as a distinct form of artful walking with historical congruence; his walking body summoned the history of walking:

> I have met with but one or two persons in the course of my life who understood the art of Walking, that is, of taking walks--who had a genius, so to speak, for SAUNTERING, which word is beautifully derived 'from idle people who roved about the country, in the Middle Ages, and asked charity, under pretense of going a la Sainte Terre'. (3)

In light of Thoreau's style of walking, *gestural walking* describes open-sensory, curiosity-based ambling. Gestural walking is in contradistinction to a closed-sensory style of fast-track walking through natural reserves in search of visually satisfying prospects, panoramas or vertiginous plummets (see chapter 2). For Thoreau, walking created continuous bodily tracing of habitats that helped to structure his written works on plant life (see chapters 1 and 3). Amato (2004,

p. 143) argues that "walking determined the form of his books, which were structured by the succession of what he observed rather than logical argumentation."

Walking is also stylistically and energetically varied. Walking may be solitary or collective. Edensor (2001, p. 98) again comments that "technical notions, which propose that particular forms of walking can be learnt and developed, stem from and engender appropriate ways of interacting with the countryside." Kay and Moxham classify recreational walking into two types. The first kind comprises walking (sauntering, strolling, ambling, plodding, promenading, wandering and roaming) and rambling (tramping and striding). The second is specialised, strenuous and demanding and includes hiking (trekking, marching and trail-walking) and backpacking (hill-walking, yomping, fell-walking and peak-bagging). In addition to recreational bipedality, a walkabout is a foot journey taken by Aboriginal peoples between periods of wage employment, or traditionally as a journey of spiritual alignment and economic exchange (Benterrak, Muecke, & Roe, 1996, p. 248). Day hiking is a long walk without overnight camping. A section hiker walks segments of a track discontinuously over many seasons, whereas a thru-hiker walks the full length continuously from beginning to end. Lastly, there are walks in search of particularities. For example, a plant walk is intensively focused on identifying or appreciating flora, whereas birding is walking in search of avian species.

The difference between active speed and passive speed illuminates further theoretical varieties of walking. As a technique of "active rest," walking is movement of the body which induces calmness and restfulness, invigorating physical and mental energies (Mauss, 1935/2006, p. 89). In contrast, the "passive speed" of cars, trains, planes, television, video games and the cinema induces anxiety and exhaustion without actual physical exertion (Giblett, 2008b, p. 105). In *Speed and Politics*, Paul Virilio (cited in Armitage, 2000, p. 74) proposes *dromology* as "the study and analysis of the impact of increasing speed of transport and communications on the development of land-use." For Virilio, the increasing anxiety about being passive is related to the rise of the contemporary city, whose streets are circuits for the rapid transmission of information between "dromomaniacs" or speed-obsessed postmoderns. In such terms, passive speed is linked to cars, planes and trains, as well as virtual networks where speed is thought to indicate productivity, development and progress.

Basing his idea in a bodily interpretation of speed and passivity, Mauss (1935/2006, p. 89) describes active rest as "not simply a matter of aesthetics, but also of bodily games." Hence active rest is not idle gazing or immobile transfixation on a scene, but rather a capacity of the body to rejuvenate through restful activities. As active rest, walking engages the tendency of the body towards homeostatic balance when granted restorative conditions of fresh air and water, purposeful movement and sensory stimulation. As expressed by the FABC, walking, as either active rest or active speed, can be a nature cure that alleviates the effects of urban trauma. However, walking as active speed replicates the conditions of highways, railways, flightways and other linear avenues of movement that prioritise efficiency and rapidity. Instead, walking as gesturally open active rest invites the curative properties of immersion through sense-based outreach of the body to flora: touching, listening, tasting, bending towards, sitting with, closely observing and tossing away (see chapter 2).

Distance and Contact: Early South-West Walkers and Botanisers

The Nyoongar have a history of engaging flora through eating roots, tasting flower nectars, drinking plant infusions, touching fibres and walking in habitats (see chapter 3). Nyoongar country is not a touristic scene, but a life-sustaining habitat for people and plants. The environment becomes body through the ingestion, mastication and digestion of wild foods and, as a living place, land is a body. Rather than determined deductively, the value of the land has been intrinsic to the experience of it. Alternately depicting the South-West landscape as fecund and barren in his journal, explorer George Grey (1841, p. v) "traversed extensive regions unknown to the European traveller, and probably never before trodden by the foot of civilized man." His Nyoongar guide Kaiber impressed Grey with knowledge of the edibility of local plants, especially the potentially poisonous zamia cycad (see chapter 3). For Grey, the surveyor's mode of perception was aligned to colonial exploits and mixed with an attitude towards the land as sterile and barren, hence negatively valued and to be avoided (see chapter 6). *Terra nullius* comprised "places where no man has trod before." The land through which Grey traversed was a virgin territory, an uninhabited zone of free enterprise, in his view.

Expeditions into the bush by early botanists and plant collectors, such as James Drummond and Georgiana Molloy, galvanised the formation of Western Australian identity by invoking the genteel tradition of the walking collector. Traversing the terrain furnished Drummond the perspicacity required to identify new species, as well as distinguish changes in the vegetation and soil character: "After passing 5 or 6 miles of open forest the luxuriance of the vegetation very ill agreeing with the barren appearance of the Soil we arrived at the river we were in search of running in an extensive valley" (Drummond, 1829/2004, p. 80). As with Grey, Drummond's evaluation of the land strongly emphasised the contrasts between picturesque and barren affordances. Similarly, Georgiana Molloy contributed to nineteenth century understandings of South-West plants through specimen gathering that necessarily occurred on foot (see Lines, 1994). Harper (2007, p. 10) observes that Molloy's "collecting walks at the isolated settlements of Augusta and Vasse in Western

Australia [helped her overcome] her initial sense of alienation from the landscape to embrace being in the bush as 'one of the most delightful states of existence'." Unlike Drummond, Molloy's domestic obligations required proximity to her family home. Rather than the major longitudinal traverses of colonial exploration, her botanising was locally focused, designed circularly to begin and end at home, and relied on Aboriginal knowledges. For both Drummond and Molloy, the sensory appreciation and scientific understanding of plants began in walking.

Walking engenders an embodied sense of place as part of a participatory aesthetics. Through walking, place becomes an experience, rather than an appearance. For South-West colonists, some of whom were accustomed to the green pastures and forests of the northern hemisphere, sense of place engaged detachment, disinterestedness and pictoriality. However, for twentieth century composer and obsessive long-distance walker Percy Grainger (1985), the theme of embracing wildflowers through sensuous walking exemplifies a post-colonial aesthetics of flora in which the body is the locus of curiosity, particularly involving the sense of smell:

> The W.A. wildflowers are ravishing; such fresh intense colors, & queer jolly forms. The old gum trees too are so altogether ownish: (characteristic) surely there has never been a *flat* land more uniquely itself & only itself! My heart yearns for the embraces of the glorious lovely scenery that I know will be unfolded to me on this tour. And the *scent* of everything at Perth...the woods, the very rooms all were fragrant with wild flower & 'wattle' smells, that smell crisp & fresh as wild berries in Northern Europe [italics in original]. (226)

Grainger expresses appreciation for the flora that transgresses distanced apprehension of "glorious lovely scenery." For Marcus Clarke (1876/1993, pp. 45-46), gum trees appeared "haggard" and the forests "funereal, secret, stern," as well as stifling, desolate, sullen and black. For Grainger, gum trees and the wattle smells characterise the place; they are peculiarly intriguing rather than peculiarly funereal. Just as the land is "uniquely itself & only itself," Grainger's sensuality disrupts the detachment of visual morphologies. His "heart yearns for the embraces" of intense colours and jolly forms. The "lovely scenery" is more than a painting to stand away from and admire. Instead, the scenery embraces him; the wildflowers ravish; the *seen/scene* offers sensuous encounter rather than disinterested perspective. Instead of burning gum leaves, Grainger (1985) tasted them, absorbed them into his body, claiming eloquently that their flavours:

> Tell to the palate what the birdcalls preach to the ear; they deftly sum up all the influences of the land ... What is the prevailing taste in gum? I find no used expressions to fit it. It's not salt, sweet, bitter, sour. Is it acid? Sharp it is, but that's not its overweighing quality. (238)

Grainger's physical interaction with flora embodies a call by environmental historian William Lines (1998, p. 180) for sensuous curiosity as pivotal to an ethic of care for the natural world: "Only when Australians learn to look sensuously at the continent will they learn to care for it. And not only to look at it, but to touch it, smell it, taste it, drink it, walk it." For both Grainger and Lines, bipedal movement engenders sensuous appreciation of plants.

Walking the Flowering Landscape: Metavisual Appreciation of Plants

Cooperation between track managers, bushwalking clubs, tourism initiatives and community members would promote deeper physical engagement with flora and wildflowers (see chapter 7). A push towards physical interaction with botanical diversity would counterpoise the inheritance of picturesque detachment (see chapter 5). Since walking tracks often traverse areas of botanical importance, the possibility exists for synergy. Such cooperation would broaden opportunities for both "wildflower enthusiasts" and "nature lovers," as well as local groups interested in short organised strolls or other embodied activities. More ecologically nuanced, multi-sensorial appreciation of plants is a possible outcome of a collaborative ethos.

Walking encourages appreciation of flora through all the senses, including sight. This point is summarised by nineteenth century naturalist John Burroughs, a pivotal figure in the history of the North American conservation, who speaks of the corporeality that bipedality affords:

> [The walker's] pores are all open, his circulation is active, his digestion good... He knows the ground is alive; he feels the pulses of the wind and reads the mute language of things. His sympathies are all aroused; his senses are continually reporting messages to his mind. Wind, frost, rain, heat, cold are something to him. He is *not merely a spectator of the panorama* of nature, but a participator in it. He experiences the country he passes through, —tastes it, feels it, absorbs it; the traveler in his fine carriage sees it merely [italics added]. (Burroughs, 1875/2007, p. 19)

Despite Burroughs's mechanical metaphor of "reporting messages" and his problematic notion of mute nature needing translation, he does suggest the sensory merits of walking. Participation as such holds greater potential for connection to flora, arousing sympathies and participatory involvements through taste, touch, smell and sound. Counterbalancing the visual emphasis of wildflower driving routes marketed by tourism commissions, walking tracks are pivotal avenues for offering corporeal experience. Bush tracks like the 1000-kilometere Bibbulmun Track in Western Australia ensure the longevity of multi-sensorial involvement as part of an ethos of flora. As an openness to pausing along the way, extending the body out to reach plants and embrace sensory diversity—rather than speeding along as on a superhighway—gestural walking is a *habitus* amenable to appreciating flora in intimate ways. Walking "articulates a relationship between pedestrian and place, a relationship which is a complex imbrication of the material organization and shape of the landscape, its symbolic meaning, and the ongoing sensual perception and experience of moving through space" (Edensor, 2001, p. 82). In other words, walking creates an embodied sense of place, rich with the sensual particularities of flora.

Not exclusively dictated by the view from the carriage, wildflower tourism incorporating foot travel, or based on a combination of walking and driving, augments visual appreciation (see chapter 7). This seemingly simple point was emphasised repeatedly by local botanists and plant enthusiasts in the South-West (see chapters 2, 7, and 8). Merle Bennett (pers. comm., September 13, 2009) points to the need for wildflower tourists, particularly in the southern half of the region, to set off on foot in order to have a richer experience: "On the whole, people need to stop and get out of their vehicles, to walk rather than to just expect to see the walls of everlastings that occur up north." In Bennett's view, whereas the carpets of everlastings in northern areas can be visually appreciated from a vehicle, wildflowers in the southern part of the state require close engagement (see chapter 7). The appreciator needs to bend down to inspect the plant, smell it, listen to it, taste the droplets of nectar on its small leaves, get his or her hands dirty and then return to the plant at a different time of the year. In a similar vein, Lyn Alcock (pers. comm., September 8, 2009) comments in an interview on the differences in depth and quality between walking and gazing at plants from a distance: "Some people want to go on wildflower tours purely to sit in the bus and to see what's out there. They don't want to get out, they don't want to find out much about it."

The perspectives of Bennett and Alcock are important to consider as wildflower tourism infrastructure, and ecotourism in general, increase in the South-West (Western Australian Tourism Commission, 2003). Over 2,000 plant species, nearly one-quarter of the South-West's botanical tally, grow along the Bibbulmun Track (Ireland & Macey, 2003, p. ii). As a transect of the diverse karri and jarrah forest communities between Perth and Albany, the Track offers bipeds an opportunity for metavisual interaction. Known as "The Showgrounds," an open heath and grassland near Albany along the Bibbulmun stands out prominently from the generally forested character of other sections. Hence, bipedality and wildflower tourism converge. Brampton and Maher (1998, p. 294) point to the depth of association afforded by walking: "The walker witnesses firsthand the unfolding vegetation patterns which result from millions of years of evolution and natural processes and from fewer, but nonetheless influential years of human activity." The interstices between movement on foot and botanical appreciation are recognised by pre-eminent regional botanists as well. Neville Marchant (cited in Ireland & Macey, 2003) in *Wildflowers of the Northern Bibbulmun Track* supports bipedal travel in relation to the experience of floristic diversity:

> There is no doubt that the best way to appreciate the varied plant wealth of Australia's South West is by walking. The incredible diversity of *plant form, flower types, colours, leaf shapes and textures* of the flora is internationally renowned, but these elements need to be examined close up to appreciate their character and intricacies [italics added]. (back cover)

To add to the appreciation of appearances, there are also olfactory aspects, varying between pungency and sweetness, as well as auditory varieties—the sounds of wind, rain and animals interacting with the stiff leaves of sclerophylls. A slight sip of banksia nectar would complete an experience of flora. On the sensual diversity of walking, Edensor (2001, p. 100) is again perspicacious here: "The different distribution of sensory stimuli—the smells, the sounds, the sights, the feelings and the tastes of the countryside—are also part of the ever changing panoply of experience which walking produces." Through sensoriality, walking disrupts the hierarchies intrinsic to sight and central to the metaphor of the *flâneur*.

Kevin Collins is certainly not a bush *flâneur*, speculatively reserved about his interactions with the plants he propagates (see chapters 7 and 8). As we walk around Banksia Farm, Collins calls on all the senses in his presentation on their history, ecology and conservation challenges. We touch bristled cones, taste the slightly sweet sap, and take in the fragrant aroma of the crushed leaves of other hardy related species, all the while viewing the astonishingly variegated colours of the world's only complete living collection of banksia. The flowers of banksia are notably long-lasting and robust, and are thus part of a lucrative international cut-flower industry. Yet, Collins is embodied in his approach to plants, giving a well-rounded appraisal of their survival strategies to enhance my touristic experience of their diverse qualities. In the cultivated garden space of Banksia Farm, we are "far away from the hustle and bustle of the modern city" but certainly not "in places where no man has trod before." And although we are not in the wilderness of the bush, we still "appreciate the strange, peaceful charm" of Western Australian plants. Collins exemplifies gestural walking as a *habitus* of bipedality with sense implications: foot movement around banksia specimens, with interludes of reaching towards, photographing, sketching, reading about, touching and smelling, contemplating, sitting down alongside and viewing the plants from a panoramic distance.

Interlude XXVI: Peripateia

Walking infuses consciousness as proprioceptive orientation to place. Nevertheless, walking often is considered a profane act of getting somewhere. Although not referring specifically to walking, Susan Griffin (1995) connects consciousness, embodiment and earth:

> If human consciousness can be rejoined not only with the human body but with the body of earth, what seems incipient in the reunion is the recovery of meaning within existence that will infuse every kind of meeting between self and the universe, even in the most daily acts, with an eros, a palpable love, that is also sacred. (9)

The onomatopoeic "Walking the Waterwheel" recounts an episode of sacred walking in the Anstey-Keane Damplands (see chapter 2). Rather than an easily decipherable or linear movement, walking gains sacred dimensions, especially with the growing ire of hunger and thirst and the disorientation of getting lost. The metaphor of a whirling waterwheel, spraying sheets of water as it rotates, is preferable to the more static image of a mandala. There is a spatially transformative element of walking that contests exact geometries with transmogrifications of shape: circles transmuting into lines, spheres comprising patterns of squares.

Peripateia
Anstey-Keane Damplands

buried water arcs the ambit
of the wheel that birls
shadows of the paperbarks;
I have walked this wheel
before through frozen berry bogs
    squooshswashsquash

not the mandala of a monk
but a whirling waterwheel;
I round its outer limits
tracing hard lines to the axle,
in the centre and circumference
    squelchsploshslosh

thirsty tho' have taken drink
hungry tho' have taken food
sun singing mug and nape
bread soaked in jagertee
brewing whorls of spirulina
    splashplashsquish

I walk the water wheel
ambling its gambit
a circle made of lines
triangles making spheres
straightforward as the crow
splishspleeshscrash

Bodily Interaction and Embodied Appreciation

I end here with an open-ended paradox. Bushwalking both imperils, through the spread of plant diseases, and ensures, through a deeper degree of cultural appreciation, an embodied aesthetic of flora. Considering this, I reflect on the definition of technique as "an action which is effective and traditional" (Mauss, 1935/2006, p. 82). Gesturally open to plants, walking is a habitus uniquely apposite to reducing the two cultures divide between appreciators and flora. Yet, walking that is not trained in an ethos of care for botanical realities, as part of the cultural transmission of style and concern between generations, can lead to devastating consequences, or quarantine status for areas of plant diversity, as evident in parts of the Fitzgerald River National Park. Indeed, a state initiative to construct a 60 kilometre walk trail through the national park has been met with adamant criticism by activists who wish to prevent the incursion of dieback into the largely unaffected reserve (Hammond, 2011). East of the Fitzgerald area, on the Bibbulmun Track, the bushwalker encounters a familiar sight: boot scrubbing stations positioned at the entrance to most segments of the route. In contrast to Drummond, Molloy and Grainger, contemporary bushwalkers are part of a new environmental sensitivity that connects the body of the human appreciator to the future of botanical bodies. Chapter 13 pursues a theoretical framework for these kinds of corporeal proximities between people and plants through the idea of *aesthesis*.

References

Amato, J. (2004). *On foot: A history of walking*. New York, NY: New York University Press.

Armitage, J. (2000). *Paul Virilio: From modernism to hypermodernism and beyond.* London, England: Sage.

Baker, J. (2010). *The Bibbulmun Track: Its history, its beauty, its walkers* (1 ed.). Perth, Australia: Bibbulmun Track Foundation.

Benterrak, K., Muecke, S., & Roe, P. (1996). *Reading the country: Introduction to nomadology*. Fremantle, Australia: Fremantle Arts Centre Press.

Brampton, J., & Maher, M. (1998). *Guide to the Bibbulmun Track: Southern half* (1 ed.). Como, Australia Department of Conservation and Land Management.

Burroughs, J. (1875/2007). *Winter sunshine*. Middlesex, England: The Echo Library.

Caffin, R. (2002). What is bushwalking? Retrieved April 14, 2011, from http://www.bushwalking.org.au/FAQ/FAQ_General.htm

Clark, N. (2001). "Botanizing on the asphalt?" The complex life of cosmopolitan bodies. In P. Macnaghten & J. Urry (Eds.), *Bodies of nature* (pp. 12-33). London, England: Sage.

Clarke, M. (1876/1993). Preface to Adam Lindsey Gordon's Sea spray and smoke drift. In M. Ackland (Ed.), *The penguin book of 19th century Australian literature* (pp. 43-46). Ringwood, Australia: Penguin Books.

Drummond, J. (1829/2004). Journal of Mr James Drummond on an Expedition on the Main under the direction of M.J. Currie Esq. In J. Shoobert (Ed.), *Western Australian exploration Volume One December 1826-December 1835: The letters, reports & journals of exploration and discovery in Western Australia* (Vol. 1, pp. 79-80). Carlisle, Australia: Hesperian Press.

Edensor, T. (2001). Walking in the British countryside: Reflexivity, embodied practices and ways to escape. In P. Macnaghten & J. Urry (Eds.), *Bodies of nature* (pp. 59-81). London, England: Sage.

Giblett, R. (2008a). *The body of nature and culture*. Basingstoke, England: Palgrave Macmillan.

Giblett, R. (2008b). *Sublime communication technologies*. Basingstoke, England: Palgrave Macmillan.

Giblett, R. (2009). *Landscapes of culture and nature*. Basingstoke, England: Palgrave MacMillian.

Gillbank, L. (2007). The University of Melbourne Herbarium, from McCoy to MELU: A broken paper-trail. University of Melbourne Collections, *1* Retrieved from http://www.unimelb.edu.au/culturalcollections/research/collections1/gillbank.pdf

Grainger, P. (1985). *The farthest north of humanness: Letters of Percy Grainger 1901-14* (K. Dreyfus, Trans.). South Melbourne, Australia: The Macmillan Company.

Grey, G. (1841). *Journals of two expeditions of discovery in North-west and Western Australia during the years 1837, 38, and 39*. (Vol. 1). London, England: T. and W. Boone.

Griffin, S. (1995). *The eros of everyday life: Essays on ecology, gender and society*. New York, NY: Doubleday.

Hammond, J. (2011, March 23). Tourism walk blasted over dieback concern, *The West Australian* Retrieved from http://au.news.yahoo.com/thewest/a/-/breaking/9058184/tourist-walk-blasted-over-dieback-concern/

Harper, M. (2007). *The ways of the bushwalker: On foot in Australia.* Sydney, Australia: University of New South Wales Press.

Ireland, A., & Macey, K. (2003). *Wildflowers of the northern Bibbulmun Track & jarrah forests: A walker's field guide.* Perth, Australia: The Bibbulmun Track Foundation.

Kay, G., & Moxham, N. (1996). Paths for whom? Countryside access for recreational walking. *Leisure Studies, 15*(3), 171-183.

Leggo, C. (2006). Attending to winter: A poetics of research. In W. Ashton & D. Denton (Eds.), *Spirituality, ethnography, and teaching: Stories from within* (pp. 140-155). New York, NY: Peter Lang.

Lines, W. (1994). *An all consuming passion: Origins, modernity, and the Australian life of Georgiana Molloy.* St. Leonards, Australia: Allen & Unwin.

Lines, W. (1998). *A long walk in the Australian bush.* Sydney, Australia: University of New South Wales Press.

Mabey, R. (2010). *A brush with nature: 25 years of personal reflections on the natural world.* London, England: BBC Books.

Mauss, M. (1935/2006). Techniques of the body In N. Schlanger (Ed.), *Techniques, technology and civilisation* (pp. 77-95). Oxford, England: Durkheim Press.

Ross, K. (2008). *Going bush: New Zealanders and nature in the twentieth century.* Auckland, New Zealand: Auckland University Press.

Ryan, J. (*in press*). John Charles Ryan. In S.-P. Mitchell (Ed.), *Fremantle poets 3: Performance poets.* Fremantle, Australia: Fremantle Press.

Solnit, R. (2001). *Wanderlust: A history of walking.* London, England: Verso.

Thoreau, H. D. (1862/2007). *Walking.* LaVergne, TN: Filiquarian Publishing.

Trollope, A. (1873/1967). *Australia* (P. D. Edwards & R. B. Joyce, Trans.). St Lucia, Australia: University of Queensland Press.

Western Australian Tourism Commission. (2003). *Research into the wildflower tourism industry: An exploration of consumer and industry perceptions.* Perth, Australia: Western Australian Tourism Commission.

13

Aesthesis

Abstract: The final chapter theorises people and plants intermeshing and fluxing in close contact as bodies in becoming. The concept of *aesthesis* enfolds change through corporeal intersections with the world. An aesthesis of flora adumbrates the heterogeneous points of contact—exchange, nourishment and healing—between plants and people. Floraesthesis is embodied engagement with plants through eating, touching, tasting, leaning towards and incorporating into. As such, aesthesis encompasses the mutability of flora through cycles of flowering, fruiting, seeding and dormancy. Moreover, an aesthesis of plants allows a framework for considering the rejuvenation and destruction of botanical bodies in colonial and contemporary contexts. An idea with ancient precedents, this concept puts the locus of perception in the contact between bodies. It is a floral aesthetics of sense immanence.

The Bodily Senses and Flora

Green Sense has suggested the need for an aesthetics of flora, based in multi-sensorial experience, to augment the visual appreciation of plants. As argued, the implications of visual biases for human experience are considerable. A plant can fall outside of aesthetic desirability, or it can be objectified by ocular appetite (see chapter 5). Botanical appreciation evolves when the bodily interdependencies between plants and people become measures of impact. This chapter urges the botanophilists of the future to move towards aesthesis through the conceptualisation of bodies as complex sites of engagement. *Floraesthesis* deepens the perception of plants as immutable imagery. In addition to the extension of the eye, gesture implies reaching out the hand intimately through touch. Hence, floraesthesis is gestural contact between bodies in living fields in which the categories of appreciation and sustenance, perception and experience, vision and the senses, and culture and nature become entangled. Through floraesthesis, perception recognises plants as experiential sources of enjoyment. Yet, the appreciator also acknowledges that plants form material interdependencies with people that are co-evolutionary and pan-sensory. As embodied appreciation, floraesthesis broadens human relationships to wild flora through sensory immediacy.

This account of floraesthesis draws from select expositions of aesthesis by Aristotle (1961, 2009), Heidegger (1971), Mules (2006, 2008) and Nancy (1997). Although he did not theorise *aesthesis*, Thoreau practiced floraesthesis in the intimacies he invited with plants through taste, smell and touch (see chapters 2 and 3). Vigilant about setting up an aesthesis-aesthetics binary, this discussion maintains that floraesthesis can augment the visual appreciation of plants (see chapter 5). Following Serres and others, the chapter goes on to theorise the *metasenses*: topaesthesia, temporaesthesia, kinaesthesia and coenesthesia. The Heideggerean explication of "the Open" is again relevant as the field of contact (see chapters 2 and 9). Floraesthesis conceptualises the integration of plants into the flesh and bones of human bodies, invigorating experiences of liminality, curiosity, intrigue and connectivity. The end of the chapter argues that floraesthesis relates to the longevity, sustenance and wellbeing of bodies, evidenced by ancient traditions of bush tucker and contemporary developments in essential oils and flower essences from South-West species. An aesthesis of plants proffers conservation consciousness oriented towards life quality and ecosystem integrity.

Interlude XXVII: At A Bend in the Track, I Help a Marri Return to Earth

Although speaking of North American places, Leggo (2001) advocates synaesthesiac rumination as a practice of enquiry. His language includes a gestural aesthetics of contact with the Earth towards a:

> Poetics of research in long walks on the dike where I listen to light, smell the line of a heron startled into slow motion by my presence, taste the screeches of eagles and hawks, poke with the roots of alders and aspens into the black earth, see the scent of the seasons. (177)

"At a Bend in the Track" was written in Jarrahdale, WA, during the Edith Cowan University School of Communications and Arts Writing Retreat of 2010. Helping a marri return to earth entails participation its processes of decline and decay. Sinking a finger into the dead flesh of an old tree is a curious gesture:

> I could burrow to this tree's pith
> with fingertips, through cambium
> once hard as concrete, now rotting
> in its sleep... (ll. 1 – 4)

Bodily tropes abound in the first stanza along with multi-sensorial interaction with the body of the tree; I could have chosen to stay on the track and bypass the encounter (to read "At a Bend in the Track" in full, see Hickman & Ryan, *in press*).

Aesthesis: Perception and Sensation

Aesthesis is the root of *aesthetics*, practiced with strong visual alignments for the last 250 years. Aesthesis is "embodied sense" (Mules, 2006, p. 2). The recent revival of aesthesis in critical theory indicates a renewed interest in sensation as the content of aesthetic experience (Mules, 2006, 2008; Nancy, 1994, 1997). Aesthesis restores sensation to the bodily domain. Whereas aesthesis signifies open sensory apprehension, aesthetics requires judgements of taste derived largely through sight and exemplified by the picturesque, sublime and beautiful modes of appreciation (see chapter 5). Aesthesis is liminally receptive experience and connectivity between bodies. However, the objectification of plants as "horticultural objects" (see chapter 4) or the taxonomic constructs of classical science is contingent upon closure to indeterminate sensoriality (see chapters 1 and 12).

Aristotle (2009) gives an early account of aesthesis in the *De Anima* or *On the Soul* written around 350 BCE. Book II, chapters 5-11 provide the most concerted discussion of sensation. He defines aesthesis as change encompassing perception and sensation (Hamlyn, 1959, pp. 6-8). Aristotelian aesthesis centralises touch as intrinsic to sensation. Aesthesis is immanent sense independent of positivist mediation. Aristotle refutes the clear division between perception and sensation, and the mediation of sense experience by reason. In this Aristotelian framework, sense is unified perception and sensation. As exemplified by touch, perception is sensation because it produces a bodily response, such as nausea or relaxation, without a kind of cognitive judgement. Aristotle (2009, Bk II Pt 7) claims that "the primary form of sense is touch, which belongs to all animals" and describes it as "the most indispensable" of the senses. As immediate sense, touch "means the absence of any intervening body" (Aristotle, 2009, Bk II Pt 10). In other words, Aristotle "assimilates perception to sensation," aligning to sense immanence, bodily immediacies and visceral engagements (Hamlyn, 1959, p. 7).

Indeed, there are critics of Aristotle's muddling of the separation between perception and sensation. Hamlyn (1959, p. 6) claims that "to perceive something, however, is not merely to have sensations (indeed we do not speak of having sensations when we are perceiving)." Hamlyn's position suggests the valorisation of sight as the primary sense. The mind and the body seem polarised in a Cartesian dualism; sensation is not immanent but mediated by the mind's faculties. Moreover, Hamlyn (1959, p. 11) stresses that sight permits a differentiation between perception and sensation because it requires an extension over a landscape, rather than an internal integration of sensation: "We speak in this way of hearing a sound, tasting a taste, and smelling a smell; but there is nothing internal in this way to seeing (it will not do to invoke sights here, for sights are not emitted)." Sounds, tastes and smells are emitted and enter the body, producing response, but vision entails perceptual interposition. As such, vision is aligned to distance and speculation. Yet, in conjunction with the autocentric senses, visual encounters with nature can induce sensation.

Wolfgang Welsch (1997) affirms the Aristotelian assimilation of perception to sensation as intrinsic to aesthesis (alternately, *aisthesis*). The double sense of aisthesis is foundational to ancient terms for perception: "On the one hand, aisthesis means perception, on the other hand, sensation. This double sense was already present in the Greek and is to be found in most other languages" (Welsch, 1997, p. 62). As immanent sense, aisthesis is sensation. Welsch differentiates between cognitive aisthesis and sensation-bound aisthesis. In Welsch's view (1997, pp. 62-63), categorical predeterminations are associated with cognitive aisthesis. In cognitive aisthesis, "primary sensations" based on vital criteria, such as pleasant or rancid flavours, are replaced by analytical descriptors: beautiful, harmonic, repulsive or ugly. Certainly, the autocentric senses entail judgements of taste. Yet, assessments of sensation, as pleasant or repugnant, are body-engaged.

In *Metaphysics* from 350 BCE, Aristotle further defines aesthesis as amalgamated perception and sensation, or generalised sensory apprehension belonging intrinsically to life (McNeill, 1999, p. 24). However, Aristotle (1961, p. 52) demarcates the boundaries between the senses according to cognitivist mediation: "Wisdom, again, is not to be identified with sense-perception which, though it is our primary source of knowledge of particulars, can never tell us why anything is so (e.g. why fire is hot) but only that it is so." The bodily senses do not lead to knowledge about the world. Aristotle's interpretation of aesthesis falls outside of epistemology. Perception within sensation is not thought-driven or mentally mediated, but immanent and experiential. Although aesthesis is an attribute of living beings, vision is allied to cognition because "sight is the principal source of knowledge and reveals many differences between one object and another" (Aristotle, 1961, p. 51). In *Metaphysics*, Aristotle's account of aesthesis emphasises visual acuity in discerning between objects taxonomically.

Aristotle describes vision as a distinctive sense but without proliferating a hierarchy of the senses. According to an interpretation by McNeill (1999, p. 39), *Nicomachean Ethics*, written around the same time as *Metaphysics* and *De Anima*, posits "practical aesthesis" as more than sense perception. Practical aesthesis signifies that the object is apprehended as a whole amongst all its particularities. Apprehension of the whole occurs through a state of openness. For example, Lindley's idea of the horticultural object exemplifies a closed model of perception based in the predeterminations of science and visual aesthetics (see chapter 4). Practical aesthesis for Aristotle, however, signifies bodily integration of sense experience. As aesthesis, sight links to the other senses through the assimilation of perception to sensation (Hamlyn, 1959). Aristotle tends to describe the non-visual senses as bodily faculties with vision alone elevating the human mind. Although Aristotle seems to privilege vision, "seeing" in an open-whole context includes sight and the autocentric senses. Perception assimilated to sensation is aesthesis, whereas mental knowing is particular to human beings as wisdom or *sophia.*

How does aesthesis relate to the idea of the Open? Heidegger interprets practical aesthesis as the perception of sensory particulars. He translates the term *boule* as "open resolve" or "resolute openness" with openness being the disclosedness of the sense world to the body and the integration of the body to perception (McNeill, 1999, p. 41). For Heidegger, aesthesis is "circumspective seeing" that is inclusive and not speculative "mere looking" or solely visual apprehension (McNeill, 1999, p. 39). The horticultural object is seen within its context as a whole—its environment, ecology, seasonal or climatic nuances—and within its particularity, as an individual living being in relation to other sensing beings. The habitat as a whole—the plant within its circumstances—constitutes the field of practical aesthesis rather than the sense-object itself extruded from its context. As practical aesthesis, perception is the making sense of a phenomenon through sensation.

For Heidegger, open-whole sense perception is the foundation of experience (*empeiria*) and wisdom (*sophia*) but not exclusively that which is attained through the mind. Aesthesis occurs in an open field of contact, represented in *Green Sense* by the application of botanic field aesthetics at three South-West sites (see chapter 2). Heidegger (1971, p. 124) characterises the Open "as the pure forces serried, boundlessly flowing into one another and thus acting toward one another. The widest orbit is the wholeness of the whole draft of attraction." Moreover, "objectification ... blocks us off against the Open" (Heidegger, 1971, p. 120). As aesthesis, sense perception is unfolding, as well as indeterminate, "open, at stake, yet to be decided," hence mutable (McNeill, 1999, p. 36). Seeing involves situations, such that we are "present to ourselves immediately, without any contemplative distance or objectification" (McNeill, 1999, p. 36). The situation of something within a milieu is so unique that it can only be apprehended by practical aesthesis (McNeill, 1999, p. 43).

Phenomenological seeing is a pre-dualistic disposition towards the world of living things based in curiosity and a sensory placement between bodies; it occurs in the "open-whole" but differs to seeing as one of the five senses. In Section 36 of *Being and Time* (1953/1996, pp. 159-162), Heidegger distinguishes between seeing as visual perception and "seeing" as phenomenological openness or seeing-thinking. He approaches this distinction through the concept of curiosity as "characteristically not limited to seeing and [expressing] the tendency toward a peculiar way of letting the world be encountered in perception" (pp. 159-160). By "letting the world be encountered," Heidegger implies a pre-subjective state of open sense receptivity—curiosity—liberated from the deterministic arrangement of perceived object and perceiving subject that is characteristic of taxonomic thinking, for example. In early Greek philosophy, according to Heidegger's reading, cognition was conceptualised as the "desire to see" (p. 160). For Augustine in particular, seeing was integral to *concupiscentia*, the root of the English term "concupiscence" for sensual longing or lust. In a phenomenological context, "seeing" is a metaphor for the acquisition of knowledge—"a knowing something"—through sense wholeness: "'We also say: see how that sounds, see how it smells, see how it tastes, see how hard that is'" (Augustine quoted in Heidegger 1953/1996, p. 160). Seeing is an open mode of experience in which the senses entwine with a thing for the purpose of knowing, through curiosity, about the thing. Moreover, in *History of the Concept of Time: Prolegomena*, Heidegger (1992) extends his interpretation of seeing again through a translation of Augustine:"'We use this word "seeing" also for the other senses when we take them in their cognitive performance'" (quoted in Heidegger, 1992, p. 275). Seeing is body sense in the open-whole and is aligned to aesthesis as sensory occupation of place; through seeing, cognition opens itself to possibilities normally excluded from dualistic, object-making thinking. Heidegger (1992, p. 275) concludes that phenomenological seeing is the directedness of the body senses at the thing being perceived: "It becomes clear here that seeing has a pre-eminence in apprehending and that the sense of seeing is therefore not restricted alone to perceiving with the eyes."

Warwick Mules theorises aesthesis and open-wholeness in contemporary critical theory. Rather than an aesthetics based in disinterested speculation, Mules (2006, p. 3) argues for an aesthetics of sensation; aesthesis is immanent sense, or "the gathering of the senses in contact with the absolute as an open-whole (i.e. the immanence of sense to itself within an open field)." Moreover, sense is unified in a field of contact—the open-whole—gathered through seeing, hearing, tasting, touching and feeling. Aesthesis is contact or "the openness of the body to the outside, the gesture that makes contact with the world" (Mules, 2006, p. 6). Multi-sensorial experience reaches a resolution, unified despite prejudgements and epistemological interpositions, and embedded within irresolute gesture. Bodily aesthesis signifies open-wholeness (Mules, 2008). Sensation emerges from the senses in contact with the substance of the Earth.

As aesthesis, creativity engages the content of aesthetics, allowing more complex explorations of sensation. Mules (2006, p. 1) argues for the broader inflection of aesthesis: "To become more fully engaged, critical theory needs to re-address art works [and nature, I would add] in terms of aesthesis: as immanent sense." Aesthesis is "contact with the earth" and gesture towards the earth as "the incessant rematerialisation of signification within technologically entangled milieu" (Mules, 2006, p. 3). Both substance and location, both the field of contact and the edge of intersection, sense is the fertile place where interaction occurs and where:

> Aesthesis runs through bodies in their exposure to the outside as immanent sense. As aesthesis, sense becomes available as 'material to work with', to create new forms or objective modalities for conditions and situations that have not yet been experienced, that are yet to come. (Mules, 2006, p. 12)

As a perceptual mode linking creativity to sense encounters with the Earth's materiality, aesthesis is "an overflowing of sense within the experience itself" (Mules 2006, p. 4). Mules (2006, p. 4) goes on to say that "an aesthetics based on immanence invokes the opening of sense to experience as open-whole, as a potential to make new sensory connections and modes of embodiment through experimentation and creativity." For Mules, aesthesis signifies the potential to forge innovative connections in the context of open-wholeness, rather than be limited by reifications, reductionisms and predeterminations (see chapter 5).

For Jean-Luc Nancy, aesthesis marks the reinvention of sense and the demise of "rationalistic forgetfulness of the multiplicity of sense experience" (Singer, 2003, p. 187). In Nancy's terms, aesthesis represents an amelioration of the cognitivist appropriation of perception in which experience of the world "does not get co-opted by conceptual understanding" (Singer, 2003, p. 188). In such works as *The Muses*, Nancy (1994, p. 17) emphasises touch as the primary bodily sense, invoking Lucretius who believed that the tactile faculty engages "the sense of the body in its entirety." Drawing upon Aristotle's account, Nancy's characterisation of aesthesis is pre-cognitive, non-determinate and positioned in the entangled field between perception and sensation. Similarly, perception is assimilated to sensation towards sense immanence. As perceptual complexity, aesthesis constitutes immanent sound, touch, taste and smell in conjunction with sight to induce experience without conceptual predeterminations.

Floraesthesis and the Metasenses

Aesthesis marks an aesthetics of bodily sensation and offers a framework for describing corporeal engagements between people and plants. If aesthesis is embodied sense, then floraesthesis is immanent sense for plants, invigorating sensual experience of the botanical world (see chapters 1 and 12). Floraesthesis is integrated sense imparted through embodied contact with flora. As sensory gesture, the practice of floraesthesis bears distinctive qualities or tendencies: diachrony, seasonality, successionality, narrative awareness and participatory involvement. The locus of attention is in the contact between human and plant bodies, reflexively integrated to visual perception. As suggested by Aristotle, Heidegger, Mules and Nancy, sensation stems from the heterogeneity of the senses. Floraesthesis neither culminates in pseudo-scientific epistemologies nor asserts that corporeal apperception is the foundation for wisdom. It is a conceptual framework that seeks to articulate the bodily content—not the intellectual outcome only—of experience of plants: how one feels in the presence of this plant in addition to what one can know objectively.

An aesthesis of plants in which perception is assimilated to sensation complements visual aesthetics. Visual appreciation privileges the powers of the subject to isolate surface features of plants within the plate of botanical illustration, the bounds of the camera shot, or the cursory gaze of the eye over a carpet of wildflowers (see chapter 7). Embodied sense of fruits dehiscing, seeds dispersing, leaves rattling, bark shedding and sap sticking coheres in the Open as sense itself. Sense for plants becomes intersubjective, geoautobiographical, non-hierarchical and place-based (see chapters 1 and 2). Considering the need to broaden visual judgements of beauty, floraesthesis could be described as a counter-aesthetic of plants. Giblett (2011, p. 72) calls for a multi-sensorial aesthetic of conservation that engages the body and which "rather than appealing to aesthetics and so predominantly to the sense of sight, a conservation counter-aesthetic would appeal to all the senses." Hence, floraesthesis is sense experience of plants in relation to the open-whole: the environment, the habitat, the ecosystem and the biome. An aesthesis of flora leads to extensions of contact in an open and "meaningful perceptual whole" (Jonas, 1966, p. 153).

The field of contact at the liminal edge between the bodies of people and plants obviates the subject-to-object power structures residual in the Western notion of aesthetics and intrinsic to Welsch's concept of cognitively bound aisthesis. The tradition of flower appreciation that visual artist Gregory Pryor responds to is the determinate discourse of visual beauty (see chapter 11). *Black Solander* moves beyond the imagistic mediation of plant presence towards visceral reception of gallery stimuli that invigorate bodily and ethical sensibilities. Sense as immanence is embedded in the artistic medium of contact: the gallery space turned crypt. Pryor's exhibition induces felt immediacy rather than sense collapsed upon reason. As floraesthesis, the openness of bodily sensation is pivotal to Pryor's project to infuse the gallery experience with a sense of the imperilled status of many South-West plant species. *Black Solander* suggests the botanic field where floraesthesis can range from artistic settings to habitats.

Floraesthesis furthermore opens appreciation to the possibility of senses beyond the conventionally accepted ones. In addition to sensations produced by the five senses, floraesthesis considers the metasenses: topaesthesia, coenesthesia, kinaesthesia and temporaesthesia. Although Aristotle claimed the possibility of only five senses in *De Anima*, Socrates in Plato's *Theaetetus* purported the unboundedness of the senses, "a great number which have names, an infinite number which have not" (cited in Connor, 2008, p. 2). The unboundedness of the senses refers to the metasenses, each of which is brought to a particular significance with respect to the appreciation of plants.

Topaesthesia comprises sense of place, or the experience of one's body within a place inducing instances of memory and emotion (see chapter 8) as well as embodied spatiality (see chapter 12). Multi-sensorial experience invokes the formation of sense of a place, or the topaesthesiac faculty. This is resonant for the experience of regional character, including the tactile, olfactory, gustatory, visual and auditory qualities of a place. Hence, topaesthesia is a hybrid sense composed of sensory perception, cultural values and the preferences of the percipient. A city-dweller's sense of place will differ for someone from the country. Rebecca Solnit (1994, p. 203) characterises topaesthesia as a metasense comprising spatial awareness and remembrance: "Sense of place is the sixth sense, an internal compass and map made by memory and spatial perception together." For Solnit, topaesthesia is the internal orientation of the body consisting of the recollections of the past, the direct sensory perception of the present, and the cartographic possibility of the future. Sense of place moreover locates the human body amongst the other bodies of the world: vegetative, animal, fungal and mineral.

For Stephen Hopper, who has spent the last three decades studying the flora and fauna of rock outcrops, topaesthesia is partly a scientific sense emanating from the diachronic diligence of revisiting outcrops, recording information as words and pressed specimens, and noting field impressions. Hopper suggests the nodes between memory, writing, flora and the formation of topaesthesia through a "new form of journal taking:"

> The written notes are the same as travel journals written by people in notebooks forever, but I combine that with these collections of plants and to me it's quite a compelling way of bringing me right back to the space and place [of the rock outcrop]. (pers. comm., S. Hopper, September 9, 2009)

Hopper positions his work in the tradition of travel journals written by John Eyre (1845/1964), the Gregory brothers (1884/1981), James Backhouse (1843), James Drummond (1848) and May Vivienne (1901), who all commented on the flora of the South-West. Hopper's field notebook, however, combines written commentary with dried specimens as both a scientific and expository instrument, bringing him "right back to the space and place" of the rock outcrop. At the crossroads of botany and art, Hopper's notebooks are "a combination of science and a bit of scrap-booky art work." His practice demonstrates the diverse scales of sense of place—from a region, habitat or ecosystem to a rock outcrop—and the relationship between topaesthesia and plants. Seddon's *Sense of Place* (1972) also discusses plant taxa as intrinsic to regional character (see chapter 1). Building upon topaesthesia as practiced by Hopper and theorised by Seddon and Solnit, floratopaesthesia is the sense of place evoked through the successional and seasonal experience of the plants co-evolved with that place.

In addition to the sense of a place, there is the sense of bodies in place and space. As a seventh sense, coenesthesia is the internal sense of one's body occupying space, such that, as Serres notes, "if I close my eyes, I have a sense of my own body" (Serres & Zournazi, 2002, p. 199). The *Oxford English Dictionary* defines coenesthesia as "the general sense or feeling of existence arising from the sum of bodily impressions, as distinct from the definite sensations of the special senses; the vital sense" (cited in Sheets-Johnstone, 1992, p. 152). The vital sense, echoing Welsch's concept of vital aisthesis, is the antonym of anaesthesia in which sensation is nullified. Sheets-Johnstone (1992, p. 152) maintains that coenesthesia entered into usage in the 1800s when anaesthetics such as chloroform were introduced into medicine, but the usage of the term declined because "what has no organ—no identifiable structure—has no reality in Western medicine." Rather than being within an organ, coenesthesia has a metaphysical connate in the ancient principle of *pneuma*, which can be translated as metabolism or the life-generating heat of the body (see "Wheatbelt Pneuma," Interlude XX).

Asenjo (1988, p. 12) argues that coenesthesia, as a totality of sensation for our physiological bodies, has been considered an obscure form of perception because of "visual prejudice" in phenomenological research. Coenesthesia imparts "physiological depth" and extension to the body such that "without any regard for inner or outer limits, we incorporate the world" (Asenjo, 1988, pp. 12-14). The coenesthesiac sense entails some degree of absorption into our environments and that conversely our surrounds incorporate us, which is a radically different conceptualisation of perception. Serres (2008, p. 164) employs the evocative expression "mingled bodies" to describe the incorporation of bodies and environments through sensation: "Deep, pungent, dense, black aromas, underground, in graves ... compost and soil are mixtures of bodies and plants, flora and fauna." Serres (1995, p. 80) comments that poetic awareness is especially amenable to the exploration of coenesthesia: "A poet is generally sensitive to his own coenesthesia and is subtly aware of the internal and secret rapport within the body itself" (see chapters 1 and 2).

Closely related to coenesthesia, kinaesthesia, as an eighth sense, points to self-generated movements, such as walking, leaping or turning through space (Serres & Zournazi, 2002, p. 22). Proposed in the 1880s by the English neurologist Henry Charlton Bastian, kinaesthesia is "the sensation of movement of body and limbs relating to sensations originating in muscles, tendons and joints" (Paterson, 2007, p. ix). Like coenesthesia, kinaesthesia is an integrative sense that challenges the Aristotelian notion of separation between discrete senses and the uniform association of the faculties to organs of perception, such as vision to the eyes. Florakinaesthesia, therefore, refers to the sense of moving through plant communities through walking, running, crawling or bounding. Self-trained botanist and coordinator of the Ravensthorpe Wildflower Show, Merle Bennett (pers. comm., September 13, 2009) calls for the engagement the kinaesthesiac sense towards regional appreciation of the flora of the Fitzgerald River area. When I asked her what mode of perception she felt is needed to deeply appreciate the plants of the region, Bennett poses a response at the margin between flora and kinaesthesia, "to walk rather than to just expect to see...walls of everlastings" (see chapter 12). Bennett poses the need for embodied appreciation through kinaesthesia because the "orchid effect" necessitates the proximity afforded by walking, bending over or squatting down (see chapter 7).

As the sense of the seasonal passage of time, a ninth sense might be described as temporaesthesia. Floratemporaesthesia relates temporal experience of flora based in a sense of the cycles of time and the successional experience of plants in a place. Barbara York Main's *Between Wodjil and Tor* (1967) shows the conjunction of temporaesthesia and flora by tracing the symphonic progression of bushland plants through the seasons in a parcel of remnant Wheatbelt vegetation (see chapter 11). A sensibility for South-West plants in Seddon's writings expresses sense of place through protracted contact with flora (Seddon, 1972, 2005). Merle Bennett has forged a long-standing sense for the Fitzgerald River area. Her interview alludes to the somatic metasenses, suggested by Solnit and Serres. Bennett's topaesthesia comprises precise knowledge of where to locate species and when they will bloom. When asked of the seasonal cycles of local flora and the potential for botanical tourism outside of the spring flush of flowers, Bennett charts the calendrical progression of orchid blossoms. She invokes temporaesthesia here:

> In the summer you get the sun flowers, the *Thelymitras*, which are always later. They won't open unless the sun is shining. They're late. Then some of the *Diuris* (donkey orchids) are quite late as well. But then you get the *Caladenia*, the leafless orchid, which is a winter flowering orchid. Then you have the *Eriochilus*. They're probably the first ones in about March, April, and the hare orchids about the same time. Then you get into the *Pterostylis*: the snails, the midgets and the greenhoods generally. (pers. comm., M. Bennett, September 13, 2009)

Orchids flower throughout the year near Ravensthorpe, along with wattles and banksia. In a comparable vein, at Anstey-Keane Damplands near Armadale, David James expresses floratemporaesthesia by synchronising the temporal progression of the landscape to its flowering plants:

> So the peak of flowering is October. By Christmas Day, the Christmas tree will flower as well. A lot of the flowering will be over, but there's still a few more. By January, when it's really hot, almost 99% of the flowers are gone. But come March, the Sand Plain Bottlebrush will be in flower. (pers. comm., D. James, September 23, 2009)

Floraesthesis theorises the somatic interdependencies between plants and people in a place, arising through the five senses but also through the metasenses of topaesthesia, coenesthesia, kinaesthesia and temporaesthesia. As a model of embodied sense, it assimilates perception to sensation, opening the body for integration to what has been construed as the outside, the external environment, the inanimate landscape or the framed plantscape of a picturesque scene.

Jason Barrow is a former employee of the Department of Environment and Conservation and presently an Education Officer at Kurongkurl Katitjin, which houses the Centre for Indigenous Australian Education and Research at Edith Cowan University in Mount Lawley, Western Australia. Jason points to floratemporaesthesia in his narrative account of the plantings around the centre. As an educational display, the Kurongkurl Katitjin garden emphasises the intersections between flowers, ecological rhythms and the cultural patterns of Nyoongar movement and time. Jason explains the arrangement of three species—*balga (Xanthorrhea presseii), mimidi (Xanthorrhoea brunonis)* and *mudja (Nuytsia floribunda*—for their cultural significance during the fire season or *birok*:

> In conjunction with those six *balgas* around the two *Nuytsias*, we've got three other sorts, three individuals of a different species of *balga* called *mimidi*. They flower the same, but as a much skinnier or more slender flower. The *mimidi* is the drill for your fire lighting stick that goes down into the *balga borna*, which is the wood from the *balga*. The fire lighting kit is planted around those two trees that are flowering in *birok* season and that tells you that's the fire season. (pers. comm., J. Barrow, July 21, 2010)

Birok is signified by the flowering of the *balga* and the *mimidi*, which together constitute a fire technology perceived through the kinaesthetic experience of drilling, the smell of smoke and the sensing of warmth. As a unit of time, the season of birok is intrinsic to flowering and cultural livelihood. As chapter 2 suggested, the six Nyoongar seasons differ from the Western concept of a season as fitting within specific dates regardless of surrounding ecological rhythms such as flowering. Nyoongar seasons trace flowering cycles. As Jason relates, the Kurongkurl Katitjin garden exemplifies this interrelation: "You can tell the seasons by looking at the flowers. If something is flowering early, you know that is out of character when you know and understand the old way." Subtle changes in plants indicate broader shifts and possible long-term weather outcomes. As successional, seasonal and diachronic perception of habitats, the temporal awareness offered by Jason's account spans cycles over years:

> If flowers, particularly the yams, are flowering early, you know straight away you are going to have an early break to the season with nothing in the middle and rain at the end. When you look at the flowering and you know how it changes, it gives you a perspective to observe larger patterns through the years. (pers. comm., J. Barrow, July 21, 2010)

The succession of blossoming points to broader correspondences between flowers and Nyoongar culture. When the flower of the common sheoak (*Allocasuarina fraseriana*) turned yellowish-brown, kangaroos were at their most corpulent and ready to hunt (Daw, Walley, & Keighery, 1997). Even when not consumed directly as bush tucker, flowers invoke aesthesis through the bearing of the metasenses of time and place on bodily orientation.

Body Futures: Healing, Wellbeing and Flora

Stemming from the autocentric senses and metasenses, floraesthesis involves entanglement between plants and people. This concept can expand the detached appreciation of aesthetics (see chapter 5). This section further conceptualises floraesthesis in practical terms of health and healing relating indigenous plants to human wellbeing. The longevity of plants connects symbiotically to people. David James recognises the interrelationship between botanical conservation efforts at Anstey-Keane Damplands and human wellbeing:

Figure 25 The Kurongkurl Katitjin Gardens. The plantings around Kurongkurl Katitjin Centre for Indigenous Australian Education and Research at Edith Cowan University, Western Australia, reflect a succession of year-round flowering. *Image courtesy of the author.*

> If every suburban town in Perth had an appealing park, you could walk there on a Saturday morning with a wild flower book, look at birds and animals. It would be good for your body. It would be good for your mind. (pers. comm., D. James, September 23, 2009)

David suggests that the experience of wildflowers—involving physical participation in their habitats—imparts a favourable influence on minds and bodies. Direct engagement with botanical communities through the act of gestural walking summons bodily integrations and enables wellbeing through time spent in the bush.

The wildflowers that are gazed upon by international tourists today as items of visual consumption have been the life blood of Aboriginal peoples for thousands of years. Some of the plants of the South-West have been consumed as roots, fruits, seeds, tubers and flowers by Nyoongar people for over 50,000 years (see chapter 3). "Bush tucker" represents an interpenetration between plants and people for sustenance and pleasure. Traditional bush tucker epitomises floraesthesis as corporeal interaction with flora. Interdependencies between plants and people are also evident in traditional practices of firing the landscape to promote its fecundity as a food resource (for example, see Hallam, 1975). During a walk around Banksia Farm, Kevin Collins explains how some Aboriginal peoples used banksia for food, fire and health:

> Flower heads, when they are predominantly open, were placed into water in a container made from a soft wood or paper bark. This produced a nectar drink. Moreover, large *B. grandis* cones in WA were used as *nulla nullas* (clubs for fighting). The same cones have bracts that smoulder. They were lit, wrapped in fire resistant paperbark (melaleuca bark) and transported in kangaroo hide bags to carry fire. Once the fire was lit by placing twigs and grass on the cone and blowing it, the cone was retrieved, rewrapped and used as a water bottle. Banksia flower spikes open on the warm north face first in the southern hemisphere in an arc following the sun. A great indication of direction. And Northern Territory Aboriginal peoples placed *B. dentata* cones on a camp fire producing a lot of smoke for awhile. Pants were lowered and their bare butts exposed to the smoke as a cure for diarrhoea. (pers. comm., K. Collins, June 4, 2012)

For Collins, the assimilation of consuming edible parts of banksias is vital to instilling in the public a means for appreciating the genus. Physical interaction deepens the ways in which people come to know about their specialised ecologies. Sucking the nectar and eating the seeds engenders a nuanced bodily sense of the plant that classical science or visual speculation might not afford (see chapter 8).

Floraesthesis is also exemplified in plant-based micro-industries that have appeared historically and have reappeared in contemporary South-West contexts (see Loneragan, 1990). A prime example is the company Mt. Romance near Albany, an operation that sustainably harvests Australian sandalwood (*Santalum spicatum*). Known for its aromatic essential oil, Australian sandalwood occurs throughout Western Australia but is most concentrated in the South-West (Department of Environment and Conservation, 2011). Western Australia contains about 161 million hectares of Australian sandalwood, the densest distribution of the tree in the world (Mt Romance Australia, 2011b, para. 1). Australian sandalwood was first distilled in 1875, but the export of unprocessed sandalwood to China as an incense wood for ceremonial purposes dates back to the 1840s (Mt Romance Australia, 2011a, para. 1). In 1916, the commercial extraction of sandalwood oil began, shortly before legislation was prompted in the 1920s to control gathering, processing and harvesting (Mt Romance Australia, 2011a, para. 1-2). During this time, sandalwood oil was recognised in Europe and Asia as a herbal treatment for venereal diseases.

With the introduction of antibiotics in the middle of the twentieth century, the production of Australian sandalwood oil ceased. In the 1990s, interest in sandalwood oil as a treatment for a wide range of skin conditions reignited the production of oil for cosmetics, fragrance and aromatherapy markets (Mt Romance Australia, 2011a, para. 7). The development of techniques for sustainably harvesting sandalwood still goes on. For every sandalwood tree harvested, Mt. Romance plants twelve seedlings to ensure the long-term survival of the slow-growing trees. Dead wood is collected in the field, and 400 seeds are planted per tonne of wood. Although the focus of the commercial enterprise is singly on smell, the educational experience at the centre near Albany brings the appreciator into the domain of aesthesis through touch. Visitors are invited to hear, see, smell and even taste the tree, considering that the factory itself is outside the natural distribution range of the species.

An aesthesis of plants proffers a more promising outlook for botanical and cultural integrations through sustainable enterprises based on bodily experience of plants like sandalwood. However, floraesthesis also offers the healing of the human body through botanical substances. Aromatherapy involves the incorporation of essential oils, such as Australian sandalwood, into the body for therapeutic purposes through three main pathways: ingestion, inhalation and absorption through the skin (Price, 2007, p. 139). Similar to the tea tree, fragonia essential oil derived from *Agonis fragrans* is being cultivated near Harvey, Western Australia for aromatherapy. Its smell has been described as "a pleasant fresh cineolic odour with a hint of a citrus note...mixed with a slight spicy cinnamon tonality and sweet balsamic undertones" (Burfield cited in The Paperbark Co, 2008, Fragonia Oil, para. 5). Recent scientific analysis of fragonia suggests anti-microbial, anti-inflammatory and pain-relieving qualities. Aromatherapists report the effectiveness of fragonia for the relief of muscle pain, respiratory problems, digestive disorders, inflammatory conditions, yeast allergies, infectious diseases, stress-related problems and mental illness (Aromatherapy College of Australia, *n.d.*).

Another noteworthy regional example of floraesthesis entails the shift away from plants as imagery to flowers as subtle substances for healing. Flower essence therapy is an example of mind-body integration. Physical ailments are approached through pathways opened by a wildflower essence. Vasudeva and Kadambii Barnao (1997) in *Australian Flower Essences for the 21st Century* enumerate the subtle qualities of South-West wildflowers. As intimated by David James, botanically rich reserves protecting wildflower habitats are not only for biodiversity preservation, community recreation or wildflower tourism but also for health maintenance, personal rejuvenation and spiritual sustenance. Barnao and Barnao (1997, p. 13) concur that "the gardens, parks and wildernesses that we seek to replenish our spirit and clear our minds and to invigorate ourselves, these are the settings for our interactions with flowers, and are precious to us." In a form comparable to a homeopathic dosage, wildflower essences are assimilated into one's body. The integration between flower, body and mind occurs at a subtle level. The authors claim to derive their approach to wildflower essences from Aboriginal healing practices:

> The Mobarn or Wadinyoongahri, as leaders and healers of the tribe, would conduct a healing ceremony where a person was laid in a pit of earth-covered coals, over which was sprinkled water and flowers, with a kangaroo skin laid on top in a type of sauna. It was considered that the person would get a new spirit from the flowers. (Barnao & Barnao, 1997, pp. 13-14)

Barnao and Barnao attribute the therapeutic agency to the flowers themselves rather than the flower essence practitioner: "The wildflowers offer this wide range of healing, and combined with diagnostic techniques, also help to uncover the many layers upon which suffering is built" (Barnao & Barnao, 1997, p. 17). Rather than images, wildflowers are agents for alleviating suffering.

A technique called Flower Affinity Diagnosis works on the premise that we are attracted to flowers for particular bodily reasons. The first step is for the patient to concentrate on a problem, attempting to visualise—and re-experience—the difficult situation. The patient then looks through a stack of images and takes note of positive or negative bodily responses. A positive reaction means that the flower conveys serenity; a neutral response signifies that the flower offers only moderate potential; a negative response suggests the ability of the flower to rouse strong emotions and hence heal embodied traumas. The practitioner prescribes an oral dose of flower essence. As an example, the Christmas Tree focuses on the resolution of irresponsibility, selfishness, inconsistency and immaturity, and helps to orient someone struggling with the responsibilities of family life (Barnao & Barnao, 1997, p. 82). Grippingly, this therapy focuses on the appearance of flowers, but not for aesthetic reasons only; the image bears upon the body. Through essences, sight is integrated to the senses. Flower essence therapy is yielding beneficial results in South-West hospitals as a complementary medicine. A study indicates the use of Western Australian bush essences in nine Perth area hospitals for stress and pain management. Safe and consistent results were produced without interfering with modern Western procedures (Balinski, 1998).

Interlude XXVIII: Smokebush Coda

Common Smokebush (*Conospermum stoechadis*) is distributed widely throughout the South-West, between Geraldton and Esperance. The genus *Conospermum* was among the earliest identified by botanists in Western Australia. Smokebush has been a traditional healing agent to Nyoongar people. In the 1980s, Smokebush was identified out of a pool of seven-thousand species as containing significant levels of conocurovone for treating HIV, and thus became the subject of international pharmaceutical speculation (Biber-Klemm, Cottier, Cullet, & Berglas, 2006, p. 97). The poem "Smokebush Coda" is a somewhat enigmatic take on the subtle healing attributes of the endemic. In 2010, I visited a local flower essence practitioner to broaden my understanding of the therapeutic qualities of South-West plants. I was instructed to apply drops of Common Smokebush essence to the top of my head where, the year before, I had a cancerous lesion excised, resulting in a permanent crater in my epidermis:

> the crown of my skull
> has a shallow crater
> carved out by the surgeon's
>
> falchion (ll. 1–4)

The poem considers the connectivity between local plants and human wellbeing. Importantly, bodily and earthly metaphors involve actual interdependencies between plants and people that long-standing traditions of plants as medicines confirm (to read "Smokebush Coda" in full, see Hickman & Ryan, *in press*).

Aesthetics and Aesthesis: Theorising Bodily Appreciation of Flora

The visual appreciation of plants can reinforce the subject-to-object binary positioning of plants as separate from people and culture. The creation of an aestheticised "horticultural" object is made possible through visual regimes. As such, plants are expected to put on a show, to erupt in colour and form and to flower in a brilliant array. This is a performative and horticultural expectation with the audience poised for passive entertainment. An aestheticised plant is a pleasing, even compelling, one: a good performer or a good work of art (see chapter 5). Yet, embodied engagements open sense appreciation to the interdependencies between plants, habitat realities and somatic implications. As the most captivating phase in the cycle of plants, flowering might be no more compelling than seed dispersion, or haptic and gustatory experiences.

In their becoming, plants evade synchronic representation. As Prigogine (2003, p. 39) comments, "we are led from a world of 'being' to a world of 'becoming'." But, as *Green Sense* has suggested, our becoming connects to our being in an ongoing interplay with the natural world. An aesthesis of plants comprises seasonal, successional and narrative awareness. Perception assimilated to sensation entails dialogical exchange between plants and people over time. It is multi-sensorial interpenetration in which the "rational subject" and the "perceptual object" find contact in a liminal field of unresolved openness and sensation. Floraesthesis entails the sense of what is yet to come: the seeding ensuing (and ensuring) the flowering. Instead of taxonomic focus on single species, an aesthesis might also consider the succession of flowering between species over various temporal and spatial dimensions, as has been suggested by Merle Bennett, David James and Jason Barrow.

Regional micro-enterprises incorporating the taste of flora and educational tourism centred on embodied experiences reinvigorate sensuous histories that also risk being lost before being fully appreciated. Diversified forms of appreciation, such as wildflower tourism in tandem with bush tucker education and essential oil micro-enterprises, are in urgent need to reconnect human wellbeing to the fate of plants. Former director of the Western Australian Herbarium Neville Marchant (2005, p. 20) comments that "a sombre thought is that diversity of WA plants will decline significantly before its incredible range in plant form, flower structure, pollinators and survival strategies are fully appreciated in Western Australia." Species diversity is sensory diversity, which also risks decline: smells and tastes that no longer exist, or foods and fibres whose traditions of are gone. Where once the body of the wild plant tangled with the body of the human, our postmodern means of sustenance might make such relationships seem no longer germane.

References

Aristotle. (1961). *Metaphysics* (J. Warrington, Trans.). London, England: J. M. Dent & Sons.

Aristotle. (2009). *On the soul* [Internet Classics Archive]. (J.A. Smith, Trans.) Retrieved from http://classics.mit.edu/Aristotle/soul.2.ii.html

Aromatherapy College of Australia. (n.d.). Essential oil fragonia Retrieved April 17, 2011, from http://www.acoa.com.au/site/acoa/FRAGONIA.pdf

Asenjo, F. (1988). *In-between: An essay on categories*. Lanham, MD: University Press of America.

Backhouse, J. (1843). *A narrative of a visit to the Australian colonies*. London, England: Hamilton, Adams and Co.

Balinski, A. (1998). Use of Western Australian flower essences in the management of pain and stress in the hospital setting. *Complementary Therapies in Nursing and Midwifery*, 4(4), 111-117. doi: 10.1016/S1353-6117(98)80048-6

Barnao, V., & Barnao, K. (1997). *Australian flower essences for the 21st century*. Perth, Australia: Australasian Flower Essence Academy.

Biber-Klemm, S., Cottier, T., Cullet, P., & Berglas, D. S. (2006). The current law of plant genetic resources and traditional knowledge. In S. Biber-Klemm & T. Cottier (Eds.), *Rights to plant genetic resources and traditional knowledge: Basic issues and perspectives* (pp. 56-111). Oxfordshire, England: CABI.

Connor, S. (2008). Introduction. In M. Serres (Ed.), *The five senses: A philosophy of mingled bodies* (pp. 1-16). London, England: Continuum.

Daw, B., Walley, T., & Keighery, G. (1997). *Bush tucker plants of the South-West*. Kensington, Australia: Department of Environment and Conservation.

Department of Environment and Conservation. (2011). Santalum spicatum. *FloraBase* Retrieved April 17, 2011, from http://florabase.calm.wa.gov.au/browse/map/2359

Drummond, J. (1848). Notice of Mr. Drummond's discovery of three remarkable plants in South-West Australia. *Companion to Curtis's botanical magazine, 74*, 1-3. Retrieved from http://books.google.com

Eyre, E. J. (1845/1964). *Journals of expeditions of discovery into Central Australia, and overland from Adelaide to King George's Sound, in the years 1840-1*. Adelaide, Australia: Libraries Board of South Australia.

Giblett, R. (2011). *People and places of nature and culture*. Bristol, England: Intellect.

Gregory, A. C., & Gregory, F. T. (1884/1981). *Journals of Australian explorations*. Victoria Park, Australia: Hesperian Press.

Hallam, S. J. (1975). *Fire and hearth: A study of Aboriginal usage and European usurpation in south-western Australia*. Canberra, Australia: Australian Institute of Aboriginal Studies.

Hamlyn, D. (1959). Aristotle's account of aesthesis in the "De Anima". *The Classical Quarterly*, 9(1), 6-16.

Heidegger, M. (1953/1996). *Being and time: A translation of* Sein und Zeit (J. Stambaugh, Trans.). Albany, NY: State University of New York Press.

Heidegger, M. (1971). *Poetry, language, thought* (A. Hofstadter, Trans.). New York, NY: HarperCollins.

Heidegger, M. (1992). *History of the concept of time: Prolegomena* (T. Kisiel, Trans.). Bloomington, IN: Indiana University Press.

Hickman, E., & Ryan, J. (*in press*). *Two with nature: Ellen Hickman and John Ryan*. Fremantle, Australia: Fremantle Press.

Jonas, H. (1966). *The phenomenon of life: Toward a philosophical biology*. Westport, CT: Greenwood Press.

Leggo, C. (2001). Research as poetic rumination: Twenty-six ways of listening to light. In L. Neilsen, A. Cole & J. G. Knowles (Eds.), *The art of writing inquiry* (pp. 173-195). Halifax, Canada: Backalong Books.

Loneragan, O. (1990). *Historical review of sandalwood (Santalum spicatum) research in Western Australia*. Como, Australia: Department of Conservation and Land Management.

Main, B. Y. (1967). *Between wodjil and tor*. Melbourne, Australia: Jacaranda.

Marchant, N. G. (2005). Documenting the diverse Western Australian flora. In *Black solander, Gregory Pryor*. Perth, Australia: Perth Institute of Contemporary Arts.

McNeill, W. (1999). *The glance of the eye: Heidegger, Aristotle, and the ends of theory*. Albany, NY: State University of New York Press.

Mt Romance Australia. (2011a). Our story Retrieved April 17, 2011, from www.mtromance.com.au/our-story/history/

Mt Romance Australia. (2011b). Sustainable sandalwood Retrieved April 17, 2011, from http://www.mtromance.com.au/our-ethics/

Mules, W. (2006). Contact aesthetics and digital arts: At the threshold of the Earth. *Fibreculture, 9* Retrieved from http://nine.fibreculturejournal.org/fcj-058-contact-aesthetics-and-digital-arts-at-the-threshold-of-the-earth/

Mules, W. (2008). Open country: Towards a material environmental aesthetics. *Continuum, 22*(2), 201-212.

Nancy, J.-L. (1994). *The muses* (P. Kamuf, Trans.). Stanford, CA: Stanford University Press.

Nancy, J.-L. (1997). *The sense of the world* (J. Librett, Trans.). Minneapolis, MN: University of Minnesota Press.

Paterson, M. (2007). *The senses of touch: Haptics, affects and technologies*. Oxford, England: Berg.

Price, L. (2007). How essential oils enter the body. In S. Price & L. Price (Eds.), *Aromatherapy for health professionals* (3 ed., pp. 139-154). Philadelphia, PA: Elsevier.

Prigogine, I. (2003). *Is future given?* Toh Tuck Link, Singapore: World Scientific Publishing.

Seddon, G. (1972). *Sense of place: A response to an environment.* Perth, Australia: University of Western Australia Press.
Seddon, G. (2005). *The old country: Australian landscapes, plants and people.* Cambridge, England: Cambridge University Press.
Serres, M. (2008). *The five senses: A philosophy of mingled bodies.* London, England: Continuum.
Serres, M., & Latour, B. (1995). *Conversations on science, culture, and time* (R. Lapidus, Trans.). Ann Arbor, MI: University of Michigan Press.
Serres, M., & Zournazi, M. (2002). The art of living: A conversation with Michel Serres. In M. Zournazi (Ed.), *Hope: New philosophies for change* (pp. 192-209). New York, NY: Routledge.
Sheets-Johnstone, M. (1992). The materialization of the body: A history of Western medicine, a history in process. In M. Sheets-Johnstone (Ed.), *Giving the body its due* (pp. 132-158). Albany, NY: State University of New York Press.
Singer, A. (2003). *Aesthetic reason: Artworks and the deliberative ethos.* University Park, PA: Pennsylvania State University Press.
Solnit, R. (1994). *Savage dreams: A journey into the landscape wars of the American West.* Berkeley, CA: University of California Press.
The Paperbark Co. (2008). Essential oils Retrieved April 17, 2011, from www.paperbarkoils.com.au/essential-oils.html#oil_agnois
Vivienne, M. (1901). *Travels in Western Australia: Being a description of the various cities and towns, goldfields, and agricultural districts of that state.* London, England: William Heinemann.
Welsch, W. (1997). *Undoing aesthetics.* London, England: Sage.

Coda

Prospects

> *When I grew up, having a patch of bush in the backyard or close enough within walking distance to the city was normal. Increasingly, as areas of bushland diminish and the city expands, that experience is on-offer less and less. So, I think this is where we have to be vigilant and continue to celebrate Western Australia, what's special about it, how we should live with it, into the arts, into broader society. When we start making plays and celebrating the landscape and wildflowers in literature, it shows that the society is maturing.* (*pers. comm.*, S. Hopper, September 9, 2009)

Botanist Stephen Hopper points to possibilities of exchange between the arts and the sciences towards a "two cultures" resolution. His collaborations with the botanical artist Philippa Nikulinsky, published as *Life on the Rocks* (1999) and *Soul of the Desert* (2005), complement science with art—and art with science—through images and words but with a conservation ethos that marks a contemporary evolution from Lindley's treatise, discussed in chapter 4. Nikulinsky and Hopper's works are examples of the forms of cross-disciplinary partnership that can increase public appreciation of flora. As chapter 1 described, this kind of arts/sciences conversation is a tenet of cultural botany. *Green Sense* has outlined the context of cultural botany and the approach of botanic field aesthetics to illustrate the diverse cultures of South-West flora. Moreover, as bodily sense for plants, *floraesthesis* has provided a prism through which to "read" the cultures of flora as represented in regional sources: colonial-era documents (see chapters 4 and 6); wildflower tourism ethnographies (see chapter 7); the memory narratives of botanists, environmentalists and tourists (see chapter 8); embodied language and place-based writing (see chapters 9 and 10); and the body cultures of South-West plants (see chapters 11, 12 and 13).

Indeed, the cultures of flora in the South-West will continue to evolve according to popular trends, environmental conditions, horticultural innovations, global economic patterns and media coverage of Australian biodiversity, for example. As Western Australian plants become better-known internationally, there will be emerging cultures of flora established through local and global means. The continued use of the arts and humanities to research plants will offer opportunities for increasing appreciation of flora and the understanding of the threats to botanical communities. It is a comprehension of botanical aesthetics that is key to an understanding of cultures of flora. In *Green Sense*, cultural botany has provided a context for the study of aesthetics. In my fieldwork in the South-West from 2008-11, I aimed to produce an expansive account of South-West botanical aesthetics through analytical and autopoietic methods. I wanted to assess written and spoken resources to understand the aesthetic values and modes underpinning representations of flora. I also endeavoured to contribute to the appreciation and celebration of South-West plants through my poetic interludes instigated in the field.

The Coda will set out to consider the implications of my questions for broader environmental research and practice in the South-West and beyond the region. Three questions were posed in the Prologue: How have aesthetic values towards South-West plants been expressed in written and spoken language? How is somatic engagement with plants different from sensory engagement with artistic objects? As *aesthesis*, how can aesthetics be a theoretical and reflective praxis based in bodily experience of plants? While summarising the concepts that have been developed thus far, the Coda will also speculate about the application of these philosophies and practices to Australian ecohumanities research. The concepts and practices—although developed here as a project of cultural botany—should be extended to humanities-based studies of nature generally, both within Australia and abroad, in accordance with the cultural and ecological particularities of places.

Philosophical Implications: Questions of Intimacy and Union

A corporeal aesthetics of flora—one which encourages and engages the senses—is a pressing post-colonial task that raises important philosophical questions about human experience of the natural world. In post-Kantian aesthetic philosophies of landscape, distance—afforded by sight—sustains the desirability and controllability of what is seen. *Green Sense* has argued that a significant inadequacy of landscape aesthetics is the conflation of living plants and inanimate objects of art. For example, Carlson's cognitive "natural environmental" model analogises living plants and non-living objects (see chapter 5). Unlike the approach of much research in environmental aesthetics, *Green Sense* has taken a linguistic turn (see chapter 9). The written and spoken sources—when read in the contexts of aesthetics and aesthesis—underscore a cultural tendency to objectify plants through scientific and artistic media. Additionally, in contemporary practices, the semantic tendencies surrounding wildflower tourism also suggest the primacy of visual appreciation of natural "objects" (see chapter 7).

Through a Thoreau-Heidegger framework, chapters 9 and 10 theorised that language is integral to the cultivation of embodied place consciousness. However, smell, taste and touch—in combination with ocular and aural stimuli—produce experience that can be invoked in images or words, but not reproduced pictorially. To say it differently, the contact moment between bodies—the immanence of sensory embodiment—cannot be extended to describe contact between an appreciator and an object of visual art. Multi-sensorial appreciation requires the immediacy of bodies through *gestures to*, *extensions into* and *exchanges with* the natural world. An aesthetics of nature incorporating the active tenses of tasting, smelling and touching places a human "subject" in proximity to living plants. Embodied encounters with wild flora take place in the field; sense immediacy makes possible more complex appreciation through physical interaction (see chapter 2). For example, the smells of sap and other exudates at particular times of year bring about all-enveloping sensations in contrast to the spatial orderliness of beauty or picturesqueness of iconic flowers, such as everlastings. In this light, Part V, "Plant Futures," argued that the return of sensation to floral aesthetics can be developed through conceptual premises (aesthesis and the metasenses) used to elucidate cultural and interpersonal practices (e.g., mourning and walking) and creative outputs (e.g., visual and literary).

Nevertheless, why should proximity between a "subject" and an "object" be of concern? Does closeness axiomatically enhance the pleasure of aesthetic experience of the natural world? The enduring question is one of intimacy. Heidegger argues that intimacy "obtains only where the intimate—world and thing—divides itself cleanly and remains separated. In the midst of the two, in the between of world and thing, in their *inter*, division prevails: a *dif-ference*" (Heidegger, 1971, p. 199). Whereas intimacy exists as *dif-ference*, "union" implies identity loss through unmediated sensory embodiment. Heidegger's notion, instead, suggests a dialogical position between "subject" and "object" in which each living body moulds the other through sense reciprocity. I eat a wild quandong fruit. I walk away. I spit out its pit. A new quandong takes root somewhere else. An array of possibility extends from the initial sensory encounter. Entering my body and assimilating into my flesh, the quandong affords sweetness, piquancy and sustenance through its substance. The nut remains a *dif-ference*, a figure of the prevailing separation through which the world proliferates, through which ecological processes happen, through which seeds disperse. This is the stuff—the flesh, the material, the matter—of environmental networks. In terms of appreciation and aesthetics, the pleasure of eating the quandong complements the visual impression of the fruit's ruddiness.

In summary, an embodied ecological aesthetics can make experience more beneficial and more pleasant than distanced contemplation of forms and colours. Yet, the abolition of distance advocated here—while contributing to aesthetic experience in complex ecological ways—merits further philosophical evaluation. Sensory appreciation of flora is not axiomatically more desirable in all circumstances. Corporeal intermeshing—the fruit enters me but I extend the fruit through physical gestures—has conceptual implications that are best considered when experience is linked to place as *sense of place*. Indeed, some plants are repulsive to taste, but pleasing to touch. Others bear—simultaneously—drastically toxic leaves and highly delicious flowers. The fruits of other species are dangerous and unpalatable during phases of the year or when not processed. The zamia palm produces a nut that sustained Aboriginal peoples throughout Australia for the last 50,000 years. But before consumption the nuts required detoxification by long-standing practices. Hence, contact and sensoriality should be moderated by place-based knowledges, although not diminished or made contingent upon those knowledges, as Carlson's model holds.

Rather than a situation of union in which the "subject" and "object" division collapses, *dif-ference* allows proximity in which bodies can reach contact and attain intimacy through sensory entwinements. Undeniably, there come to pass different interferences with aesthetic enjoyment when all the senses are activated. The zamia nut was consumed by explorers and colonists—often with disastrous consequences—yet relished by Aboriginal peoples who cached nuts for retrieval during seasonal journeys. The act of stashing detoxified the nut, while aiding the longevity of the zamia species—a relationship of co-modification, in which both "subject" and "object" benefitted through nutriment and distribution, respectively. This profound historical reality underscores a pressing concern for ecophilosophy. Sensory embodiment in nature unfolds a raft of contemporary philosophical problems that should be theorised and researched through various disciplines as thoroughly as visual distance has been theorised since the modern inception of Western aesthetics and its operative term: disinterestedness (see chapter 5).

Past Into Present: Researching Cultures of Flora

There are three historical phases of aesthetics that converge in the present discussion: traditional Aboriginal embodied interdependencies predating European contact; pre-colonial and colonial scientific representations; and post-colonial appreciation based in bodily futures. This latter phase involves a contemporary adaptation of the physical intimacies between plants and people during the human history of South-West biota. Tracing historical materials, chapters 3, 4 and 6 focused on the accounts of settlers, explorers and naturalists. A close reading of Lindley's *Sketch* in chapter 4 demonstrates that scientific documents can be infused with aesthetic evaluations and cultural preferences. This third phase is a movement from a visually privileged colonial landscape aesthetics to a post-colonial aesthesis of botanical conservation and body engagement. As exemplified by the pre-colonial practice of wild yam cultivation, Nyoongar people enacted, on a regular basis, extensive embodied relationships with nature in which plants were cultivated and consumed, thus becoming *acculturated*. In contrast, the aesthetic modes of the beautiful, picturesque and sublime fused visual assessment and scientific object-making. Through the aesthetic outputs of writing, painting and illustration, the sensory manifold of plants became marginalised into curious economic prospect by a distant audience. In these historical tracings, *Green Sense* argues that a form of aesthetic-scientific hybridity was at work. It was expressed culturally by settlement-era documents and scientifically by the "synoptic tableau" of taxonomy (see chapter 1).

Hopper remarks that "science provides a perspective and an approach that are global in application and know no boundaries, political or cultural. This is not to say that science is value-free or lacks a cultural context" (Nikulinsky & Hopper, 2005, p. 14). In expanding Hopper's position, I suggest here that aesthetic modes—the beautiful, picturesque and sublime—also know no boundaries and are not confined to discrete historical or cultural moments. Rather, aesthetic attitudes towards plants exhibit a variety of visual, sensory, cognitive and emotional tones through settlement era to our contemporary context of tourism and conservation science. These tones, in turn, are shaped by the preferences, moods and historical circumstances of the appreciators—the subjects—in connection to the milieu of the appreciated—the "horticultural objects," to invoke Lindley's term. As the interviews with botanists, conservationists and tourists reveal, an individual shifts naturally between aesthetic inclinations within a conversation; hence experiences of plants—and the communication of experiences—are best described through a syncretic aesthetics (see chapters 7 and 8). The critique of visuality central to Heideggerean thought and the cultural botany of Henry David Thoreau present perspectives for assessing the interchanges between images and language—and images *in* language—with respect to flora (see chapters 1 and 9).

Drawing from archives and other sources, potential also exists for related research into the medicinal attributes of South-West plants (see chapter 13). Despite the scarcity of historical documentation of colonial adoption of plants as medicine, there are promising potentials and ethical pitfalls related to contemporary therapeutic uptake of South-West species. This is shown by current research into common smokebush as a treatment for cancer and H.I.V. (see "SmokeBush Coda," Interlude XXVIII). Another example is *Stirlingia latifolia*, known as blueboy or rust flower. The common name *blueboy* refers to the fact that wall plaster—made with sand from where the species grows—turns blue. In 1918, essential oil of *Stirlingia latifolia* examined at the Imperial Institute in the UK was found to consist almost entirely of acetophenone, which has soporific properties and was employed in medicine during that era. Yet another example is *Chamelaucium uncinatum* or

Geraldton wax which was researched in the 1940s to yield a medicine comparable to penicillin. Moreover, *Anigozanthos humilis* contains haemocorin, responsible for the red colour of the root and presently being researched for antitumor and antibacterial applications. Indeed, one of the promising cultures of flora is the development of medicines from South-West species. Yet, an ethical concern is the intellectual property of Aboriginal groups; these plants and their usages constitute part of Nyoongar cultural heritage.

Further considering the past through the lens of the present, the history of South-West Australian wildflower tourism, the development of the concept of botanical tourism and the analysis of the practices of each offer directions for related studies. Wildflower tourism is a culture of flora with aesthetic underpinnings. However, there are other angles from which to approach wildflower tourism. As chapter 7 concluded, further comparative research into South-West ecotourism should situate the region's plant tourism within Australian and international contexts. On a philosophical level, a return to corporeal appreciation contributes to the exploration of aesthetics as sensation, but what should this mean for the public and for non-botanical nature? How should the scent of wildflowers, the taste of sap or the singing of birds be introduced into the management of human experience of biodiversity? Despite such questions, these moments of intimacy offer possibility for increased appreciation through sensory openness; visitors have memorable experiences of nature through tasting, touching and smelling. A wildflower tourism model of embodiment and conservation would draw from the senses synergistically.

Through a model of embodiment, the Lesueur Management Plan—alluded to at the beginning of chapter 5—would be revised to incorporate visitors' sensory interactions with nature. Educational tours into the park might introduce or expand programs for tasting, touching, smelling, squatting down beside and looking closely at wild plants as vital modes of learning. On the one hand, visual models of appreciation are integral to the management of botanical heritage sites, such as Lesueur. The landscape becomes an object of visual pleasure and hence management; conversely, the *non-pleasurable visual* and *non-visual sensory* qualities tend to be excised out. On the other hand, multi-sensoriality can enhance models of land management and rehabilitation. The methodology makes possible a transformation of the "field" from one constructed by the colonial gaze and reasserted by the gridlines of ecological managerialism to an indeterminate place of participatory meaning-making. Hence, the practical implications of a sensory approach for botanical conservation should be elaborated and applied further within the region and Australia broadly. The implications of corporeality for land management should be the subject of successive theoretical or reflective research.

In further research on cultures of nature, memory and mourning studies, presented in chapters 8 and 11 respectively, should expand upon existing literature in memory research and emotional geography. The ongoing clearing of bushland in the metropolitan Perth area, and in most parts of Australia, will continue to affect adversely the emotional and psychological states of activist communities and others for whom access to nature is an essential part of wellbeing and economy. In related research, it will be important to consider the intrinsic connections between memory and mourning in post-colonial ecological contexts. Moreover, documentation of botanical memories—stories, narratives and anecdotes about plants—will contribute to a fuller portrayal of species change in the region (see chapter 8). A central archive of recorded interviews with residents and visitors from a range of locales—and a series of community-based publications about collective memories of flora, including mournful memories—would be very useful contributions to the cultural-botanical record. These materials could be compiled in an open-access repository made available online and affiliated with *FloraBase*—the digital archive of Western Australian plant taxonomy. This archive should encompass the three historical phases of botanical aesthetics: Aboriginal interdependencies; pre-colonial and colonial scientific representations; and post-colonial conservation, tourism and bodily appreciation.

Memory studies of nature in the region and elsewhere will benefit from an interrogation of ethnography's relationship to environmental memory, as well as the connections between aesthetic experiences of nature and human mourning of biodiversity loss (see chapter 11). Subsequent research into cultural botany might focus on ethnography as the primary research method. However, the course of *Green Sense* has been polyvocal because of its blend of ethnography, poetic enquiry and analysis of historical and literary sources. Ethnographic information reflects cultural values towards flora that expand the personal quality of poetry (see chapters 7 and 8). Indeed, ethnography has been an anchor point to a sense of community. Although the interviews were balanced between botanical experts and novices, there are limitations to ethnography that should be considered. The process of selecting interviewees is inherently biased and will often depend on the availability or interest of the respondents, as well as the good fortune of the ethnographer in finding people to interview. The narratives emerging from ethnography are always partial accounts that should be reflexively positioned within a theoretical framework. The interplay between theory and transcript should work to augment the voice of interviewees by drawing out their positions more clearly rather than producing a value judgment about the stories or opinions presented in their transcripts. The selection of material from the transcripts, again, offers only a fractional account of their sensory memories.

Finally, as a technique of embodiment with the long-standing history and cultural relevance, walking—the primary mode of human mobility—has been overshadowed with the advent of motorised transport only since the late 1800s (see chapter 12). Thus, walking should be researched more fully as fundamental to a participatory aesthetics of the natural world, in Australia and internationally. As the third branch of enquiry, gestural walking is practical aesthesis in the botanic field where plants develop and decay (see chapter 2). Extending the work of Solnit (2001) and Edensor (2001), the cultural history of walking provides an embodied history of culture. In the context of ecocultural studies of Australian environments, walking merits further critical consideration as a somatic practice performed in the bush or the city, for the appreciation of flora and nature broadly. How can our national parks become more walkable? What kind of facilities will be required for a rebirth of walking? Paul Virilio (quoted in Armitage, 2000, p. 74) proposes *dromology* as the study of speed and its implications for hypermodern cities; with the term *dromomaniacs* referring to a speed-obsessed populace. Yet, there is an opportunity in cultural studies and critical theory for the emergence of the complementary study of two-footed movement: walking as *peripatetica* or *bipedology* (see chapters 2 and 12). Situated in a transdisciplinary milieu, bipedology would draw from a diverse framework of historical, phenomenological and creative methods. It would underscore the potential for walking to be a premise of ecocultural change, just as dromology has been used as a lens of cultural critique in postmodern theory.

Implications for Creativity: Environmental Practices and Processes

The concepts developed and cultural patterns identified in *Green Sense* have implications for creative practices and scholarly criticism of those practices. For instance, related research into Western Australian environmental aesthetics should develop comparative analyses of visual productions and artists, including early botanical illustrators, such as William Westall (1781-1850) and Georgiana Leake (arrived in Fremantle, 1833; died 1869), and contemporary artists, such as Philippa Nikulinsky. A limitation of this language-based project of aesthetics is the small emphasis given to the visual art of South-West landscapes. Indeed, contemporary artists, such as Gregory Pryor, are part of a movement towards physically immersive, ecologically conscious representations of flora (see chapter 11). In this vein, Stephen Hopper (pers. comm., September 9, 2009) notes an evolution in botanical aesthetics in the region. Colonial artists "applied considerable artistic licence to the form and shape" so that audiences in Europe would recognise the antipodean species as plants (see chapters 4 and 5). As we have seen, illustrations of the Swan River Colony's flora may have reflected aesthetic values developed outside Australia (see chapter 3). Nevertheless, according to Hopper, "the celebration of South-West plants in all forms of art is becoming increasingly more common. Instead of trying to shoehorn plants into familiar forms from elsewhere, people are celebrating the uniqueness of the plants of South-West Australia." The maturing of Western Australian society, as Hopper speculates, entails a reconfiguration of aesthetic ideas through the language we use to communicate about the natural world. The development of regional aesthetic and scientific vocabularies to reflect what is special about South-West plants is exemplified by Alex George's neologism *diallagy* (see chapter 6).

Creative practices of landscape should consider the related principle of process and its implications for aesthetics. The plant morphologist Rolf Sattler (1994) identifies how stasis—stillness, inertia, synchronicity—figures into understandings of plants. He transfigures the binary between stasis and movement by characterising plant structures themselves as processes:

> Structure tends to be considered static, whereas process is dynamic. If we mistake the map for the territory, we conclude that plants consist of structures within which processes occur. On closer inspection we learn, however, that what appears static is in fact also dynamic. (Sattler, 1994, p. 451)

Extending Sattler's ideas here, one could say that plants exist dynamically at the margins between being and becoming; stasis and process; synchronicity and diachrony; visual aesthetics and bodily aesthesis. Sustained sensory awareness of flora in a place implies that the map—the static appearance of greenness, for example—is not the territory—the field of bodily interaction or what has been referred to as botanic field aesthetics. In contrast to an atomistic philosophy of nature as an aggregation of stable things, process elicits "temporality, historicity, change, and passage as fundamental facts" (Rescher, 2000, p. 3). Rather than isolated extracts or individual parts, plants should be defined by their connectivity (see chapter 11). A creative practice involving flora should consider process as an underlying principle. As plants change before our senses, so too do our bodies in relation to plants. Process as principle taps the capacity of language—visual, written or spoken—to express liminality and non-duality.

Indeed, the decay process figures into some works of art and further identifies a post-colonial aesthetics of flora in practice amongst regional artists. Hopper characterises Nikulinsky's botanical illustrations as unidealised and ecological, and hence epitomising a movement from plants as *objets d'art* to plants as dynamic processes:

> She portrays plants and animals in natural situations with warts and all. She doesn't go for the perfect flower or leaf. If it's been a bit 'moth-eaten' or chewed over by insects or mammals, it gets featured as much as a perfect leaf and flower would be. I guess traditional botanical artists ignored all the blemishes, aiming to illustrate in fine detail the structure of organisms. (pers. comm., S. Hopper, September 9, 2009)

A post-colonial aesthetics of decomposition engages with the ecological changeability of flora. Plate 72 of *Life on the Rocks* depicts a honey possum (*Tarsipes rostratus*) perched atop the flower of *Banksia verticillata*, endemic to the Walpole-Albany area, five hours south of Perth by car. The foliage of the banksia is "moth-eaten" and partially browned. The yellow inflorescence is obscured by the foraging honey possum and the unidealised—though ecologically realistic—foliage. There is attention in this representation to natural symmetries between the possum and the banksia flower. Similarly, Plate 67 illustrates sandalwood with gnawed and truncated leaves, warped and aged bark on the lower limb and dead weather-worn matter at the right side of the composition. The composition is *decomposition,* and of habitat processes. Hopper differentiates between the idealised images of early artists and the ecological approach of Nikulinsky. "Warts" and blemishes intimate the plant's individuality, rather than an idealised species. Indeed, Hopper's distinction between idealised and ecological renderings could be said to parallel the distinction between landscape poetry and habitat poetry, set out in chapter 10. Habitat poetry has the capacity to induce corporeality through sensory language (see chapters 2 and 10).

Other process-conscious contemporary environmental art depends on interactions between the creator, viewer and the object in its milieu. These creative works problematise the extraction of an "object" from a "subject" in its context. For example, the artist Andy Goldsworthy sympathises with environmental processes. His work exhibits an awareness of surface-depth interplay:

> *Movement, change, light, growth and decay* are the lifeblood of nature, the energies I try to tap through my work. I need *the shock of touch*, the resistance of place, materials and weather, the earth as my surface. I want to *get under the surface*. When I work with a leaf, rock, stick, it is not just the material itself, it is opening into the *processes of life within and around it* [italics added]. (Mabey, 2010, p. 154)

In many ways, an aesthetics of nature summons the things of now through the satiation of visual desire. Aesthetics as embodied process, however, summons now, but with respect to a common future in a place. Certainly in these works we note Goldsworthy's interest in reinventing nature as an object of art. Nevertheless, the interdependencies between natural things and their habitats, strong currents in his place-based installations, point to a basic difference between objects of visual art and living plants, argued for in chapter 5. This kind of creative contact with place through process, epitomised by Nikulinsky and Goldsworthy, is connected to natural conservation and human sustenance (see chapter 5).

Interlude XXIX: Qualup Bell

Cultural botany contextualises wildflowers in interrelationships. In the spring of 2009, I visited Hopetoun, WA, where the eastern edge on the Fitzgerald River National Park meets the Southern Ocean. Hopetoun is home to Merle Bennett (see chapters 7 and 13). A mood of disappointment, uncertainty and deflation shrouded the little town, as the Broken Hill Proprietary corporation had halted further construction of the newly established Ravensthorpe nickel mine due to the "global financial crisis." The repercussion of the pull-out had been a local metaphysical crisis. The icons of industrial zealousness were left behind. A massive truck wash, built for oversized mining vehicles, provided testimony to an inescapable loss reflected by Hopetoun residents. The novelty of the mining infrastructure and the fickleness of its corporate profiteers contrasted to the delicacy, timelessness and constancy of the Qualup Bell (*Pimelea physodes*) in flower. The poem shows the often acute angles between human culture and botanical nature, between contemporary financial behemoths and wild flora preserved in national parks, between industrial and primordial senses of time, and between the gargantuan and the diminutive. Endemic to coastal areas between Albany and Esperance, the Qualup Bell symbolises to me that there is still hope in Hopetoun.

Qualup Bell

when nickel prices plunged
BHP reneged on promises
of dazzling wealth to Hopetoun—

now half-hatched façades of
slapdash boom burbs riddle
the hinterland between Veal Street
and Fitzgerald River heath;

north side of town,
a colossal mining truck wash
stands sentinel, belittling all small
tokens of catharsis—

in the saddened hodgepodge,
retailers peddle couches and caffeine,
with that expectant look of refugees;

but then I found a Qualup Bell
pendulating a fuchsia flower trio
each veined-heart bract
poised breathless at its heights

as the Southern Ocean beat alabaster sand,
it would soon bellow into the ground,
fomenting the white-capped sea.

Another key creative practice of place, poetic enquiry should be further articulated as a qualitative approach to landscape research. Extending its theorisation in the social sciences, poetic enquiry should be deployed further in ecocultural studies, particularly in relation to the mourning of habitat loss (see chapter 12). Arts-based, practice-led approaches to landscape research, although promising, are presently unrealised in Australia where poetry as a qualitative "method" is the domain of creative writing departments. Yet, autopoietic methods like writing should be extended to the study of flora, fauna and inanimate "objects" in various disciplines, beginning with the field of cultural studies of nature. Sensory experience of land(scape) through gestural walking and poetry cultivate a kind of creative practice of corporeal place (or a *corporeal practice of creative place* or a *creative corporeality of place practice*). As botanic field aesthetics has endeavoured to show, poetic methods should happen in tandem with other qualitative approaches—ethnography and walking are examples used here—to produce a polyphony in research. By reconstituting the botanic field as a site of creative enquiry, poetry might become a more accepted mode of ecocultural research alongside and in collaboration with quantitative research as, for example, conservation science. There is no doubt that the terrain produced by innovative, transdisciplinary research into ecology will be murky. Yet the challenge to writers of poetry about place is to develop intimacy through touch, smell and taste in order to ground and concretise their works. As Heidegger and Thoreau propound, language is experience: a bodily hexis that centres activity. We feel Thoreau's submersion in the swamp. We re-enact a personal memory of being inside a tree. We taste—as a kind of sense re-enactment—the tang of a wild apple through words. My body responds; I am enthralled by language expressed.

Poetic place practice is even more pressing today when we consider that nature writing is poised to settle on South-West soil as a literary tradition with important possibilities for bioregional consciousness. As chapter 10 maintained, nature writing incorporating body sense should be thought of as habitat writing, in contrast to landscape writing which tends to focus on the visual "scape" of a place. Habitat writing—in which the diverse senses are mobilised—is unmistakable in the works of contemporary British nature writer Richard Mabey (2010), in the tradition of naturalist-writers Gilbert White (1720-1793), Henry David Thoreau and others. This mode of nature writing expresses human-plant intimacies through the sensory immersion and curiosity of the writer. For Heidegger, as we saw in chapter 13, curiosity is phenomenological seeing as body sense in the Open rather than seeing as the single sense of sight. Thus, the biodiversity of South-West

Australia, while the subject of sustained scientific investigation, has the potential to be celebrated, mourned and explored phenomenologically through language. An intriguing combination of philosophy, place consciousness and poetry can be found in the works of South-West writers of local landscapes. These include poets John Kinsella and Glen Phillips, writing about the Wheatbelt region, and poet and essayist Annamaria Weldon, writing about Lake Clifton between Perth and Mandurah. Their works embody Stephen Hopper's hope for "the celebration of South-West plants in all forms of art." These writers could be added to Hopper's vision that Australian environments will continue to be acknowledged and celebrated through different forms of art.

Conclusion: Possibilities for Ecocultural Studies and the Humanities

For there to be a higher degree of transdisciplinarity in ecocultural studies, continued dialogue between the humanities and the sciences is paramount. Stephen Hopper (pers. comm., September 9, 2009) establishes the need for such dialogue when he explains that "combining science and art as different ways of perceiving and learning about the world is a very powerful way of communicating with a broader audience." On the collaboration between disciplines, Hopper adds that "it's part of the maturing of the settler society coming to terms with what is special, unusual and valuable in relation to the West Australian environment." Hopper intimates a regional, post-colonial aesthetics of flora—an evolution away from aesthetic values formed in other places. Collaborations between artists and scientists are pivotal to a postmodern coming to terms with Western Australian sense of place and aesthetic identity. This also has practical consequences for conservation, as Hopper suggests: "I came to realise that science alone doesn't win hearts and minds. You can do the best science in the world as a conservation biologist but if the people who are making political decisions are unaware, it's like fiddling as Rome burns." This pivotal exchange will engage visual artists, writers, scientists and phenomenologists in common projects.

Defining cultural botany within the field of ecocultural studies, this discussion of aesthetics has been situated where the humanities and the sciences intersect (see chapter 1). In the spirit of Goethe whose epigraph to *Life on the Rocks* calls for sympathies between artists and scientists, Hopper is a botanist with collaborative interests in aesthetics and science; in other words, he is a cultural botanist. The arrangement of images and narrative vignettes by Hopper and Nikulinsky in *Life on the Rocks* adds a sense of ecology and change to aesthetic appreciation. In the spirit of both Goethe and Hopper, interchange between the arts and the sciences characterises the poetic interludes appearing throughout *Green Sense.* Chemistry between knowledge forms generates a wider range of appreciative possibilities. To this purpose, *Green Sense* began with the notion of transdisciplinary research into flora that interrogates the foundations of aesthetics and botanical science. The ways in which the arts and the sciences cross over has been central to questioning specialist forms of knowledge-making. But how does transdisciplinarity result in boundary-defying research approaches? How should qualitative and quantitative methodologies converge or diverge? What are the outcomes of transdisciplinary research in environmental contexts? How can approaches developed in one region be used in other contexts, places, ecologies and communities? These are crucial questions agitated by critical theory and postmodern ecocultural studies. They are also central questions for creative research into place.

Through the cultural botany of Thoreau and the phenomenology of Heidegger (see chapters 1 and 9), *Green Sense* has argued for an embodied aesthetics of plants within a place-based, transdisciplinary context. The assemblage of methods reflects Thoreau's field practices; he was a paragon of cultural botany. Thoreau dealt extensively with local flora in participatory ways, prefiguring Heidegger's critique of pictorialisation; his theorisation of "the Open" as the metaphysical basis for the botanic field; emphasis on poetic thought; critique of science; and theorisation of *poiēsis*. For both theorists poetic language plays a role in body openness. The eclecticism of Thoreau suits the development of aesthetic concepts which go beyond philosophical aesthetics—in the vein of Carlson and Berleant—but contextualised in the history, literature and praxis of place; hence the study of botanical aesthetics as part of cultural and creative research. A contribution to ecocultural studies developed here is the study of aesthetic theory within cultural botany and through poetic enquiry, ethnographic approaches and gestural walking (see chapters 1 and 2). Simply put, aesthetic perception is a major mode through which people come to appreciate and value places. It warrants ongoing ecocritical attention through transdisciplinary means.

By studying the aesthetic aspects of language, botanic field aesthetics has provided a practical approach for describing floral aesthetics theoretically and experientially. The approach allows personal reflections through poetry; collective evocations through ethnography; and bodily expressions through walking. These methods suggest that creative, ethnographic and experiential representations of the natural world should be deepened through sensory embodiment in place. Through this configuration of methods, the botanic field has been transfigured from a space of speculation—demarcated by taxonomic and geographical grids—to a phenomenological "field" of contact between plants and people. The botanic field is generated by curiosity, perceptual openness and sensory embodiment; its conceptualisation seeks to reconcile scientific and aesthetic understandings of place with bodily experience. As "the Open," the botanic field is a liminal space of bodily exertion and participatory immersion. Through sensory embodiment in place, the dualisms of otherness—our otherness in nature and the otherness of an unfamiliar nature—recede into phenomenological intimacy. As aesthesis, aesthetics mobilises the senses; meaning is invoked through contact in the Open. Hence, the approach broadens the conditions for aesthetic experience. A regional context re-localises knowledge of the natural world as a balance to the species logos at the heart of the technicised plant. Arts-based, practice-led ecocultural researchers will confront the question of how to approach objectivist ways of knowing. Transdisciplinarity encourages critical dialogue between disciplines, stressing the validity of partial accounts and fostering the development of hybrid methodologies.

An embodied aesthetics of nature depends on more than redefining cultural theory or environmental philosophy. There will remain aspects to consider, such as the personal tendencies of human subjects, the nuances of place and the misuse of Aboriginal knowledges in research contexts, including in creative practices. Hence, any ecocultural approach should comprise diverse methods from the humanities to begin adressing these complexities. Although the senses of taste, touch and smell enhance the appreciation of nature, corporeality is individualistic and multifaceted, yet also part of larger cultural values. Sensory embodiment and its theorisation exact exertion; in contrast, visual appreciation makes possible the detached, passive, even relaxing "taking in of a scene" at the centre of our post-Kantian notions of landscape aesthetics. The intention of *Green Sense* has been to unfold these potential implications of sensory embodiment for artistic, literary, cultural and touristic interactions with nature and place. The research identifies the possibility for sensory embodiment to be introduced to arts-based, practice-led methodologies of landscape research, historical research into Australian environments, nature writing, ecotourism and contemporary conservation agendas. These themes should be explored through the continued application of humanities-based approaches to ecocultural research in South-Western Australia and elsewhere.

References

Abbott, I. (1983). *Aboriginal names for plant species in south-western Australia.* Perth, Australia: Forests Department of Western Australia.

Abraham, T. (2006). *Mountains so sublime: Nineteenth-century British travellers and the lure of the Rocky Mountain West.* Calgary, Canada: University of Calgary Press.

Adelaide Bushwalkers. (2009). Club Information. Retrieved April 17, 2011, from http://www.adelaidebushwalkers.org/general_info.html

Adorno, T. W. (1984). The beauty of nature (C. Lenhardt, Trans.). In G. Adorno & R. Tiedemann (Eds.), *Aesthetic theory* (pp. 91-115). London, England: Routledge and Kegan Paul.

Albrecht, G. (2010). Solastalgia and the creation of new ways of living. In S. Pilgrim & J. Pretty (Eds.), *Nature and culture: Rebuilding lost connections* (pp. 217-234). London, England: Earthscan.

Alexander, A. (1979). Nuytsia Floribunda. In B. Bennett & W. Grono (Eds.), *Wide domain: Western Australian themes & images.* London, England: Angus & Robertson.

Almenberg, G. (2010). *Notes on participatory art.* Central Milton Keynes, England: AuthorHouse.

Amato, J. (2004). *On foot: A history of walking.* New York, NY: New York University Press.

American Psychological Association. (2010). *Publication manual of the American Psychological Association* (6 ed.). Washington, DC: American Psychological Association.

Andrews, M. (1989). *The search for the picturesque: Landscape aesthetics and tourism in Britain, 1760-1800.* Aldershot, England: Scolar.

Arber, A. (1912/1938). *Herbals: Their origin and evolution* (3 ed.). Cambridge, England: Cambridge University Press.

Arber, A. (1950). *The natural philosophy of plant form.* Cambridge, England: Cambridge University Press.

Aristotle. (1961). *Metaphysics* (J. Warrington, Trans.). London, England: J. M. Dent & Sons.

Aristotle. (2009). *On the soul* [Internet Classics Archive]. (J.A. Smith, Trans.). Retrieved from http://classics.mit.edu/Aristotle/soul.2.ii.html

Armitage, J. (2000). *Paul Virilio: From modernism to hypermodernism and beyond.* London, England: Sage.

Armstrong, F. (1836/1979). Manners and habits of the Aborigines of Western Australia, from information collected by Mr F. Armstrong, Interpreter, 1836. In N. Green (Ed.), *Nyungar—The people: Aboriginal customs in the southwest of Australia* (pp. 186-206). North Perth, Australia: Creative Research.

Armstrong, P. (1985). *Charles Darwin in Western Australia: A young scientist's perception of an environment.* Nedlands, Australia: University of Western Australia Press.

Arnold, M. (1882/2011). Literature and science. Retrieved April 4, 2011, from http://homes.chass.utoronto.ca/~ian/arnold.htm

Aromatherapy College of Australia. (n.d.). Essential oil fragonia. Retrieved April 17, 2011, from http://www.acoa.com.au/site/acoa/FRAGONIA.pdf

Arthur, J. M. (2003). *The default country: A lexical cartography of twentieth century Australia.* Sydney, Australia: UNSW Press.

Asenjo, F. (1988). *In-between: An essay on categories.* Lanham, MD: University Press of America.

Aspects of Kings Park. (2007). *Kings Park and Botanic Garden, Western Australia postcard.* Perth, Australia: Kings Park.

Austin, R. (1856). Report by Assistant Surveyor Robert Austin, of an expedition to explore the interior of Western Australia. *Journal of the Royal Geographic Society of London, 26,* 235-274. doi: 10.2307/1798360

Australian Broadcasting Corporation. (2011). Transcript: Capturing the beauty of wildflowers in photos, broadcast 09/09/2005. Retrieved April 10, 2011, from http://www.abc.net.au/stateline/wa/content/2005/s1458437.htm

Australian Television Information Archive. (2001). Bush tucker man. Retrieved January 26, 2011, from http://www.australiantelevision.net/btm.html

Averill, J., Stanat, P., & More, T. (1998). Aesthetics and the environment. *Review of General Psychology, 2*(2), 153-174.

Bachelard, G. (1971). *The poetics of reverie: Childhood, language and the cosmos.* Boston, MA: Beacon Press.

Backhouse, J. (1843). *A narrative of a visit to the Australian colonies.* London, England: Hamilton, Adams and Co.

Bailey, J. (n.d.). Power struggle: Unions' role in a conservation campaign. Retrieved April 14, 2011, from http://www98.griffith.edu.au/dspace/bitstream/10072/1574/1/20646.pdf

Bain, M. A. (1975). *Ancient landmarks: A social and economic history of the Victoria District of Western Australia 1839-1894.* Nedlands, Australia: University of Western Australia Press.

Baisya, R. K., & Das, G. G. (2008). *Aesthetics in marketing*. New Delhi, India: SAGE.

Baker, J. (2010). *The Bibbulmun Track: Its history, its beauty, its walkers* (1 ed.). Perth, Australia: Bibbulmun Track Foundation.

Balinski, A. (1998). Use of Western Australian flower essences in the management of pain and stress in the hospital setting. *Complementary Therapies in Nursing and Midwifery, 4*(4), 111-117. doi: 10.1016/S1353-6117(98)80048-6

Baltzly, D. (2008). Stoicism. In E. N. Zalta (Ed.), *The Stanford encyclopedia of philosophy* (Spring 2008 ed.). Retrieved from http://plato.stanford.edu/archives/spr2008/entries/stoicism/.

Bandt, R. (2008). Place as acoustic space: Hearing Australian identity. In F. Vanclay, M. Higgins & A. Blackshaw (Eds.), *Making sense of place: Exploring concepts and expressions of place through different senses and lenses* (pp. 95-102). Canberra, Australia: National Museum of Australia Press.

Banfield, E. (1968). *The confessions of a beachcomber: Scenes and incidents in the career of an unprofessional beachcomber in tropical Queensland*. Sydney, Australia: Angus & Robertson.

Barker, C. (2008). *Cultural studies: Theory and practice* (3 ed.). London, England: Sage.

Barnao, V., & Barnao, K. (1997). *Australian flower essences for the 21st century*. Perth, Australia: Australasian Flower Essence Academy.

Barrett-Lennard, J. (2005). Outline and absence. In *Black solander, Gregory Pryor* (pp. 3-7). Perth, Australia: PICA Press.

Barthes, R. (1977). *Image, music, text* (S. Heath, Trans.). London, England: Fontana Press.

Basford, K. (1978). *The Green Man*. Ipswich, England: D.S. Brewer.

Bates, D. (1930/1992). *Aboriginal Perth and Bibbulmun biographies and legends*. Victoria Park, Australia: Hesperian Press.

Baudin, N. (1974). *The journal of post captain Nicolas Baudin Commander-in-Chief of the Corvettes GEOGRAPHE and NATURALISTE* (C. Cornell, Trans.). Adelaide, Australia: Libraries Board of South Australia.

Beard, J. (1979a). Phytogeographic regions. In J. Gentilli (Ed.), *Western landscapes* (pp. 107-121). Nedlands, Australia: University of Western Australia Press.

Beard, J. (1979b). *The vegetation of the Pinjarra area, Western Australia: Map and explanatory memoir*. Perth, Australia: Vegmap Publications.

Beard, J., & Pate, J. (1984). Foreword. In J. S. Beard & J. S. Pate (Eds.), *Kwongan, plant life of the sandplain: Biology of a south-west Australian shrubland ecosystem* (pp. xvii-xxi). Nedlands, Australia: University of Western Australia Press.

Belich, J. (2009). *Replenishing the Earth: The settler revolution and the rise of the Anglo-world, 1783-1939*. New York, NY: Oxford University Press.

Benterrak, K., Muecke, S., & Roe, P. (1996). *Reading the country: Introduction to nomadology*. Fremantle, Australia: Fremantle Arts Centre Press.

Beresford, Q., Bekle, H., Phillips, H., & Mulcock, J. (2001). *The salinity crisis: Landscape, communities and politics*. Crawley, Australia: University of Western Australia Press.

Berg, L. (2008). *Introductory botany: Plants, people, and the environment* (2 ed.). Belmont, CA: Thomson Learning, Inc.

Berleant, A. (1993). The aesthetics of art and nature. In S. Kemal & I. Gaskell (Eds.), *Landscape, natural beauty and the arts* (pp. 228-243). Cambridge, England: Cambridge University Press.

Berleant, A. (2005). *Aesthetics and environment: Variations on a theme*. Burlington, VT: Ashgate Publishing.

Berlin, I. (1979). *Against the current: Essays in the history of ideas*. London, England: Hogarth Press.

Berthold, D. (2004). Aldo Leopold: In search of a poetic science. *Human Ecology Review, 11*(3), 205-214. Retrieved from http://www.humanecologyreview.org/pastissues/her113/berthold.pdf

Biber-Klemm, S., Cottier, T., Cullet, P., & Berglas, D. S. (2006). The current law of plant genetic resources and traditional knowledge. In S. Biber-Klemm & T. Cottier (Eds.), *Rights to plant genetic resources and traditional knowledge: Basic issues and perspectives* (pp. 56-111). Oxfordshire, England: CABI.

Bickerton, E. (1905, February 18). Along the Midland railway, Perth to Geraldton, a lady's journey [Australia Trove]. *Western Mail*, 39-40. Retrieved from http://trove.nla.gov.au/

Blamey, R. (2001). Principles of ecotourism. In D. Weaver (Ed.), *The encyclopedia of ecotourism* (pp. 5-22). New York, NY: CABI.

Blunt, W. (2004). *Linnaeus: The compleat naturalist*. London, England: Francis Lincoln Limited.

Bockemühl, J. (1981). *In partnership with nature* (J. Meeks, Trans.). Wyoming, RI: Bio-Dynamic Literature.

Boland, D. J., Brooker, M. H., Chippendale, G. M., Hall, N., Hyland, B. M., Johnston, R. D., et al. (2006). *Forest trees of Australia* (5 ed.). Collingwood, Australia: CSIRO Publishing.

Bondi, L., Davidson, J., & Smith, M. (2005). Introduction: Geography's "emotional turn." In J. Davidson, L. Bondi & M. Smith (Eds.), *Emotional geographies* (pp. 1-16). Aldershot, England: Ashgate.

Botanic Gardens and Parks Authority. (2009). Kings Park Festival over all too soon! Retrieved May 15, 2009, from http://www.bgpa.wa.gov.au/o/content/section/14/74/

Boyer, P., & Wertsch, J. (2009). *Memory in mind and culture*. Cambridge, England: Cambridge University Press.

Brady, E. (2003). *Aesthetics of the natural environment.* Tuscaloosa, AL: University of Alabama Press.

Brampton, J., & Maher, M. (1998). *Guide to the Bibbulmun Track: Southern half* (1 ed.). Como, Australia Department of Conservation and Land Management.

Breeden, S., & Breeden, K. (2010). *Wildflower country: Discovering biodiversity in Australia's Southwest.* Fremantle, Australia: Fremantle Press.

Brewer, J. D. (2000). *Ethnography.* Buckingham, England: Open University Press.

Bropho, R. (1998). Interview 17: Robert Bopho. Swan Valley Aboriginal Community, Circle of Elders. 9:30am, 17:06:98. In T. McCabe (Ed.), *Nyoongar views on logging old growth forests* (pp. 30-32). West Perth, Australia: Wilderness Society.

Brown, A., Batty, A., Brundrett, M., & Dixon, K. (2003). Underground orchid (Rhizanthella gardneri) Interim Recovery Plan 2003-2008. Retrieved from http://www.environment.gov.au/biodiversity/threatened/publications/recovery/r-gardneri/background.html

Brown, R. (1832). General view of the botany of the vicinity of Swan River. *Journal of the Royal Geographic Society*, 17-21. Retrieved from http://books.google.com

Browning, R. (1855/2008). One more word. In C. Warner (Ed.), *A library of the world's best literature* (Vol. VI: Brantome-Bulwer, pp. 2589-2592). New York, NY: Cosimo Publications.

Buckley, R., Pickering, C., & Weaver, D. (2003). *Nature-based tourism, environment, and land management, Volume 2001.* Wallingford, England: CABI.

Buddhist Council of Western Australia. (2009). Jade Buddha for Universal Peace exhibition Perth. Retrieved April 8, 2011, from http://www.wa.buddhistcouncil.org.au/home/jade-buddha-for-universal-peace-exhibition-perth/

Burchell, J. (2002). *The evolution of green politics: Development and change within European green parties.* London, England: Earthscan Publications.

Burke, E. (1756/1990). *A philosophical enquiry into the origin of our ideas of the sublime and beautiful* (5 ed.). Oxford, England: Oxford University Press.

Burroughs, J. (1875/2007). *Winter sunshine.* Middlesex, England: The Echo Library.

Bychkov, O., & Sheppard, A. (2010). Introduction. In O. Bychkov & A. Sheppard (Eds.), *Greek and Roman aesthetics* (pp. xi-xxx). Cambridge, England: Cambridge University Press.

Caffin, R. (2002). What is bushwalking? Retrieved April 14, 2011, from http://www.bushwalking.org.au/FAQ/FAQ_General.htm

Cameron, J. (2005). Place, Goethe and phenomenology: A theoretic journey. *Janus Head, 8*(1), 174-198. Retrieved from http://www.janushead.org/8-1/cameron.pdf

Carlson, A. (1993). Appreciating art and appreciating nature. In S. Kemal & I. Gaskell (Eds.), *Landscape, natural beauty and the arts* (pp. 199-227). Cambridge, England: Cambridge University Press.

Carlson, A. (2000). *Aesthetics and the environment: The appreciation of nature, art, and architecture*. New York, NY: Routledge.

Carr, D. J., & Carr, S. G. M. (1981). The botany of the first Australians. In D. J. Carr & S. G. M. Carr (Eds.), *People and plants in Australia* (pp. 3-44). Sydney, Australia: Academic Press.

Carson, R. (1962/1982). *Silent spring*. Harmondsworth, England: Penguin in association with Hamilton.

Carter, P. (2010). *The road to Botany Bay: An exploration of landscape and history*. Minneapolis, MN: University of Minnesota Press.

Casey, E. (1998). *The fate of place: A philosophical history*. Berkeley, CA: University of California Press.

Casey, E. (2000). *Remembering: A phenomenological study*. Bloomington, IN: Indiana University Press.

Cavell, S. (2000). Night and day: Heidegger and Thoreau. In J. E. Faulconer & M. A. Wrathall (Eds.), *Appropriating Heidegger* (pp. 30-49). Cambridge, England: Cambridge University Press.

Caygill, H. (1995). *A Kant dictionary*. Malden, MA: Blackwell Publishing.

Chawla, L. (1994). *In the first country of places: Nature, poetry, and childhood memory*. Albany, NY: State University of New York Press.

Choate, A. (1978). *Gifts upon the water*. Fremantle, Australia: Fremantle Arts Centre Press.

Choate, A. (1986). *A marking of fire*. Fremantle, Australia: Fremantle Arts Centre Press.

Choate, A. (1995). *Mind in need of a desert*. Fremantle, Australia: Fremantle Arts Centre Press.

Clarfield, G. H., & Wiecek, W. W. (1984). *Nuclear America: Military and civilian nuclear power in the United States 1940-1980*. New York, NY: Harper & Row.

Clark, N. (2001). "Botanizing on the asphalt?" The complex life of cosmopolitan bodies. In P. Macnaghten & J. Urry (Eds.), *Bodies of nature* (pp. 12-33). London, England: Sage.

Clarke, M. (1876/1993). Preface to Adam Lindsay Gordon's *Sea spray and smoke drift*. In M. Ackland (Ed.), *The penguin book of 19th century Australian literature* (pp. 43-46). Ringwood, Australia: Penguin Books.

Clarke, P. (2007). *Aboriginal people and their plants*. Dural Delivery Centre, Australia: Rosenberg Publishing.

Clarke, P. (2008). *Aboriginal plant collectors: Botanists and Australian Aboriginal people in the nineteenth century*. Dural Delivery Centre, Australia: Rosenberg Publishing

Classen, C. (1997). Foundations for an anthropology of the senses. *International Social Science Journal, 153*, 401-412.

Clewell, T. (2002). Mourning beyond melancholia: Freud's psychoanalysis of loss. *Journal of the American Psychoanalytic Association 52*(1), 43-67. Retrieved from http://www.apsa.org/portals/1/docs/japa/521/clewell.pdf

Clifford, J. (1986). Introduction: Partial truths. In J. Clifford & G. Marcus (Eds.), *Writing culture: The poetics and politics of ethnography* (pp. 1-26). Berkeley, CA: University of California Press.

Colbung, K. (1998). Interview 30: Ken Colbung. Elder. Joondalup. 9:30am. 10:07:98. In T. McCabe (Ed.), *Nyoongar views on logging old growth forests* (pp. 51-56). West Perth, Australia: Wilderness Society.

Cole, A., & Knowles, J. G. (2007). Arts-informed research. In J. G. Knowles & A. Cole (Eds.), *Handbook of the arts in qualitative research: Perspectives, methodologies, examples, and issues* (pp. 55-70). Thousand Oaks, CA: Sage.

Coleman, H. R. (2008). Eremophila R. Br. *FloraBase*. Retrieved from http://florabase.calm.wa.gov.au/browse/profile/22098

Collard, D. (1998). Interview 19: Dorothy Collard. Actress and great grandmother. In T. McCabe (Ed.), *Nyoongar views on logging old growth forests* (pp. 32-35). West Perth, Australia: Wilderness Society.

Collins, K., Collins, K., & George, A. (2008). *Banksias*. Melbourne, Australia: Blooming Books.

Conley, V. A. (1993). Eco-Subjects. In V. A. Conley (Ed.), *Rethinking technologies* (pp. 77-91). Minneapolis, MN: University of Minnesota Press.

Connerton, P. (1989). *How societies remember*. Cambridge, England: Cambridge University Press.

Connor, S. (2008). Introduction. In M. Serres (Ed.), *The five senses: A philosophy of mingled bodies* (pp. 1-16). London, England: Continuum.

Conservation International. (2007). Southwest Australia. Retrieved May 5, 2011, from http://www.biodiversityhotspots.org/xp/hotspots/australia/Pages/default.aspx

Conservation International. (2008). Biological diversity in Southwest Australia. In K. Caley (Ed.), *Encyclopedia of Earth*. Retrieved from http://www.eoearth.org/article/Biological_diversity_in_Southwest_Australia

Corrick, M. G., & Fuhrer, B. A. (2002). *Wildflowers of Southern Western Australia*. Noble Park, Australia: The Five Mile Press.

Cosgrove, D. (1998). *Social formation and symbolic landscape*. Madison, WI: University of Wisconsin Press.

Cotton, C. M. (1996). *Ethnobotany: Principles and applications.* Chichester, England: John Wiley & Sons.

Cowling, R. M., Richardson, D. M., & Pierce, S. M. (1997). *Vegetation of Southern Africa.* Cambridge, England: Cambridge University Press.

Crary, J. (1990). *Techniques of the observer: On vision and modernity in the nineteenth century.* Cambridge, MA: MIT Press.

Culpeper, N. (1653/1981). *Culpeper's complete herbal and English physician.* Manchester, England: J. Gleave and Son.

Cunningham, I. (2005). *The land of flowers: An Australian environment on the brink.* Caringbah, Australia: Otford Press.

Curry, P. (2010). Grizzly man and the spiritual life. *Journal for the Study of Religion, Nature and Culture, 4*(3), 206-219. doi: 10.1558/jsrnc.v4i3.206

Daniels, S., & Cosgrove, D. (1988). Introduction: Iconography and landscape. In D. Cosgrove & S. Daniels (Eds.), *The iconography of landscape* (pp. 1-10). Cambridge, England: Cambridge University Press.

Dark Side Chocolates. (2011). Dark side chocolates. Retrieved January 26, 2011, from http://www.darksidechocolates.com.au/

Daw, B., Walley, T., & Keighery, G. (1997). *Bush tucker plants of the South-West.* Kensington, Australia: Department of Environment and Conservation.

De Vries, A. (1974). *Dictionary of symbols and imagery.* Amsterdam, Netherlands: North-Holland Publishing Company.

Dean, B. P. (2000). Introduction. In B. P. Dean (Ed.), *Wild fruits: Thoreau's rediscovered last manuscript* (pp. ix-xvii). New York, NY: W.W. Norton.

Deleuze, G., & Guattari, F. (1987). *A thousand plateaus.* London, England: Continuum.

Department of Conservation and Land Management. (1995). *Management plan: Lesueur National Park and Coomallo Nature Reserve 1995-2005.* Perth, Australia: Department of Conservation and Land Management.

Department of Environment and Conservation. (2011). Santalum spicatum. *FloraBase.* Retrieved April 17, 2011, from http://florabase.calm.wa.gov.au/browse/map/2359

Department of Environment and Conservation. (n.d.). An online visitor's guide to Western Australia's parks, reserves and other recreation areas: Fitzgerald River National Park. Retrieved May 5, 2011, from http://www.naturebase.net/component/option,com_hotproperty/task,view/id,105/Itemid,755/

Department of Environment and Conservation, Shire of Dalwallinu, & Australian Tree Seed Centre. (2011). Wattle Day (1 September). Retrieved April 9, 2011, from http://www.worldwidewattle.com/infogallery/symbolic/wattleday.php

Dewey, J. (1934). *Art as experience.* New York, NY: Capricorn Books.

Dewey, J. (1967). *John Dewey: The early works 1882-1898* (Vol. 2). (H. Schneider, Ed.). Carbondale, IL: Southern Illonois University Press.

Dixon, R. (1995). *The Baumgarten corruption: From sense to nonsense in art and philosophy*. East Haven, CT: Pluto Press.

Don, D. (1835). Anigozanthos manglesii: Mr. Mangles's Anigozanthos Vol. 3. In R. Sweet (Ed.), *The British flower garden* (pp. 265-266). Retrieved from http://books.google.com

Driver, F. (2004). Distance and disturbance: Travel, exploration and knowledge in the nineteenth century. *Transactions of the RHS, 14*, 73-92.

Drummond, J. (1829/2004). Journal of Mr James Drummond on an Expedition on the Main under the direction of M.J. Currie Esq. In J. Shoobert (Ed.), *Western Australian exploration Volume One December 1826-December 1835: The letters, reports & journals of exploration and discovery in Western Australia* (pp. 79-80). Carlisle, Australia: Hesperian Press.

Drummond, J. (1848). Notice of Mr. Drummond's discovery of three remarkable plants in South-West Australia. *Companion to Curtis's botanical magazine, 74*, 1-3. Retrieved from http://books.google.com

Eagleton, T. (1990). *The ideology of the aesthetic*. Cambridge, England: Basil Blackwell.

Eaton, M. (1989). *Aesthetics and the good life*. Cranbury, NJ: Associated University Presses.

Edensor, T. (2001). Walking in the British countryside: Reflexivity, embodied practices and ways to escape. In P. Macnaghten & J. Urry (Eds.), *Bodies of nature* (pp. 59-81). London, England: Sage.

Ednie-Brown, J. (1899). *The forests of Western Australia and their development, with plan and illustrations*. Perth, Australia: Perth Printing Works.

Elliott, B. (1967). *The landscape of Australian poetry*. Melbourne, Australia: F.W. Cheshire.

Engen, T. (1982). *The perception of odors*. New York, NY: Academic Press.

Engen, T. (1991). *Odor sensation and memory*. New York, NY: Praager.

Environmental Humanities Graduate Program. (2010). About the EH graduate program. Retrieved April 4, 2011, from http://www.hum.utah.edu/eh/?pageId=1083

Eyre, E. J. (1845/1964). *Journals of expeditions of discovery into Central Australia, and overland from Adelaide to King George's Sound, in the years 1840-1*. Adelaide, Australia: Libraries Board of South Australia.

Fabian, J. (2007). *Memory against culture: Arguments and reminders*. Durham, NC: Duke University Press.

Farber, P. (2000). *Finding order in nature: The naturalist tradition from Linnaeus to E.O. Wilson*. Baltimore, MD: The Johns Hopkins University Press.

Featherstone, M. (1992). Postmodernism and the aestheticization of everyday life. In S. Lash & J. Friedman (Eds.), *Modernity and identity* (pp. 265-290). Oxford, UK: Blackwell.

Feenberg, A. (2005). *Heidegger and Marcuse: The catastrophe and redemption of history*. New York, NY: Routledge.

Firestein, P. (2009). *Crisis of character: Building corporate reputation in an age of skepticism*. New York, NY: Sterling Publishing.

Fitzgerald River National Park Advisory Committee. (1987). *An introduction to the Fitzgerald River National Park*. Perth, Australia: The Committee.

Flinders, M. (1814). *A voyage to Terra Australis*. Retrieved from http://books.google.com

Florida Wildflower Foundation. (2011). Florida's native wildflowers. Retrieved April 9, 2011, from http://flawildflowers.org/

Foucault, M. (1966/2002). *The order of things: An archaeology of the human sciences*. London, England: Routledge.

Fraser, C. (1830). Remarks on the botany, &c. of the banks of Swan River, Isle of Buache, Baie Geographie, and Cape Naturaliste. *Botanical miscellany, 1*, 221-236. Retrieved from http://books.google.com

Freud, S. (1917/2009). Mourning and melancholia. In L. G. Fiorini, T. Bokanowski & S. Lewkowicz (Eds.), *On Freud's* Mourning and melancholia (pp. 19-34). London, England: Karnac Books.

Freud, S. (1997). The antithetical meaning of primal words. In *Writings on art and literature* (pp. 94-100). Stanford, CA: Stanford University Press.

Gage, J. (1993). *Color and culture: Practice and meaning from antiquity to abstraction*. Berkeley, CA: University of California Press.

Garber, F. (1991). *Thoreau's fable of inscribing*. Princeton, NJ: Princeton University Press.

Gardham, J. (2004). Curtis's botanical magazine. Retrieved April 7, 2011, from http://special.lib.gla.ac.uk/exhibns/month/oct2004.html

Gardner, C. A. (1959/1981). *Wildflowers of Western Australia* (14 ed.). Perth, Australia: St. George Books.

Garrard, G. (2004). *Ecocriticism*. Oxfordshire, England: Routledge.

Gascoigne, J. (1996). The ordering of nature and the ordering of empire. In D. Miller & P. Reill (Eds.), *Visions of empire: Voyages, botany, and representations of nature* (pp. 107-116). Cambridge, England: Cambridge University Press.

Geertz, C. (1988). *Works and lives: The anthropologist as author*. Stanford, CA: Stanford University Press.

George, A. (2002a). *The long dry: Bush colours of summer and autumn in South-western Australia*. Kardinya, Australia: Four Gables Press.

George, A. (2002b). The south-western Australian flora in Autumn: 2001 presidential address. *Journal of the Royal Society of Western Australia, 85*, 1-15.

Giblett, R. (1996). *Postmodern wetlands: Culture, history, ecology*. Edinburgh, Scotland: Edinburgh University Press.

Giblett, R. (2004). *Living with the Earth: Mastery to mutuality*. Cambridge, England: Salt.

Giblett, R. (2006). *Forrestdale: People and place*. Bassendean, Australia: Access Press.

Giblett, R. (2008a). *The body of nature and culture*. Basingstoke, England: Palgrave Macmillan.

Giblett, R. (2008b). *Sublime communication technologies*. Basingstoke, England: Palgrave Macmillan.

Giblett, R. (2009). *Landscapes of culture and nature*. Basingstoke, England: Palgrave MacMillian.

Giblett, R. (2011). *People and places of nature and culture*. Bristol, England: Intellect.

Giblett, R., & James, D. (2009). Anstey-Keane botanical jewel. *Landscope, 24*(4), 42-44.

Gillbank, L. (2007). The University of Melbourne Herbarium, from McCoy to MELU: A broken paper-trail. *University of Melbourne Collections, 1*. Retrieved from http://www.unimelb.edu.au/culturalcollections/research/collections1/gillbank.pdf

Gilpin, W. (1786). *Observations relative chiefly to picturesque beauty, made in the year 1772*. Retrieved from http://books.google.com/

Ginsberg, A. (1994). C'mon pigs of western civilization eat more grease. *Literal Latté* Retrieved from http://www.literal-latte.com/1994/11/cmon-pigs-of-western-civilization-eat-more-grease/

Glazebrook, T. (2000). *Heidegger's philosophy of science*. New York, NY: Fordham University Press.

Goethe, J. W. V. (1790/2009). *The metamorphosis of plants*. Cambridge, MA: MIT Press.

Goethe, J. W. V. (1828/2010). *Faust*. (G. De Nerval, Trans.). Retrieved from http://books.google.com/

Goldgar, A. (2007). *Tulipmania: Money, honor, and knowledge in the Dutch Golden Age*. Chicago, IL: University of Chicago Press.

Goody, J. (1993). *The culture of flowers*. Cambridge, England: Cambridge University Press.

Graham, D. (Writer). (1990). Nyoongar bush tucker with Ken Colbung [VHS]. The Institute of Applied Aboriginal Studies Mt. Lawley (Producer).

Graham, M. (2008). Some thoughts about the philosophical underpinnings of Aboriginal worldviews. *Australian Humanities Review*, 45, unpaginated Retrieved from www.australianhumanitiesreview.org/archive/Issue-November-2008/graham.html

Grainger, P. (1985). *The farthest north of humanness: Letters of Percy Grainger 1901-14* (K. Dreyfus, Trans.). South Melbourne, Australia: The Macmillan Company.

Gregory, A. C., & Gregory, F. T. (1884/1981). *Journals of Australian explorations.* Victoria Park, Australia: Hesperian Press.

Grey, G. (1841a). *Journals of two expeditions of discovery in North-west and Western Australia during the years 1837, 38, and 39.* (Vol. 1). London, England: T. and W. Boone.

Grey, G. (1841b). *Journals of two expeditions of discovery in North-West and Western Australia during the years 1837, 38, and 39.* (Vol. 2). London, England: T. and W. Boone.

Griffin, S. (1995). *The eros of everyday life: Essays on ecology, gender and society*. New York, NY: Doubleday.

Griffiths, T. (1996). *Hunters and collectors: The antiquarian imagination in Australia.* Cambridge, England: Cambridge University Press.

Griffiths, T. (2007). The humanities and an environmentally sustainable Australia. *Australian Humanities Review,* (43). Retrieved from http://www.australianhumanitiesreview.org/archive/Issue-December-2007/EcoHumanities/EcoGriffiths.html

Grimwade, R. (1920). *An anthography of the eucalypts*. Sydney, Australia: Angus & Robertson.

Groser, T. (1927). *The lure of the golden West: Experiences and adventures in a bush brotherhood of Western Australia: Early problems and conquests: All about group settlements of the West: The land of sunshine and opportunity*. London, England: Alexander-Ouseley.

Hall, C., & Page, S. (2006). *The geography of tourism and recreation: Environment, place and space* (3 ed.). New York, NY: Routledge.

Hallam, S. J. (1975). *Fire and hearth: A study of Aboriginal usage and European usurpation in south-western Australia*. Canberra, Australia: Australian Institute of Aboriginal Studies.

Ham, J., & Senior, M. (1997). Introduction. In J. Ham & M. Senior (Eds.), *Animal acts: Configuring the human in Western history* (pp. 1-8). New York, NY: Routledge.

Hamilakis, Y. (2010). Re-collecting the fragments: Archaeology as mnemonic practice. In K. Lillios & V. Tsamis (Eds.), *Material mnemonics: Everyday memory in prehistoric Europe* (pp. 188-199). Oxford, England: Oxbow Books.

Hamlyn, D. (1959). Aristotle's account of aesthesis in the De Anima. *The Classical Quarterly, 9*(1), 6-16.

Hammersley, M., & Atkinson, P. (1995). *Ethnography: Principles in practice* (2 ed.). New York, NY: Routledge.

Hammersley, M., & Atkinson, P. (2007). *Ethnography: Principles in practice* (3 ed.). New York, NY: Routledge.

Hammond, J. (2011, March 23). Tourism walk blasted over dieback concern, *The West Australian* Retrieved from http://au.news.yahoo.com/thewest/a/-/breaking/9058184/tourist-walk-blasted-over-dieback-concern/

Harari, J., & Bell, D. (1982). Introduction. In M. Serres, J. Harari & D. Bell (Eds.), *Hermes: Literature, science, philosophy* (pp. ix-xl). Baltimore, MD: Johns Hopkins University Press.

Haraway, D. J. (1991). Situated knowledges: The science question in feminism and the privilege of partial perspective. In *Simians, cyborgs, and women: The reinvention of nature* (pp. 183-201). New York, NY: Routledge.

Harper, D. (2010). green. *Online Etymology Dictionary*. Retrieved from http://www.etymonline.com/index.php?search=green&searchmode=none

Harper, M. (2007). *The ways of the bushwalker: On foot in Australia*. Sydney, Australia: University of New South Wales Press.

Hassell, E. (1975). *My dusky friends*. Dalkeith, Australia: C.W. Hassell.

Hayles, K. (1990). *Chaos bound: Orderly disorder in contemporary literature and science*. Ithaca, NY: Cornell University Press.

Hayward, C. (2009). *Engagement and partnership*. Paper presented at the CREATEC Special Workshop, Edith Cowan University, Mount Lawley, Australia.

Head, L. (2007). Cultural ecology: The problematic human and the terms of engagement. *Progress in Human Geography, 31*(6), 837-846.

Head, L., & Atchison, J. (2009). Cultural ecology: Emerging human-plant geographies. *Progress in Human Geography, 33*(2), 236-245.

Heidegger, M. (1934/2009). *Logic as the question concerning the essence of language* (W. Gregory & Y. Unna, Trans.). Albany, NY: State University of New York Press.

Heidegger, M. (1938/2009). The age of the world picture (J. Veith, Trans.). In G. Figal (Ed.), *The Heidegger reader* (pp. 207-223). Bloomington, IN: Indiana University Press.

Heidegger, M. (1953/1996). *Being and time: A translation of* Sein und Zeit (J. Stambaugh, Trans.). Albany, NY: State University of New York Press.

Heidegger, M. (1971). *Poetry, language, thought* (A. Hofstadter, Trans.). New York, NY: HarperCollins.

Heidegger, M. (1977). The question concerning technology. In D. Krell (Ed.), *Martin Heidegger: Basic writings* (pp. 284-317). New York, NY: Harper & Row.

Heidegger, M. (1982). *On the way to language* (P. Hertz, Trans.). San Francisco, CA: Harper & Row.

Heidegger, M. (1992). *History of the concept of time: Prolegomena* (T. Kisiel, Trans.). Bloomington, IN: Indiana University Press.

Hepburn, R. (1966/2004). Contemporary aesthetics and the neglect of natural beauty. In A. Carlson & A. Berleant (Eds.), *The aesthetics of natural environments* (pp. 43-62). Ontario, Canada: Broadview Press.

Herndl, C. G., & Brown, S. C. (1996). Introduction. In C. G. Herndl & S. C. Brown (Eds.), *Green culture: Environmental rhetoric in contemporary America* (pp. 3-20). Madison, WI: The University of Wisconsin Press.

Herz, R., & Cupchik, G. (1995). The emotional distinctiveness of odor-evoked memories. *Chemical Senses* 20(5), 517-528. doi: 10.1093/chemse/20.5.517

Hewson, H. (1999). *Australia: 300 years of botanical illustration*. Collingwood, Australia: CSIRO Publishing.

Hickman, E., & Ryan, J. (*in press*). *Two with nature: Ellen Hickman and John Ryan*. Fremantle, Australia: Fremantle Press.

Hill, M. (1998). Interview 7: Mike Hill. RFA Nyoongar Action Group chairperson. Public meeting. In T. McCabe (Ed.), *Nyoongar views on logging old growth forests* (pp. 18-20). West Perth, Australia: Wilderness Society.

Hillman, A. (1990). *The Hillman diaries 1877-1884: The personal diaries of Alfred James Hillman for 21 December 1877 to 24 April 1884*. Applecross, Australia: F V Bentley.

Hitchings, R. (2003). People, plants and performance: On actor network theory and the material pleasures of the private garden. *Social & Cultural Geography, 4*(1), 99-113.

Hitchings, R., & Jones, V. (2004). Living with plants and the exploration of botanical encounter within human geographic research practice. *Ethics, Place & Environment, 7*(1 & 2), 3-18.

Hobbs, R., & Hopkins, A. (1990). From frontiers to fragments: European impact on Australia's vegetation. *Proceedings of the Ecological Society of Australia, 16*, 93-114.

Hoffman, B., & Gallaher, T. (2007). Importance indices in ethnobotany. *Ethnobotany Research & Applications, 5*, 201-218.

Hope, A. D. (1977). *Collected Poems, 1930-1970* (2 ed.). London: Angus & Robertson.

Hopkins, A., Keighery, G., & Marchant, N. (1983). Species-rich uplands of south-western Australia. *Proceedings of the Ecological Society of Australia, 12*, 15-26.

Hopper, S. (1990). European Exploration. In A. A. Burbidge, S. Hopper & S. Van Leeuwen (Eds.), *Nature conservation, landscape and recreation values of the Lesueur area: A report to the Environmental Protection Authority from the Department of Conservation and Land Management* (pp. 9-14). Perth, Australia: Environmental Protection Authority.

Hopper, S. (1993). *Kangaroo paws and catspaws: A natural history and field guide*. Como, Australia: Department of Conservation and Land Management.

Hopper, S. (1998). An Australian perspective on plant conservation biology practice. In P. Fiedler & P. Kareiva (Eds.), *Conservation biology for the coming decade* (2 ed., pp. 255-278). New York, NY: Chapman Hall.

Hopper, S. (2004). Southwestern Australia, Cinderella of the world's temperate floristic regions. *Curtis's Botanical Magazine, 21*(2), 132-180.

Hopper, S. (2010). Nuytsia floribunda. *Curtis's Botanical Magazine, 26*(4), 333-368. doi: 10.1111/j.1467-8748.2009.01671.x

Howes, D. (2003). *Sensual relations: Engaging the senses in culture and social theory.* Ann Arbor: The University of Michigan Press.

Hua, A. (2009). "What we all long for:" Memory, trauma and emotional geographies. In M. Smith, J. Davidson, L. Cameron & L. Bondi (Eds.), *Emotion, place, and culture* (pp. 135-148). Farnham, England: Ashgate.

Hutchings, J. (1997). Folklore and symbolism of green. *Folklore, 108*, 55-64.

Ireland, A., & Macey, K. (2003). *Wildflowers of the northern Bibbulmun Track & jarrah forests: A walker's field guide.* Perth, Australia: The Bibbulmun Track Foundation.

Isaacs, J. (1989). *Bush food: Aboriginal food and herbal medicine.* Chatswood, Australia: New Holland Publishers.

Jay, M. (1993). *Downcast eyes: The denigration of vision in twentieth-century French thought.* Berkeley, CA: University of California Press.

Jolly, E. (2010). *Challenging the divide: Approaches to science and poetry.* Adelaide, Australia: Lythrum.

Jonas, H. (1966). *The phenomenon of life: Toward a philosophical biology.* Westport, CT: Greenwood Press.

Jones, O. (2005). An ecology of emotion, memory, self and landscape. In J. Davidson, L. Bondi & M. Smith (Eds.), *Emotional geographies* (pp. 205-218). Aldershot, England: Ashgate.

Jung, C. G. (1959). *The archetypes and the collective unconscious* (R. F. C. Hull, Trans. 2 ed.). London, England: Routledge and Kegan Paul.

Kant, I. (1790/2008). *The critique of judgement* (J.C. Meredith, Trans.). Retrieved from http://books.google.com

Kant, I. (1798/1974). *Anthropology from a pragmatic point of view.* The Hague, Netherlands: Nijhoff.

Kaplan, C. (1996/2000). *Questions of travel: Postmodern discourses of displacement.* Durham, NC: Duke University Press.

Kaptchuk, T. (2000). *The web that has no weaver: Understanding Chinese medicine.* Chicago, IL: Contemporary Books.

Kay, G., & Moxham, N. (1996). Paths for whom? Countryside access for recreational walking. *Leisure Studies, 15*(3), 171-183.

Keane, B. (2009). Pounamu – jade or greenstone - Pounamu – several names. *Te Ara: The Encyclopedia of New Zealand.* Retrieved from http://www.TeAra.govt.nz/en/pounamu-jade-or-greenstone/1

Keighery, G., & Beard, J. (1993). Plant communities. In C. Thomson, G. Hall & G. Friend (Eds.), *Mountains of mystery: A natural history of the Stirling Range* (pp. 43-54). Como, Australia: Department of Conservation and Land Management.

King, H. (2004). *The disease of virgins: Green sickness, chlorosis and the problems of puberty*. London, England: Routledge.

King, V. (2003). Embodied perceptions: Aboriginal expressions of place. In J. Cameron (Ed.), *Changing places: Re-imagining Australia* (pp. 286-295). Double Bay, Australia: Longueville Books.

Kings Park & Botanic Garden. (n.d.). *Aboriginal Life*. West Perth, Australia: Kings Park & Botanic Garden.

Kinsella, J. (1997). *Poems 1980-1994*. Fremantle, Australia: Fremantle Arts Centre Press.

Kinsella, J. (2005). *The new arcadia*. New York, NY: Norton.

Kinsella, J. (2008). *Contrary rhetoric: Lectures on landscape and language*. G. Phillips & A. Taylor (Eds.). Fremantle, Australia: Fremantle Press.

Klein, J. T. (1990). *Interdisciplinarity: History, theory, and practice*. Detroit, MI: Wayne State University Press.

Knight, P. (1986). *Flower poetics in nineteenth-century France*. Oxford, UK: Oxford University Press.

Knowles, J. G. (2001). Writing place, wondering pedagogy. In L. Neilsen, A. Cole & J. G. Knowles (Eds.), *The art of writing inquiry* (pp. 89-99). Halifax, Canada: Backalong Books.

Kurongkurl Katitjin. (2009, Kambarang). Nyoongar elder in residence at Kurongkurl Katitjin. *Our Place: Official Newsletter of Kurongkurl Katitjin.* Retrieved from http://www.kk.ecu.edu.au/data/shared/documents/newsletter/kambarang_newsletter_09_final_we.pdf

Lansdown, A. (1979). *Homecoming*. Fremantle, Australia: Fremantle Arts Centre Press.

Larkin, P. (2001). *Terrain seed scarcity: Poems from a decade*. Cambridge, England: Salt.

Lash, S., & Friedman, J. (1992). Introduction. In S. Lash & J. Friedman (Eds.), *Modernity and identity* (pp. 1-30). Oxford, UK: Blackwell.

Latour, B. (1999). Circulating reference: Sampling the soil in the Amazon Forest. In *Pandora's hope: Essays on the reality of science studies* (pp. 24-79). Cambridge, MA: Harvard University Press.

Lawrence, D. H., & Skinner, M. (1924/2002). *The boy in the bush*. P. Eggert (Ed.). Cambridge, England: Cambridge University Press.

Leakey, R., & Lewin, R. (1996). *The Sixth Extinction: Biodiversity and its survival.* London, England: Orion Books.

Leggo, C. (2001). Research as poetic rumination: Twenty-six ways of listening to light. In L. Neilsen, A. Cole & J. G. Knowles (Eds.), *The art of writing inquiry* (pp. 173-195). Halifax, Canada: Backalong Books.

Leggo, C. (2004). Tangled lines: On autobiography and poetic knowing. In A. Cole, L. Neilsen, J. G. Knowles & T. Luciani (Eds.), *Provoked by art: Theorizing arts-informed research* (Vol. 2, pp. 18-35). Halifax, Canada: Backalong Books.

Leggo, C. (2006). Attending to winter: A poetics of research. In W. Ashton & D. Denton (Eds.), *Spirituality, ethnography, and teaching: Stories from within* (pp. 140-155). New York, NY: Peter Lang.

Leggo, C. (2007). Astonishing silence: Knowing in poetry. In J. G. Knowles & A. Cole (Eds.), *Handbook of the arts in qualitative research: Perspectives, methodologies, examples, and issues* (pp. 165-174). Thousand Oaks, CA: Sage.

Lehrer, J. (2007). *Proust was a neuroscientist.* New York, NY: Houghton Mifflin Company.

Lenman, R. (2006). British photographers and tourism in the nineteenth century: Three case studies. In D. Crouch & N. Lubbren (Eds.), *Visual culture and tourism* (pp. 91-108). Oxford, England: Berg.

Leopold, A. (1949/1987). *A Sand County almanac and Sketches here and there.* Oxford, England: Oxford University Press.

Levin, D. (1993). Decline and fall: Ocularcentrism in Heidegger's reading of the history of metaphysics. In D. Levin (Ed.), *Modernity and the hegemony of vision* (pp. 186-217). Los Angeles: University of California Press.

Lindley, J. (1840). *A sketch of the vegetation of the Swan River colony.* London, England: James Ridgway.

Lindley, J. (1847). Glossary of technical terms. In *The elements of botany, structural and physiological* (pp. i-civ). London, England: Bradbury & Evans.

Lindqvist, S. (2007). *Terra nullius. A journey through no one's land* (S. Death, Trans.). London, England: Granta.

Lines, W. (1994). *An all consuming passion: Origins, modernity, and the Australian life of Georgiana Molloy.* St. Leonards, Australia: Allen & Unwin.

Lines, W. (1998). *A long walk in the Australian bush.* Sydney, Australia: University of New South Wales Press.

Lledo, P, Gheusi, G., & Vincent, J. (2005). Information processing in the mammalian olfactory system. *Physiological Reviews,* 85, 281-317. doi: 10.1152/physrev.00008.2004.

Loneragan, O. (1990). *Historical review of sandalwood* (Santalum spicatum) *research in Western Australia* Como, Australia: Department of Conservation and Land Management.

Loubster, G., Mouton, P., & Nel, J. (2001). The ecotourism potential of herpetofauna in the Namaqua National Park, South Africa. *South African Journal of Wildlife Research, 31*(1-2), 13-23.

Lovejoy, T. (1997). Biodiversity: What is it? In M. Reaka-Kudla, D. Wilson & E. Wilson (Eds.), *Biodiversity II: Understanding and protecting our biological resources* (pp. 7-14). Washington, DC: Joseph Henry Press.

Lyon, J., & Nystrom, A. D. (2002). *USA* (2 ed.). Melbourne, Australia: Lonely Planet Publications.

Mabey, R. (2010). *A brush with nature: 25 years of personal reflections on the natural world.* London, England: BBC Books.

Mahood, M. (2008). *The poet as botanist.* Cambridge, England: Cambridge University Press.

Main, B. Y. (1967). *Between wodjil and tor.* Melbourne, Australia: Jacaranda Press.

Malpas, J. E. (1999). *Place and experience: A philosophical topography.* Cambridge, England: Cambridge University Press.

Malpas, J. E. (2006). *Heidegger's topology: Being, place, world.* Cambridge, MA: MIT Press.

Marchant, N. G. (2005). Documenting the diverse Western Australian flora. In *Black solander, Gregory Pryor* (pp. 17-20). Perth, Australia: Perth Institute of Contemporary Arts.

Marchant, N. G., Wheeler, J. R., Rye, B. L., Bennett, E. M., Lander, N. S., & Macfarlane, T. D. (1987). *Flora of the Perth region: Part one.* Perth, Australia: Western Australian Herbarium, Department of Agriculture.

Martin, G. (2004). *Ethnobotany: A methods manual.* London, England: Earthscan.

Martin, S. (1993). *A new land: European perceptions of Australia 1788-1850.* St Leonards, Australia: Allen & Unwin.

Mauss, M. (1935/2006). Techniques of the body. In N. Schlanger (Ed.), *Techniques, technology and civilisation* (pp. 77-95). Oxford, England: Durkheim Press.

McCabe, T. (1998). Nyoongar kaarny Nyoongar spirituality. In T. McCabe (Ed.), *Nyoongar views on logging old growth forests* (p. 6). West Perth, Australia: Wilderness Society.

McLeod, A. L. (1980). Maturity in Australian satire: The poetry of A. D. Hope. *Modern Language Studies, 10*(2), 42-48.

McNeill, W. (1999). *The glance of the eye: Heidegger, Aristotle, and the ends of theory.* Albany, NY: State University of New York Press.

Meagher, S. J. (1974). The food resources of the Aborigines of the South-west of Western Australia. *WA Museum Records, 3*(1).

Miller, D., & Reill, P. (1996). *Visions of empire: Voyages, botany, and representations of nature.* Cambridge, England: Cambridge University Press.

Miller, G. (2009). Preface. In G. Miller (Ed.), *The metamorphosis of plants* (pp. xi-xiv). Cambridge, MA: MIT Press.

Mitchell, W. (2002). Showing seeing: A critique of visual culture. In N. Mirzoeff (Ed.), *The visual culture reader* (2 ed., pp. 86-101). London, England: Routledge.

Moore, G. F. (1846). *Descriptive vocabulary of the language in common use amongst the Aborigines of Western Australia.* Nedlands, Australia: University of Western Australia Press.

Moore, G. F. (1884/1978). *Diary of ten years eventful life of an early settler in Western Australia, and also A descriptive vocabulary of the language of the Aborigines.* Nedlands, Australia: University of Western Australia Press.

Moran, J. (2010). *Interdisciplinarity* (2 ed.). Abington, England: Routledge.

Moyal, A. (1981). Collectors and illustrators: Woman botanists of the nineteenth century. In D. J. Carr & S. G. M. Carr (Eds.), *People and plants in Australia* (pp. 333-356). Sydney, Australia: Academic Press.

Moyal, A. (1986). *A bright & savage land: The science of a new continent - Australia - where all things were* queer and opposite. Ringwood, Australia: Penguin.

Mt Romance Australia. (2011a). Our story. Retrieved April 17, 2011, from www.mtromance.com.au/our-story/history/

Mt Romance Australia. (2011b). Sustainable sandalwood. Retrieved April 17, 2011, from http://www.mtromance.com.au/our-ethics/

Mulcock, J. (2008). Planting natives: Gardening and belonging to place in Perth, Western Australia. In F. Vanclay, M. Higgins & A. Blackshaw (Eds.), *Making sense of place: Exploring concepts and expressions of place through different senses and lenses* (pp. 183-189). Canberra, Australia: National Museum of Australia Press.

Mules, W. (2006). Contact aesthetics and digital arts: At the threshold of the Earth. *Fibreculture,* (9). Retrieved from http://nine.fibreculturejournal.org/fcj-058-contact-aesthetics-and-digital-arts-at-the-threshold-of-the-earth/

Mules, W. (2008). Open country: Towards a material environmental aesthetics. *Continuum, 22*(2), 201-212.

Murray, J. (1995). *Nature writing handbook: A creative guide.* San Francisco, CA: Sierra Club Books.

Murray, L. (2007). *Selected poems.* Melbourne, Australia: Black Inc.

Nabhan, G. (1993). Foreword: Learning the language of fields and forests. In B. Dean (Ed.), *Faith in a seed: The first publication of Thoreau's last manuscript* (pp. xi-xvii). Washington, DC: Island Press.

Nancy, J.-L. (1994). *The muses* (P. Kamuf, Trans.). Stanford, CA: Stanford University Press.

Nancy, J.-L. (1997). *The sense of the world* (J. Librett, Trans.). Minneapolis, MN: University of Minnesota Press.

Neilsen, L. (2007). Lyric enquiry. In J. G. Knowles & A. Cole (Eds.), *Handbook of the arts in qualitative research: Perspectives, methodologies, examples, and issues* (pp. 93-104). Thousand Oaks, CA: Sage.

Neruda, P. (1995). *Neruda's garden: An anthology of odes* (M. Jacketti, Trans.). Pittsburgh, PA: Latin American Literary Review Press.

Newsome, D., Moore, S., & Dowling, R. (2002). *Natural area tourism: Ecology, impacts and management.* Tonawanda, NY: Channel View Publications.

Nicolescu, B. (2002). *Manifesto of transdisciplinarity* (K.-C. Voss, Trans.). Albany, NY: State University of New York Press.

Nikulinsky, P., & Hopper, S. (1999). *Life on the rocks: The art of survival.* Fremantle, Australia: Fremantle Arts Centre Press.

Nikulinsky, P., & Hopper, S. (2005). *Soul of the desert.* Fremantle, Australia: Fremantle Arts Centre Press.

Nord-Deutscher, L., & Plate, A. G. (1906). *The Lloyd guide to Australasia.* London, England: Edward Stanford.

Northover, J. (1998). Interview 21: Joe Northover. Police Officer. Collie. 4:30pm. 28:06:98. In T. McCabe (Ed.), *Nyoongar views on logging old growth forests* (pp. 35-37). West Perth, Australia: Wilderness Society.

Ogle, N. (1977). *The colony of Western Australia: A manual for emigrants 1839.* Sydney, Australia: John Ferguson.

Paczkowska, G., & Chapman, A. (2000). *The Western Australian flora: A descriptive catalogue.* Perth, Australia: Wildflower Society of Western Australia (Inc.), the Western Australian Herbarium, CALM and the Botanic Gardens & Parks Authority.

Paterson, M. (2007). *The senses of touch: Haptics, affects and technologies.* Oxford, England: Berg.

Payne, J. (1993). *Wetlands in the city of Armadale: A report to the Water Authority of Western Australia and the Environmental Protection Authority.* Leederville, Australia: Water Authority of Western Australia.

Pearman, G. (2005). Nuts, seeds, and pulses. In G. Prance & M. Nesbitt (Eds.), *The cultural history of plants* (pp. 133-152). New York, NY: Routledge.

Pelloe, E. (1921). *Wildflowers of Western Australia.* Melbourne, Australia: C.J. De Garis.

Perth Bushwalkers Club. (2007a). Constitution Retrieved April 17, 2011, from http://www.perthbushwalkers.asn.au/PBCweb/constitution.php?sid=d025076c2721e9c16af3276c80fee33f

Perth Bushwalkers Club. (2007b). Dieback Retrieved April 17, 2011, from http://www.perthbushwalkers.asn.au/PBCweb/dieback.php

Phillips, G. (1988). *Sacrificing the leaves: Poems.* Thailand: 10th World Congress of Poets.

Pink, S. (2009). *Doing sensory ethnography.* London, England: Sage.

Plate, C. (2005). *Restless spirits: The life and times of a wandering artist.* Sydney, Australia: Pan Macmillan.

Pocock, C. (2008). Reaching for the reef: Exploring place through touch. In F. Vanclay, M. Higgins & A. Blackshaw (Eds.), *Making sense of place: Exploring concepts and expressions of place through different senses and lenses* (pp. 77-85). Canberra, Australia: National Museum of Australia Press.

Porteous, J. D. (1989). *Planned to death: The annihilation of a place called Howdendyke.* Manchester, England: Manchester University Press.

Porteous, J. D. (1996). *Environmental aesthetics: Ideas, politics and planning.* London, England: Routledge.

Pratt, M. L. (1986). Fieldwork in common places. In J. Clifford & G. Marcus (Eds.), *Writing culture: The poetics and politics of ethnography* (pp. 27-50). Berkeley, CA: University of California Press.

Prendergast, M. (2009). Introduction: The phenomena of poetry in research. In M. Prendergast, C. Leggo & P. Sameshima (Eds.), *Poetic inquiry: Vibrant voices in the social sciences* (pp. xix-xlii). Rotterdam, Netherlands: Sense Publishers.

Price, L. (2007). How essential oils enter the body. In S. Price & L. Price (Eds.), *Aromatherapy for health professionals* (3 ed., pp. 139-154). Philadelphia, PA: Elsevier.

Prigogine, I. (2003). *Is future given?* Toh Tuck Link, Singapore: World Scientific Publishing.

Prigogine, I., & Stengers, I. (1984). *Order out of chaos: Man's new dialogue with nature.* London, England: Shambhala.

Priskin, J. (2003). Characteristics and perceptions of coastal and wildflower nature-based tourists in the Central Coast region of Western Australia. *Journal of Sustainable Tourism, 11*(6), 499-528.

Proust, M. (1913/2008). *Swann's way.* (C.K.S. Moncrieff, Trans.) Retrieved from http://books.google.com

Pryor, G. (2005a). Black death: Species extinction in WA. *Artlink, 25*(4). Retrieved from http://www.artlink.com.au/articles.cfm?id=2227

Pryor, G. (2005b). *Black solander, Gregory Pryor.* Perth, Australia: PICA Press.

Reich, C. A. (1970). *The greening of America.* Ringwood, Australia: Penguin Books.

Relph, E. (1976). *Place and placelessness.* London, England: Pion.

Repko, A. F. (2008). *Interdisciplinary research: Process and theory.* Los Angeles, CA: Sage.

Rescher, N. (2000). *Process philosophy: A survey of basic issues.* Pittsburgh, PA: University of Pittsburgh Press.

Robin, L. (2002). Nationalising nature: Wattle days in Australia. *Journal of Australian Studies, 73*, 5-23. Retrieved from http://www.api-network.com/jas/pdf/libby_robin.pdf

Rodaway, P. (2002). *Sensuous geographies: Body, sense and place*. Abingdon, England: Taylor & Francis.

Rose, D. (1992). *Dingo makes us human: Life and land in an Aboriginal Australian culture*. Cambridge, England: Cambridge University Press.

Rose, D., & Robin, L. (2004). The ecological humanities in action: An invitation. *Australian Humanities Review, 31-32* Retrieved from http://www.australianhumanitiesreview.org/archive/Issue-April-2004/rose.html

Rose, D. (2011). *Wild dog dreaming: Love and extinction*. Charlottesville, VA: University of Virginia Press.

Ross, K. (2008). *Going bush: New Zealanders and nature in the twentieth century*. Auckland, New Zealand: Auckland University Press.

Russo, G., & Schmitt, H. (1987). *Swan River mania: A colonial history of Western Australia*. Perth, Australia: Lynwood Enterprises.

Ryan, J. (2010a). Editorial: On the theme of "peripatetica". *Landscapes, 4*(1), 1. Retrieved from http://www.landscapeandlanguagecentre.au.com/Peripatetica/Peripatetica_Ryan_Editorial.pdf

Ryan, J. (2010b). Katoomba Incantation. *Philament*, December (Wallace, A., & Rahman, Eds.) Retrieved from http://sydney.edu.au/arts/publications/philament/issue16_pdfs/RYAN.pdf

Ryan, J. (*in press*). John Charles Ryan. In S.-P. Mitchell (Ed.), *Fremantle poets 3: Performance Poets*. Fremantle, Australia: Fremantle Press.

Ryan, S. (1996). *The cartographic eye: How explorers saw Australia*. Cambridge, England: Cambridge University Press.

Sanford, J. (1980). *The invisible partners: How the male and female in each of us affects our relationships*. New York, NY: Paulist Press.

Sarni, W. (2010). *Greening brownfields: Remediation through sustainable development* (1 ed.). New York, NY: McGraw-Hill.

Sattler, R. (1978). *Theoretical plant morphology*. Leiden, The Netherlands: Leiden University Press.

Sattler, R. (1994). Homology, homeosis, and process morphology in plants. In B. K. Hall (Ed.), *Homology: The hierarchical basis of comparative biology* (pp. 423-475). San Diego, CA: Academic Press.

Saunders, J., & O'Malley, C. (1982). *The anatomical drawings of Andreas Vesalius*. New York, NY: Bonanza Books.

Saussure, F. D. (1916/1986). *Course in general linguistics*. Retrieved from http://books.google.com

Schalow, F., & Denker, A. (2010). *Historical dictionary of Heidegger's philosophy*. Lanham, MD: Scarecrow Press.

Schiebinger, L. (2004). The private lives of plants. In *Nature's body: Gender in the making of modern science* (pp. 11-39). New Brunswick, NJ: Rutgers University Press.

Schroeder, S. (2010). *Philosophy of literature*. Chichester, England: John Wiley & Sons.

Seddon, G. (1972). *Sense of place: A response to an environment*. Perth, Australia: University of Western Australia Press.

Seddon, G. (1997). *Landprints: Reflections on place and landscape*. Cambridge, England: Cambridge University Press.

Seddon, G. (2005). *The old country: Australian landscapes, plants and people*. Cambridge, England: Cambridge University Press.

Sellers, C. (1999). Thoreau's body: Towards an embodied environmental history. *Environmental History, 4*, 487.

Seremetakis, C. N. (1994). The memory of the senses, Part I: Marks of the transitory. In C. N. Seremetakis (Ed.), *The senses stilled: Perception and memory as material culture in modernity* (pp. 1-18). Chicago, IL: University of Chicago Press.

Serres, M. (1982). *Hermes: Literature, science, philosophy* (J. Harari & D. Bell, Eds.). Baltimore, MD: Johns Hopkins University Press.

Serres, M. (1995). *The natural contract* (E. MacArthur & W. Paulson, Trans.). Ann Arbor, MI: University of Michigan Press.

Serres, M. (2008). *The five senses: A philosophy of mingled bodies*. London, England: Continuum.

Serres, M., & Latour, B. (1995). *Conversations on science, culture, and time* (R. Lapidus, Trans.). Ann Arbor, MI: University of Michigan Press.

Serres, M., & Zournazi, M. (2002). The art of living: A conversation with Michel Serres. In M. Zournazi (Ed.), *Hope: New philosophies for change* (pp. 192-209). New York, NY: Routledge.

Sharr, F. (1978). *Western Australian plant names and their meanings: A glossary*. Nedlands, Australia: University of Western Australia Press.

Sheets-Johnstone, M. (1992). The materialization of the body: A history of Western medicine, a history in process. In M. Sheets-Johnstone (Ed.), *Giving the body its due* (pp. 132-158). Albany, NY: State University of New York Press.

Sheldrake, R. (1991). *The rebirth of nature: The greening of science and god*. New York, NY: Bantam Books.

Shelton, A. (1994). Predicates of aesthetic judgement: Ontology and value in Huichol material representations. In J. Coote & A. Shelton (Eds.), *Anthropology, art and aesthetics* (pp. 209-244). Oxford, England: Clarendon Press.

Sibley, F. (2001). Some notes on ugliness. In J. Benson, B. Redfern & J. R. Cox (Eds.), *Approach to aesthetics: Collected papers on philosophical aesthetics* (pp. 190-206). Oxford, England: Oxford University Press.

Singer, A. (2003). *Aesthetic reason: Artworks and the deliberative ethos*. University Park, PA: Pennsylvania State University Press.

Skene, J. (1997). The power of naming: Women botanical collectors and the contested space of 19th century botany. In J. Gregory (Ed.), *Studies in Western Australian history: Historical traces* (Vol. 17, pp. 1-12). Nedlands, Australia: University of Western Australia.

Skinner, M. L. (1972). *The fifth sparrow: An autobiography*. Sydney, Australia: Sydney University Press.

Smith, B. (1989). *European vision and the South Pacific* (2 ed.). Oxford, England: Oxford University Press.

Smith, J. E. (1793). *A specimen of the botany of New Holland*. London, England: J. Sowerby.

Snow, C. P. (1964/1993). *The two cultures*. Cambridge, England: Cambridge University Press.

Solnit, R. (1994). *Savage dreams: A journey into the landscape wars of the American West*. Berkeley, CA: University of California Press.

Solnit, R. (2001). *Wanderlust: A history of walking*. London, England: Verso.

Soulé, M., & Press, D. (1998). What is environmental studies? *BioScience, 48*(5), 397-405.

Southern Scribes. (2007). *Wildflower country*. Ravensthorpe, Australia: Southern Scribes.

Spooner, A. (1999). Eremophila alternifolia R.Br. Poverty Bush. *FloraBase*. Retrieved from http://florabase.calm.wa.gov.au/browse/profile/7180

Spooner, A. (2003). Hakea neurophylla Meisn. *FloraBase*. Retrieved 27 March, 2011, from http://florabase.calm.wa.gov.au/browse/profile/2186

Steffen, W., Crutzen, P., & McNeill, J. (2007). The Anthropocene: Are humans now overwhelming the great forces of nature? *Ambio, 36*(8). Retrieved from http://www.allenpress.com/pdf/ambi-36-08-06_614..621.pdf

Stegner, W. (1992a). Living dry. In *Where the bluebird sings to the lemonade springs: Living and writing in the West* (pp. 57-75). New York, NY: The Modern Library.

Stegner, W. (1992b). Thoughts in a dry land. In *Where the bluebird sings to the lemonade springs: Living and writing in the West* (pp. 45-56). New York, NY: The Modern Library.

Stewart, F. (1995). *A natural history of nature writing*. Washington, DC: Shearwater Books.

Stirling, J. (1827/2009). Narrative of operations: Expedition to Swan River, 1827. In J. Shoobert (Ed.), *Western Australian Exploration Volume One December 1826 - December 1835* (Vol. 1, pp. 20-41). Carlisle, Australia: Hesperian Press.

Stoller, P. (1989). *The taste of ethnographic things: The senses in anthropology*. Philadelphia, PA: University of Pennsylvania Press.

Stoller, P. (1997). *Sensuous scholarship*. Philadelphia, PA: University of Pennsylvania Press.

Stolnitz, J. (1998). The aesthetic attitude. In C. Korsmeyer (Ed.), *Aesthetics: The big questions* (pp. 78-83). Malden, MA: Blackwell Publishing.

Stratford, E. (1997). Memory work, geography and environmental studies: Some suggestions for teaching and research. *Australian Geographical Studies, 35*(2), 206-219.

Summers, L. (2011). Wildflower season: The development of native flora protection legislation in Western Australia, 1911-1975. In A. Gaynor & J. Davis (Eds.), *Studies in Western Australian History: Environmental Exchanges, 27*, 31-44.

Süskind, P. (1985). *Perfume: The story of a murderer*. Camberwell, Australia: Penguin Books.

Synnott, A. (1991). Puzzling over the senses: From Plato to Marx. In D. Howes (Ed.), *The varieties of sensory experience: A sourcebook in the anthropology of the senses* (pp. 61-76). Toronto, Canada: University of Toronto Press.

Tennyson, A. L. (1850/2009). In memoriam A.H.H. OBIT MDCCCXXXIII: 56. Retrieved April 8, 2011, from http://rpo.library.utoronto.ca/poem/2141.html

The Paperbark Co. (2008). Essential oils. Retrieved April 17, 2011, from www.paperbarkoils.com.au/essential-oils.html#oil_agnois

The West Australian. (1893, September 2). News and notes [Australia Trove]. *The West Australian*. Retrieved from http://trove.nla.gov.au/ndp/del/home

The West Australian. (1899, August 30). News and notes [Australia Trove]. *The West Australian*. Retrieved from http://trove.nla.gov.au/ndp/del/home

Thomas, D. (1952). The force that through the green fuse drives the flower. Retrieved May 19, 2010, from http://www.poets.org/viewmedia.php/prmMID/15379

Thomas, K. (1983). *Man and the natural world: Changing attitudes in England 1500-1800*. New York, NY: Penguin Books.

Thomas, S. (2004). *Of earth and flesh and bones and breath: Landscapes of embodiment and moments of re-enactment*. Halifax, Canada: Backalong Books.

Thomas, S. (2009). Nissopoesis: Visuality and aesthetics in poetic inquiry. In M. Prendergast, C. Leggo & P. Sameshima (Eds.), *Poetic inquiry: Vibrant voices in the social sciences* (pp. 127-132). Rotterdam, The Netherlands: Sense Publishers.

Thoreau, H. D. (1854/1966). *Walden; and Civil disobedience: Authoritative texts, background, reviews, and essays in criticism*. (T. Owen, Ed.). New York, NY: Norton.

Thoreau, H. D. (1862/2007). *Walking*. LaVergne, TN: Filiquarian Publishing.

Thoreau, H. D. (1862/2010). *Wild apples*. LaVergne, TN: Filiquarian Publishing.

Thoreau, H. D. (1962). *The journal of Henry David Thoreau in fourteen volumes bound as two: Volumes VIII-XIV (November 1855-1861)*. New York, NY: Dover Publications.

Thoreau, H. D. (1993). *Faith in a seed: The dispersion of seeds and other late natural history writings*. Washington, DC: Island Press.

Thoreau, H. D. (1999). *Material faith: Thoreau on science*. (L.D. Walls, Ed.). Boston, MA: Houghton Mifflin Company.

Thoreau, H. D. (2000). *Wild fruits: Thoreau's rediscovered last manuscript*. New York, NY: W.W. Norton.

Tourism Australia. (2010). Western Australia's wildflowers. Retrieved 9 April, 2011, from http://www.australia.com/articles/wa_wildflowers.aspx

Trigger, D., & Mulcock, J. (2005). Forests as spiritually significant places: Nature, culture and "belonging" in Australia. *The Australian Journal of Anthropology, 16*(3), 306-320.

Trollope, A. (1873/1967). *Australia* (P. D. Edwards & R. B. Joyce, Trans.). St Lucia, Australia: University of Queensland Press.

Tuan, Y.-F. (1982). *Segmented worlds and self: Group life and individual consciousness*. Minneapolis, MN: University of Minnesota.

Turizem Bohinj. (2010). 5th international wildflower festival. Retrieved April 9, 2011, from http://www.bohinj.si/alpskocvetje/eng/index.php

Tyrrell, I. (2005). placing george seddon. *Environmental History, 10*(4), 752-754.

Urry, J. (2002). *The tourist gaze* (2 ed.). London, England: Sage.

Urry, J. (2005). The place of emotions within place. In J. Davidson, M. Smith & L. Bondi (Eds.), *Emotional geographies* (pp. 77-86). Aldershot, England: Ashgate.

Valley of the Giants. (n.d.). Denmark Walpole wilderness forests and ocean coastline scenic tours. Retrieved April 9, 2011, from http://valleyofthegiants.com.au/information.php

Van Dooren, T. (2010). Pain of extinction: The death of a vulture. *Cultural Studies Review, 16*(2), 271-289. Retrieved from http://epress.lib.uts.edu.au/ojs/index.php/csrj/article/viewFile/1702/1834

Vanclay, F. (2008). Place matters. In F. Vanclay, M. Higgins & A. Blackshaw (Eds.), *Making sense of place: Exploring concepts and expressions of place through different senses and lenses* (pp. 3-11). Canberra, Australia: National Museum of Australia Press.

Visit Western Australia Online. (2011). Fitzgerald River National Park. Retrieved April 18, 2011, from http://visitwa.com.au/australias-south-west-region/60-fitzgerald-river-national-park-wa.html

Visvanathan, S. (1996). Footnotes to Vavilov: An essay on gene diversity. In F. Apffel-Marglin & S. Marglin (Eds.), *Decolonizing knowledge: From development to dialogue* (pp. 306-339). Oxford, England: Clarendon Press.

Vivienne, M. (1901). *Travels in Western Australia: Being a description of the various cities and towns, goldfields, and agricultural districts of that state.* London, England: William Heinemann.

Vlamingh, W. (1985). *Voyage to the Great South Land* (C. De Heer, Trans.). Sydney, Australia: Royal Australian Historical Society.

Von Humboldt, A. (1850). *Cosmos: A sketch of a physical description of the universe* (E. C. Otté, Trans. Vol. 2). New York, NY: Harper & Brothers.

Wallace, A., & Rahman, S. (2009). Editorial. *Philament,* (December). Retrieved from http://sydney.edu.au/arts/publications/philament/issue15_pdfs/Editorial.pdf

Walls, L. D. (1995). *Seeing new worlds: Henry David Thoreau and nineteenth-century natural science.* Madison, WI: University of Wisconsin Press.

Waskul, D., Vannini, P., & Wilson, J. (2009). The aroma of recollection: Olfaction, nostalgia, and the shaping of the sensuous self. *Senses & Society, 4*(1), 5-22. doi: 10.2752/174589309X388546

Welsch, W. (1997). *Undoing aesthetics.* London, England: Sage.

West, M. (1984). Thoreau and the language theories of the French Enlightenment. *ELH: English Literary History, 51*(4), 747-770.

Western Australian Tourism Commission. (2003). *Research into the wildflower tourism industry: An exploration of consumer and industry perceptions.* Perth, Australia: Western Australian Tourism Commission.

Western Australian Tourism Commission. (2010). *Wildflower holiday guide Western Australia: Your guide to the largest collection of wildflowers on Earth.* Perth, Australia: Tourism Western Australia.

Whewell, W. (1840). *The philosophy of the inductive sciences: Founded upon their history* (Vol. 2). London, England: John W. Parker.

Whitman, W. (1855/1969). *Leaves of grass.* Retrieved from http://books.google.com

Wickens, G. E. (2001). *Economic botany: Principles and practices.* Dordrecht, Netherlands: Kluwer Academic.

Wildflower Society of Western Australia. (2010). Country wildflower shows. *Wildflower Society of Western Australia Newsletter, 48*(2), 41.

Wilkes, R. (1998). Interview 26: Richard Wilkes. Swan River Derbal Yerrigan elder. 1:15pm. 29:06:98. In T. McCabe (Ed.), *Nyoongar views on logging old growth forests* (pp. 46-47). West Perth, Australia: Wilderness Society.

Wilkes, T. (1998). Interview 25: Ted Wilkes. Perth Aboriginal Medical Service, Director. 12:55pm. 29:06:98. In T. McCabe (Ed.), *Nyoongar views on logging old growth forests* (pp. 44-45). West Perth, Australia: Wilderness Society.

Williams, R. (1643/1997). *A key into the language of America.* Bedford, MA: Applewood Books.

Williams, R. (1982). *Socialism and ecology*. London, England: Socialist Environment and Resources Association.

Wills, R., & Kinnear, J. (1993). Threats to the Stirling Range. In C. Thomson, G. Hall & G. Friend (Eds.), *Mountains of mystery: A natural history of the Stirling Range* (pp. 135-141). Como, Australia: Department of Conservation and Land Management.

Wilson, E. O. (1984). *Biophilia: The human bond with other species*. Cambridge, MA: Harvard University Press.

Wilson, E. O. (1997). Introduction. In M. Reaka-Kudla, D. Wilson & E. Wilson (Eds.), *Biodiversity II: Understanding and protecting our biological resources*. Washington, DC: Joseph Henry Press.

Wilson, E. O. (1998). *Consilience: The unity of knowledge*. New York, NY: Alfred A. Knopf.

Wolin, R. (1993). "OVER THE LINE:" Reflections on Heidegger and National Socialism. In R. Wolin (Ed.), *The Heidegger controversy* (pp. 1-22). Cambridge, MA: MIT Press.

Wright, J. (1994). *Judith Wright: Collected poems 1942-1985*. Sydney, Australia: Angus & Robertson.

Publication History

Earlier versions of chapters have appeared previously in these forms:

- Chapter 1 as Ryan, J. (2011). Cultural botany: Toward a model of transdisciplinary, embodied and poetic research into plants. *Nature and Culture*, *6*(2), 123-148.

- Chapter 2 as Ryan, J. (2010). Botanic field aesthetics: A methodology of embodied research into indigenous Southwest flora. *Landscapes*, *4*(1), 96-107 Retrieved from http://www.landscapeandlanguagecentre.au.com/Peripatetica/Peripatetica_Ryan_Botanic.pdf.

- Chapter 3 as Ryan, J. (2011). Leaves of a tree: Interweaving the many narratives of Southwest Australian flora. *Research Journal of the ooi Junior Academy: Transactions on Nature*, *11*(1), 45-60.

- Chapter 4 as Ryan, J. (2010). Values and evaluations: Reading for beauty in John Lindley's *A sketch of the vegetation of the Swan River Colony* (1839-40). *The New Critic*, *13*, 1-17 Retrieved from http://www.ias.uwa.edu.au/new-critic/thirteen/?a=1326211

- Chapter 5 as Ryan, J. (2011). Plants as objects: Challenges for an aesthetics of plants. *Philosophy Study*, *1*(3), 222-236.

- Chapter 6 as Ryan, J. (2010). Green tropism and the Southwest Australian flora: From the Green Man to the Ravensthorpe Woman. *The International Journal of the Humanities, 8*(2), 211-228.
- Chapter 7 as Ryan, J. (2011). Anthoethnography: Emerging research into the culture of flora, aesthetic experience of plants, and the wildflower tourism of the future. *New Scholar, 1*(1), 28-40 Retrieved from http://www.newscholar.org.au
- Chapter 8 as Ryan, J. (2012, *forthcoming*). Botanical memory: Exploring emotional recollections of native flora in the Southwest of Western Australia. *Emotion, Space and Society*.
- Chapter 9 as Ryan, J. (2011). An unlikely marriage? Theorising the corporeality of language at the crossroads of Thoreau, Heidegger and the botanical world. *Philosophy Study, 1*(2), 139-149.
- Chapter 11 as Ryan, J. (2009). Why do extinctions matter? Mourning the loss of indigenous flora in the Southwest of Western Australia. *Philament, 15*, 51-80 Retrieved from http://sydney.edu.au/arts/publications/philament/issue15_pdfs/RYAN_Why%20Do%20Extinctions%20Matter.pdf
- Chapter 12 as Ryan, J. (2010). Not a bush flâneur? The convergent topographies of recreational bushwalking, floristic appreciation and human embodiment in the Southwest of Western Australia. *Colloquy, 20*, 5-32 Retrieved from http://www.arts.monash.edu.au/ecps/colloquy/journal/issue020/ryan.pdf
- Chapter 13 as Ryan, J. (2012). Towards intimate relations: Gesture and contact between plants and people. *PAN: Philosophy, Activism, Nature*, 9, 29-36.

I'd also like to acknowledge the following publications that feature aspects of *Green Sense*:

- Ryan, J. (2009). 'Plants that perform for you?' From floral aesthetics to floraesthesis in the Southwest of Western Australia. *Australian Humanities Review, 47*, 117-140. Retrieved from http://www.australianhumanitiesreview.org/archive/Issue-November-2009/ryan.html.
- Ryan, J. (2010). Towards a corporeal aesthetics of plants: Ethnographies of embodied appreciation along the wildflower trail. *Continuum, 24*(4), 543-557.
- Ryan, J. (2010). Palm-like fingers gripping a coarse line of air: Poetry as a method of enquiry into Southwest Australian flora. *Provoking Texts*, 1-24. Paper presented at the Creative Margins: Tenth Humanities Graduate Research Conference, Curtin University, Bentley, WA Retrieved from http://hgsoconference.curtin.edu.au/local/pdf/ RYAN_poetic_enquiry.pdf

- Ryan, J. (2011). Sifting horticulture from botany: John Lindley's 'A sketch of the vegetation of the Swan River Colony'. *Australian Garden History, 23*(1), 9-14.
- Ryan, J. (2011). Plants, processes, places: Sensuous intimacy and poetic enquiry. *Axon: Creative Explorations, 1* Retrieved from http://axonjournal.com.au/issue-1/plants-processes-places-sensory-intimacy-and-poetic-enquiry

Index

Page numbers followed by f indicate figures. Those followed by t indicate tables.

N

O

T

U

www.ingramcontent.com/pod-product-compliance
Ingram Content Group UK Ltd.
Pitfield, Milton Keynes, MK11 3LW, UK
UKHW020219041026
15470UKWH00012B/474

9 780957 301702